29. Colloquium der Gesellschaft für Biologische Chemie
6.–8. April 1978 in Mosbach/Baden

Energy Conservation in Biological Membranes

Edited by
G. Schäfer and M. Klingenberg

With 160 Figures

Springer-Verlag
Berlin Heidelberg New York 1978

Prof. Dr. GÜNTER SCHÄFER
Medizinische Hochschule Hannover
Institut für Klinische Biochemie
und Physiologische Chemie
Karl-Wiechert-Allee 9
3000 Hannover 61/FRG

Prof. Dr. MARTIN KLINGENBERG
Universität München
Institut für Physiologische Chemie
Physikalische Biochemie und Zellbiologie
Goethestraße 33
8000 München 2/FRG

ISBN 3-540-09079-7 Springer-Verlag Berlin Heidelberg New York
ISBN 0-387-09079-7 Springer-Verlag New York Heidelberg Berlin

Offsetprinting and bookbinding: Brühlsche Universitätsdruckerei, Lahn-Gießen.
2131/3130-543210

*Dedicated to
Professor* Britton Chance
*on occasion of his
65th birthday*

Preface

The topic of the 29th Mosbach Colloquium *Energy Transduction in Biological Membranes* is one of the most formidable problems in biology. Its solution in molecular terms has proved to be a very difficult task for a whole generation of biochemists.

The Mosbach Colloquia had so far not yet covered this subject. In former Mosbach Colloquia some contributions were closely related, such as the lecture by E. C. Slater on the mitochondrial respiratory chain, presented 25 years ago. A broader coverage of this subject was given in the Mosbach Colloquia on *Biochemistry of Oxygen* in 1968, and on *Inhibitors: Tools in Cell Research* in 1969, which contained several lectures related to bioenergetics.

Today progress and understanding of the energy transduction in biological membranes had advanced to the stage where we can formulate reliable theories on many facets of the energy transduction process. On the other hand, the primary energy conservation steps are as controversial as ever and challenge the field for an all-out effort for resolving these burning problems. The 29th Mosbach Colloquium has given a broad and vivid picture of this situation, illustrating the progress and also the controversial problems currently debated.

The colloquium was divided into four sections dealing with (1) the electron transport, its structure and function (2) the primary events of energy conservation, in particular the conversion of redox energy and of light energy (3) the coupling ATPase, its structure and function, and (4) the control and thermodynamics of energy transduction.

Some related subjects, such as biogenesis of energy-transducing membranes, which has developed into a large field of its own, could not be included since this would exceed the scope of this colloquium.

The organizers are grateful to the Gesellschaft für Biologische Chemie and its chairman, Prof. K. Decker, for the active support of the colloquium. We are particularly pleased to acknowledge the efforts of the treasurer, Prof. E. Auhagen, to obtain the necessary funds. The colloquium was generously supported by the Deutsche Forschungsgemeinschaft and by some companies who are corporate members of the Gesellschaft für Biologische Chemie.

We certainly hope that this book fulfills its various purposes, such as serving as an introduction to the field of energy conservation for the nonspecialized scientists and for young scientists who might be encouraged to enter this subject, and also as a record and reference book on the present status of this field.

Hannover and München, October 1978   G. SCHÄFER

                                                      M. KLINGENBERG

Contents

Structure and Function of ATP-Synthesizing Systems

Control and Dynamics of Energy Conservation

Contributors

AMZEL, L.M.   Department of Biophysics, Johns Hopkins University
                     School of Medicine, 725 N. Wolfe Street, Baltimore,
                     MD 21205 / USA

AUSLÄNDER, W.   Max-Volmer-Institut, Technische Universität
                     Berlin, Straße des 17. Juni 135, 1000 Berlin 12

BÄUERLEIN, E.   Max-Planck-Institut für medizinische Forschung,
                     Abteilung Naturstoff-Chemie, Jahnstraße 29,
                     6900 Heidelberg / FRG

BALTSCHEFFSKY, H.   Department of Biochemistry, Arrhenius Laboratory,
                     University of Stockholm, 106 91 Stockholm / Sweden

BUSE, G.   RWTH Aachen, Abteilung Physiologische Chemie,
                     Melatenerstraße 211, 5100 Aachen / FRG

CHANCE, B.   Johnson Research Foundation, University of
                     Pennsylvania, Philadelphia, PA 19104 / USA

CINTRÓN, N.   Laboratory for Molecular and Cellular Bioenergetics,
                     Department of Physiological Chemistry, Johns Hop-
                     kins University, School of Medicine, 725 N. Wolfe
                     Street, Baltimore, MD 21205 / USA

DOSE, K.   Institut für Biochemie der Joh. Gutenberg-Univer-
                     sität, Joh. Joachim Becher-Weg 30, 6500 Mainz / FRG

ENGEL, W.D.   Institut für Physiologische Chemie und Physikali-
                     sche Biochemie der Universität München, Goethe-
                     straße 33, 8000 München / FRG

ERECINSKA, M.   Department of Biochemistry and Biophysics, Medical
                     School University of Pennsylvania, Philadelphia,
                     PA 19104 / USA

GALANTE, Y.M.   Department of Biochemistry, Scripps Clinic and
                     Research Foundation, La Jolla, CA 92037 / USA

GRÄBER, P.   Max-Volmer-Institut für Physikalische Chemie und
                     Molekularbiologie, Technische Universität Berlin,
                     Straße des 17. Juni 135, 1000 Berlin 12

HACKENBERG, H.   Institut für Physiologische Chemie und Physikali-
                     sche Biochemie der Universität München, Goethe-
                     straße 33, 8000 München / FRG

HARTMANN, R.   Institut für Biochemie der Universität Würzburg,
                     8700 Würzburg / FRG

HATEFI, Y.   Department of Biochemistry, Scripps Clinic and
                      Research Foundation, La Jolla, CA 92037 / USA

HESS, B.   Max-Planck-Institut für Ernährungsphysiologie,
                      Rheinlanddamm 201, 4600 Dortmund 1 / FRG

HULLIHEN, J.   Laboratory for Molecular and Cellular Bioenergetics,
                      Department of Physiological Chemistry, Johns
                      Hopkins University, School of Medicine, 725 N.
                      Wolfe Street, Baltimore, MD 21205 / USA

JENSEN, L.H.   Department of Biological Structure, University
                      of Washington, School of Medicine, Seattle, WA
                      98195 / USA

JUNGE, W.   Max-Volmer-Institut, Technische Universität
                      Berlin, Straße des 17. Juni 135, 1000 Berlin 12

KAGAWA, Y.   Jichi Medical School, Department of Biochemistry,
                      Minamikawachi, Kawachi-gun, Tochigi-ken,
                      Japan 32904

KOLB, H.J.   Institut für Diabetesforschung, Kölner Platz 1,
                      8000 München 40 / FRG

KOLLIA, J.   Max-Volmer-Institut, Technische Universität
                      Berlin, Straße des 17. Juni 135, 1000 Berlin 12

KORENSTEIN, R.   Max-Planck-Institut für Ernährungsphysiologie,
                      Rheinlanddamm 201, 4600 Dortmund 1 / FRG

KRAB, K.   Department of Medical Chemistry, University of
                      Helsinki, Siltavuorenpenger 10A, 00170 Helsinki 17/
                      Finland

KRÖGER, A.   Institut für Physiologische Chemie und Physika-
                      lische Biochemie der Universität München, Petten-
                      koferstraße 14a , 8000 München / FRG

KUSCHMITZ, D.   Max-Planck-Institut für Ernährungsphysiologie,
                      Rheinlanddamm 201, 4600 Dortmund 1 / FRG

McGEER, A.J.   Max-Volmer-Institut, Technische Universität
                      Berlin, Straße des 17. Juni 135, 1000 Berlin 12

MICHEL, H.   Institut für Biochemie der Universität Würzburg,
                      8700 Würzburg / FRG

OESTERHELT, D.   Institut für Biochemie der Universität Würzburg,
                      8700 Würzburg / FRG

ONUR, G.   Medizinische Hochschule, Abteilung für Biochemie,
                      3000 Hannover / FRG

PEDERSEN, P.L.   Laboratory for Molecular and Cellular Bioenergetics,
                      Department of Physiological Chemistry, Johns
                      Hopkins University, School of Medicine, 725 N.
                      Wolfe Street, Baltimore, MD 21205 / USA

RATHGEBER, G.   Institut für Biochemie der Joh. Gutenberg-Univer-
                      sität, Joh. Joachim Becher-Weg 30, 6500 Mainz / FRG

RENGER, G.   Max-Volmer-Institut für Physikalische Chemie und
                  Molekularbiologie, Technische Universität Berlin,
                  Straße des 17. Juni 135, 1000 Berlin 12

SARONIO, C.†   Johnson Research Foundation, University of Penn-
                  sylvania, Philadelphia, PA 19104 / USA

SCHÄFER, G.   Medizinische Hochschule, Abteilung für Biochemie,
                  3000 Hannover / FRG

SCHÄFER, H.-J.   Institut für Biochemie der Joh. Gutenberg-Univer-
                  sität, Joh. Joachim Becher-Weg 30, 6500 Mainz / FRG

SCHÄGGER, H.   Institut für Physiologische Chemie und Physikali-
                  sche Biochemie der Universität München, Goethe-
                  straße 33, 8000 München / FRG

SCHEURICH, P.   Institut für Biochemie der Joh. Gutenberg-Univer-
                  sität, Joh. Joachim Becher-Weg 30, 6500 Mainz / FRG

SEBALD, W.   Institut für Physiologische Chemie und Physikali-
                  sche Biochemie der Universität München, Goethe-
                  straße 33, 8000 München / FRG

SOPER, J.W.   Laboratory for Molecular and Cellular Bioenergetics,
                  Department of Physiological Chemistry, Johns
                  Hopkins University, School of Medicine, 725 N.
                  Wolfe Street, Baltimore, MD 21205 / USA

STROTMANN, H.   Abteilung Biochemie der Pflanzen, Botanisches
                  Institut, Tierärztliche Hochschule Hannover,
                  3000 Hannover / FRG

STUCKI, J.W.   Pharmakologisches Institut der Universität Bern /
                  Switzerland and Service de Chimie Physique II,
                  Université Libre de Bruxelles / Belgium

SUSSMAN, I.   Department of Biochemistry and Biophysics, Medical
                  School University of Pennsylvania, Philadelphia,
                  PA 19104 / USA

TIEMANN, R.   Max-Volmer-Institut für Physikalische Chemie und
                  Molekularbiologie, Technische Universität Berlin,
                  Straße des 17. Juni 135, 1000 Berlin

TRASCH, H.   Max-Planck-Institut für medizinische Forschung,
                  Abteilung Naturstoff-Chemie, Jahnstraße 29,
                  6900 Heidelberg / FRG

TREBST, A.   Abteilung für Biologie, Ruhr-Universität-Bochum,
                  Postfach 102148, 4630 Bochum-Querenburg / FRG

VON JAGOW, G.   Institut für Physiologische Chemie und Physika-
                  lische Biochemie der Universität München, Goethe-
                  straße 33, 8000 München / FRG

WACHTER, E.   Institut für Physiologische Chemie und Physika-
                  lische Biochemie der Universität München, Goethe-
                  straße 33, 8000 München / FRG

WAGNER, G.   Institut für Biochemie der Universität Würzburg,
                    8700 Würzburg / FRG

WARING, A.   Johnson Research Foundation, University of Penn-
                    sylvania, Philadelphia, PA 19104 / USA

WEISS, H.   Europäisches Laboratorium für Molekularbiologie,
                    Meyerhofstraße 1, Postfach 10 2209, 6900 Heidel-
                    berg / FRG

WIKSTRÖM, M.   Department of Medical Chemistry, University of
                    Helsinki, Siltavuorenpenger 10A, 00170 Helsinki 17/
                    Finland

WILSON, D.F.   Department of Biochemistry and Biophysics, Medical
                    School University of Pennsylvania, Philadelphia,
                    PA 19104 / USA

WITT, H.T.   Max-Volmer-Institut für Physikalische Chemie und
                    Molekularbiologie, Technische Universität Berlin,
                    Straße des 17. Juni 135, 1000 Berlin 12

Organization of Electrontransport Systems

Evolutionary Aspects of Biological Energy Conversion

H. BALTSCHEFFSKY

Introduction

It is a great pleasure and privilege for me to begin this 29th Mos-
bacher Colloquium on Energy Conversion in Biological Membranes with a
presentation of an evolutionary approach. By this approach one may
pursue the important but difficult task of seeking information primarily about how, in molecular terms, it all began and how the present
energy conversion systems of living cells evolved, by development of,
and subsequent variations on, some apparently very ancient fundamental
themes. However, I believe that the arrangers of this Colloquium share
my optimistic conviction that the study of the molecular evolution of
enzymes and pathways involved in biological energy conversion may be
of value in an additional way. That is, not only for learning more about
common molecular ancestries and paths of evolution, but also, and perhaps of particular significance in connection with this Colloquium,
for tackling some of the unsolved fundamental mechanistic problems still
existing in the area of biological energy conversion, in particular in
photosynthetic and respiratory electron transport phosphorylation.

Some General Concepts in Molecular Evolution

Evolution Along Metabolic Sequences

First I would like to discuss some current evolutionary concepts, as
many of you may be more familiar with areas of biochemistry other than
molecular evolution. This may be started off by recalling that the
Chairman of this Session, Professor Bücher, with his co-workers in the
early sixties drew attention to the nearly constant ratio of the activities of a "constant-proportion group" of five consecutive enzymes in
the Embden-Meyerhof pathway. These five, starting with triosephosphate
isomerase and ending with enolase, constitute an unbranched pathway
between two branching points, as was pointed out by Pette et al. (1962).

Although those investigations were performed with material from higher
organisms, they bring into focus not only the well-known successes of
the operon theory with prokaryote systems, but also the early works of
Horowitz and of Demerec. Horowitz (1945, 1965) first drew attention to
the possibility, very plausible in a Darwinian context due to the selective advantage conferred to any cell capable of breaking down the product of the preceding reaction, that biosynthetic pathways, for example
amino acid synthesis pathways, may have originated with the breakdown
of amino acids occurring in a primitive earth environment and evolved
stepwise in this reversed direction. And Demerec (1956) found with the
histidine and tryptophan synthesis pathways of *Salmonella* that the genes
for four consecutive enzymes were linked together and located in the
same order on the chromosome in which the corresponding enzymes were
operating in their metabolic pathways. These are early examples of our

increasingly detailed picture of a stepwise molecular evolution, where
a small change in protein structure may result in slightly changed
function.

Today it is well known that biological evolution proceeds through
gene duplication, mutation, and other transformations of the nucleic
acids, which are the genetic material providing information for protein synthesis in the living cell. Phenomena such as gene transfer,
gene overlap, and the recently discovered gene sequence omission or
deletion in nucleated cells, not found in bacteria, show that the rules
of the information game are not extremely simple. On the other hand,
additional possibilities for variability in the evolutionary process
do not seem to diminish its fundamentally stepwise and principally
traceable nature.

Some Assumptions

At this stage I would like to enlist some assumptions which are of
basic relevance in the field of biological evolution and of particular
interest for the latter part of this presentation:
1) The genetic code is universal for life on earth.
2) Gradually divergent trends in protein structure are evidence for
   a single common ancestor.
3) Secondary and tertiary structures of polypeptides tend to be more
   conserved during evolution than primary structures.
4) The β-pleated sheet structure was a very early secondary structure.

Each of these assumptions is founded on extensive support obtained in
the last one or two decades. My comments have to be limited to the
first and the fourth. Whereas no indication or any reason exists for
an alternative genetic code in any current organisms, the present code
must, on the other hand, have evolved from simpler molecular information systems (Orgel, 1968, 1972). Special attention should be given to
the suggestion by Orgel (1972) that from a primitive code with high
information content in the sequence ---AUAUAUA--- would result polypeptides with alternating hydrophobic and hydryphilic side chains.
Such polypeptides tend to form β-pleated sheets, and may be of importance as sources of early membrane-like structures as will be discussed below. These and similar considerations have experimental support (Brack and Orgel, 1975) and have led to the fourth of the
assumptions above.

Some Principles and Statements

The following four principles and statements will now be presented:
1) The principle of continuity (Orgel, 1968): molecular evolution
   occurs stepwise, giving increasing structural complexity as a
   general trend.
2) The principle of growing points (Baltscheffsky et al., 1978):
   polypeptide secondary structures tend to be distinct growing points
   in evolution.
3) A change in a gene may give a protein with retained or slightly
   or much changed function.
4) It is often difficult to distinguish between evolutionary divergence and functional convergence.

The note of caution given in the last statement reminds us of the
necessity to be aware of the great difficulty existing in many cases
in distinguishing finally between the two possibilities evolutionary
divergence (homology) and functional convergence (analogy).

An Approximate Time Scale

The sun and the earth have existed as celestial bodies for about
5.0 and 4.6 billion years, respectively, according to available evidence. If we limit our treatment of the cosmic evolutionary process
to the earth, we find that it may be divided into two phases. With
current terminology, molecular evolution on the earth includes a
chemical phase, which led in discrete steps to the origin of life,
probably well before the earth had reached its first billion years,
and a subsequent biological phase, which started when life first
emerged on earth, and which still continues (Fig. 1).

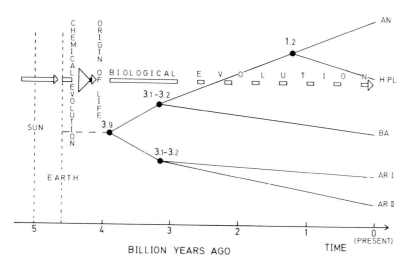

Fig. 1. Molecular evolution in a cosmic time perspective, with indicated common
ancestry between some major groups of living organisms. Approximate ages of sun,
earth and life on earth are given. *AN*, animals; *H PL*, higher plants; *BA*, bacteria;
AR I and *AR II*, the two major groups of archaebacteria (see text)

The scheme for this molecular evolution given in Figure 1 is based
on earlier investigations by Barghoorn, Schopf, Dayhoff and many others
and on very recent work by Woese and his co-workers (Fox et al., 1977;
Woese and Fox, 1977). It should be considered as only a very approximate picture of when life on earth originated and when some major
branchings in the evolution of various types of living organisms occurred. Living cells which were common ancestors to both the animals
and higher plants appear to have existed as recently as about 1.2 billion years ago. Cells, from which emerged on the one hand the animals
and the higher plants and on the other the bacterial line of descent,
appear to have existed roughly 3.1-3.2 billion years ago. Also shown
in Figure 1 are two branches, AR I and AR II, which apparently emerged
equally long ago. These represent the two main groups of methanogenic
bacteria, which according to ribosomal RNA oligonucleotide sequence
data obtained by Fox et al. (1977) are as distant from each other as
the blue-green algae are from the other major groups of bacteria. The
data also indicate that all the methanogens are extremely distant relatives of other living cells. According to Woese and Fox (1977) they
constitute an altogether different primary kingdom, "the archaebacteria".
If one uses the published dendrogram of the relationships between
methanogens and typical bacteria, and, as a first approximation,

tentatively assumes that the decrease in association coefficient is linear with time, the divergence of the methanogens and other known organisms occurred roughly 3.9 billion years ago, as is indicated in Figure 1.

We may now bring a fundamental structure in photosynthetic and respiratory energy conversion, the membrane-bound coupling factor ATPase, into the evolutionary time scale. It is well known that this complex multi-subunit polypeptide structure is remarkably similar in animals, higher plants, and both photosynthetic and nonphotosynthetic bacteria. On the other hand, it does not seem to exist in the methanogens, who have a very special metabolism, and whose energy conservation reactions are completely unknown. One may draw the tentative conclusion that the coupling factor ATPase emerged more than 3 billion years ago, before the separation of the bacterial and the (animal + higher plant) lines of descent, and was retained during the subsequent biological evolution along these lines with only relatively minor modifications. The simplest assumption with respect to the archaebacteria is that they and their ancestors never contained the intricate coupling factor ATPase system.

Studies of the primary, secondary, and tertiary structures of certain proteins functioning in energy conversion systems, with cytochrome c, ferredoxin and dehydrogenases as outstanding examples, have given rather detailed information about evolutionary trees of organisms going far back along the time scale of Figure 1 into early Precambrian time (Dickerson and Timkovich, 1975; Hall et al., 1974; Rossmann et al., 1975). Some results elucidate possible evolutionary relationships between major biological energy conversion pathways, notably between photosynthesis and respiration. But rather than discuss here such evolutionary aspects, which have been treated in great detail in recent literature, I would like to focus attention on the origin and early evolution of biological electron transport and coupled phosphorylation processes as such. In this context will be discussed also our recent work on a pathway of protein β-structure evolution and our considerations about functional concerted hydrogen bond pattern changes in proteins and their possible role in energy transfer mechanisms.

Evolution of Electron Transport and Coupled Phosphorylation

Evolution of Electron Transport

With stepwise evolution as one guiding principle and survival of the fittest as another, I proposed in 1974 a hypothesis for the origin and evolution of biological electron transport (Baltscheffsky, 1974a). Based also on the collected evidence for a reducing atmosphere when life originated and that ferredoxins are very ancient proteins and thus early electron carriers, the hypothesis envisaged biological electron transport as having originated close to the potential of the hydrogen electrode (-420 mV at pH 7), quite possibly with ferredoxin-like electron carriers, and as having evolved stepwise in suitable directions along and across the potential scale. Evidence from studies of amino acid sequences and three-dimensional structures have supported this hypothesis by indicating evolutionary relationships between different types of iron-binding (Baltscheffsky, 1974a) and nucleotide-binding (Rossmann et al., 1975) proteins and possibly even between iron-binding and nucleotide-binding proteins (Baltscheffsky, 1974a,b).

I would like to suggest that if there has existed a discernible evolutionary step from simple cellular electron transport to coupled, energy-conserving electron transport, and if this energy conservation at the oxidation-reduction level occurs as a protein conformational change, then one may assume that very similar coupling mechanisms exist at different coupling sites. Once an energy-conserving protein conformational change appeared during the evolutionary process, it was accordingly usefully elaborated upon, so that the same basic principle would have been utilized when a new coupling site appeared. In general, whether the essential aspects of energy conservation are more closely linked to protein level or to co-factor level, with or without inter- or intramembrane charge separation, one-time evolution of the coupling principle with subsequent variations on the same theme would seem to be a more attractive assumption than the alternative with several independent molecular solutions to the energy-coupling problem.

A β-Structure Evolution Pathway and the Case of Bacterial Ferredoxin

As has been pointed out by Orgel (1972) and by Carter and Kraut (1974) oligo- and polypeptides with β-structures may have been important already in connection with the evolution of the genetic code. Results obtained by Brack and Orgel (1975) with synthetic polydipeptides supported the suggestion that such polypeptides, in which hydrophobic and hydrophilic amino acids alternate, are likely to form β-sheets with one hydrophobic and one hydrophilic surface. Such sheets would stack together to form "membrane-like" aggregates or micelles. Rossmann et al. (1975) have directed attention to possible evolutionary pathways linking proteins with various kinds of β-pleated sheet structure patterns. Recently, we (Baltscheffsky et al., 1978) discussed the possibility that a long evolutionary continuity of β-pleated sheet structures in proteins may be traced. This was exemplified with proteins involved in biological electron transport chains.

The three-dimensional picture of low molecular weight bacterial 8(Fe + S)-ferredoxin did not show any such β-structures which would fit it into a β-structure evolution pathway for electron carriers (Adman et al., 1973). On the other hand, we have been able to show with secondary structure prediction methods that a high probability for β-structure formation exists in the bacterial ferredoxins in question (von Heijne et al., 1978). It thus appears that the requirement in these ferredoxin apoproteins to bind, in the inner parts of what is to become the final ferredoxin structure, two distorted cube-type 4(Fe + S) clusters prevents the polypeptide from realizing, in its final iron-sulfur-containing form, a high potential of the apoprotein for β-pleated sheet formation.

The results from our calculations showing that antiparallel β-pleated sheet structures are to be expected in bacterial ferredoxin apoprotein-type polypeptides are summarized in Figure 2. In b and c are shown the two alternatives proposed for ferredoxin apoprotein, which may be compared with the ferredoxin apoprotein half-unit in a. As was discussed in detail (von Heijne et al., 1978) the proposed temporary structure b relates well to what would be expected to give 4(Fe + S)-binding to ferredoxin, whereas the temporary structure c would be useful as a link to current rubredoxins, as shown in d and e. Such an evolutionary link would agree with the evidence obtained by Vogel et al. (1977) for phylogenetic relationships between bacterial ferredoxins and rubredoxins.

These findings were encouraging from both the electron carrier evolution and the β-structure pathway evolution points of view. In addition,

8

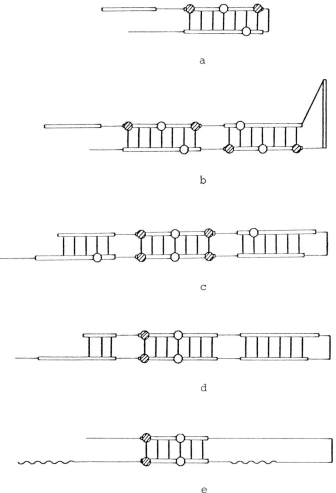

Fig. 2a-e. Proposed temporary apoprotein structures. Stretches of extended structure are represented by *heavy bars*, helices by *wavy lines* and hydrogen bonds by *thin lines*. *Filled* and *unfilled circles* represent cysresidues with sidegroups projecting out from different sides of the sheet. The amino-end of the protein is always at the *upper left-hand corner*. (a) ferredoxin halfunit; (b) ferredoxin; (c) alternative, rubredoxin-like structure; (d) rubredoxin, based on (c); (e) rubredoxin, based on secondary structure predictions for contemporary rubredoxins

a

b

c

d

e

they indicated that concerted hydrogen bond pattern changes, such as those proposed in the disappearance of antiparallel β-structures when ferredoxin was formed from apoprotein, iron and sulfur, may be of a more general relevance in connection with protein function. Before discussing this possibility in more detail I will, however, turn to the question of how the cellular electron transport coupled phosphorylation reactions may have originated and evolved.

PPi and ATP in the Evolution of Coupled Phosphorylation

The first alternative biological electron transport phosphorylation pathway was found in chromatophores from the photosynthetic bacterium *Rhodospirillum rubrum* with the discovery of PPi formation at the expense of light energy (Baltscheffsky et al., 1966). A difference between light-induced PPi formation and ATP formation was the lack of sensitivity of the former to the energy transfer inhibitor oligomycin (Baltscheffsky and von Stedingk, 1966).

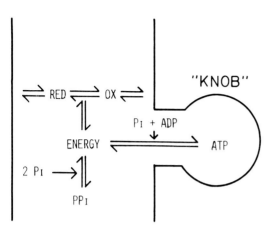

Fig. 3. Simplified scheme for the gross location of oxidation-reduction, PPi formation and ATP formation reactions in the chromatophore membrane. No attempt is made to depict all the vectorial characteristics of the system

In contrast to the coupling factor ATPase, the corresponding coupling factor PPase has been very difficult to solubilize from the chromato-phore membrane. As was shown by M. Baltscheffsky (1977), after sonic treatment practically all ATPase activity was solubilized, whereas the PPase activity remained in the membrane fraction. Very recently, how-ever, M. Baltscheffsky and Randahl (1978) have been able to solubilize PPase from chromatophores with retained activity. This appears to open up the possibility of characterizing PPase in similar molecular detail as the ATPase, which should facilitate a look from a new angle with this alternative coupling system into minimum requirements for energy coupling between electron transport and phosphorylation.

From an evolutionary point of view the reactions with PPi in the chro-matophore membrane seem to be of particular significance (Baltscheffsky, 1972). PPi may well have been a very early energy-rich phosphate com-pound, which when adenine nucleotides became available in sufficient concentrations became substituted by the more complicated and also more versatile ATP molecule. As Figure 3 shows, the Pi-PPi system may, both functionally and structurally, be more closely connected to the basic membrane containing the electron transport system. The coupling factor ATPase "knobs" performing the last step in electron transport coupled formation of ATP may be later additions to a primitive energy-coupling membrane functioning with the Pi-PPi system. One may predict that a more primitive Pi-PPi system means a simpler coupling factor PPase structure than the very complex multi-subunit coupling factor ATPase, which adds to the urgency of investigating in more detail the coupling factor PPase from chromatophore membranes.

Evolution and Mechanism in Combination

Overlaps Between Evolution, Nitrogen Fixation and Electron Transport Phosphorylation

In the last few years we have been actively interested in three dif-ferent but in part overlapping areas of biochemistry: molecular evolu-tion, nitrogen fixation, and electron transport phosphorylation. Our intention has been to use information from one area for elucidating problems in what may be regarded as the overlapping part of another. In Figure 4 an attempt is made to describe this using three area

THREE OVERLAPPING AREAS IN BIOCHEMISTRY

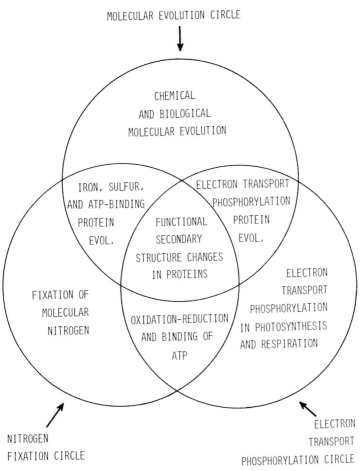

Fig. 4. Molecular evolution, nitrogen fixation, and electron transport phosphorylation as three overlapping areas in biochemistry

circles, with some common denominators given in the three regions where two area circles overlap and a possibly fundamental property of more general relevance briefly presented in the central region of overlap between all three circles.

A major challenge in this endeavor is to try to obtain by the evolutionary approach more insight into the molecular mechanisms of oxidation-reduction and phosphorylation or dephosphorylation, especially in the coupled reactions of photophosphorylation and oxidative phosphorylation. The way toward this outstanding functional problem has included the above-mentioned investigations with ferredoxin and the proposed β-structure evolution pathway in proteins.

It now seems appropriate to enter, with a new consideration on increasing evolutionary advantage, (Fig. 5), into protein function, both in general and in electron transport phosphorylation. This new

```
                          PROTEIN

 PROTEIN        PROTEIN      SECONDARY

 PRIMARY ———→ SECONDARY ———→ STRUCTURE

 STRUCTURE     STRUCTURE     FUNCTIONAL

                            CHANGE
```

Fig. 5. Increasing evolutionary advantage of some protein structural characteristics

consideration may be described as an addition to the pathway accounting for the well-known fact that protein secondary structure tends
to be better conserved during evolution than primary structure. But
if, as we have claimed, secondary structures may be regarded as "growing points" in molecular evolution (Baltscheffsky et al., 1978) an
even more pronounced "growing point" situation may exist whenever a
protein secondary structure is capable of undergoing a functional
change. This is accounted for in Figure 5, and provides a background
for our discussion of functional concerted changes in the hydrogen
bond pattern of protein secondary structures.

Functional Changes of Protein Secondary Structures

General Aspects

Our current picture of the formation of the bacterial ferredoxins
with two $4(Fe + S)$-cubes is that β-structures exist in the apoprotein
and are useful for the binding of the cysteine residues of the apoprotein to the iron in the resulting ferredoxin holoproteins. As these
predicted β-structures are not present in the final ferredoxin structure, and thus are assumed to be disappearing in the transition from
apoprotein to ferredoxin, this may be taken as an example of a functional change of a protein secondary structure. Another example of
such a change is the recent demonstration by Warrant and Kim (1978)
that protamine on binding to t-RNA changes its conformation from a
random coil to a structure containing α-helices. This further increased our attention toward the possibility that different types of
protein function may involve a concerted change in the hydrogen bond
pattern of an α- or β-structure, in such a way that this change is not
just linked to the actual mechanism, but is itself an essential and
fundamental part of the mechanism.

We may now recall the central area of Figure 4, where segments from
all the three circles overlap, and find in Figure 6 in more detail
the general fields for possible functional secondary structure changes
in proteins which will be considered.

The overlapping region of molecular evolution, nitrogen fixation, and
electron transport phosphorylation tentatively enlists four such fields
(Fig. 6): ferredoxin formation, coupled oxidation-reduction, nucleotide
binding and phosphorylation or dephosphorylation. Ferredoxin formation
was already discussed, and of the three remaining fields I would like
to turn first to nucleotide binding.

Just before leaving Stockholm for this meeting I telephoned Professor
C.-I. Bränden from the Agricultural University in Uppsala and asked
for his reaction to my attempt to generalize the secondary structural
change concept from ferredoxin formation to other functions, including

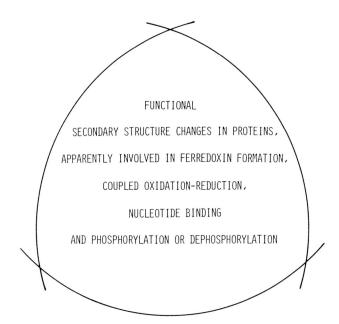

Fig. 6. Overlapping region of molecular evolution, nitrogen fixation and electron transport phosphorylation

nucleotide binding, which for example is linked to conformational changes of both nitrogenase (Zumft et al., 1974) and coupling factor ATPase (Ryrie and Jagendorff, 1972) proteins. He kindly mailed me four pages from his lecture in December 1977 at the Ciba Foundation Symposium on Molecular Interactions and Activity in Proteins (Brändén and Eklund, 1978). It was found that residues 46-55 of liver alcohol dehydrogenase apoenzyme are arranged in an α-helix and that in the NAD-containing holoenzyme structure the last residues of the helix are unwrapped and are part of the loop region. Thus Brändén and Eklund have already shown that nucleotide binding may indeed involve a secondary structural change. In this connection should be mentioned the suggestion by Eilen and Krakow (1977) that the cyclic AMP receptor protein of *E. coli* in the presence of cAMP undergoes a conformational change in which at least part of the DNA binding domain assumes a β-sheet conformation, as well as the observed decrease in α-helical content and increase in the β-form content of Bacteriophage T4 tail sheath protein during sheath contraction (Venyaminov et al., 1975). The later protein secondary structure change may be an essential element of the contraction process of this very simple contractile system.

Electron Transport Phosphorylation

The two remaining fields of the four listed above are coupled oxidation-reduction and phosphorylation or dephosphorylation. As they represent the first and the final, more or less directly interlinked parts of the electron transport phosphorylation mechanism, which is one of the focal points at this Colloquium, I shall be concerned mainly with the possible role of functional secondary structure changes in this system. However, it should be pointed out that the iron-protein of nitrogenase should not be entirely left out of the picture as (1)

ATP binds to this protein (Bui and Mortenson, 1968) and is hydrolyzed to ADP and Pi in the nitrogenase reaction; (2) in the presence of 2 mM ATP and 2 mM $MgCl_2$ the oxidation reduction potential of the iron-protein from *Clostridium* is shifted from -297 mV to -400 mV, indicating a conformational change in the protein (Zumft et al., 1974); and (3) the complete amino acid sequence of a nitrogenase iron-protein is known (Tanaka et al., 1977).

Let us now return to energy coupling. Already fifteen years ago it was pointed out by Lumry (1963) that the free energies for phosphorylation "could easily be accommodated by minor distortion of proteins without any major refolding process". In an attempt to give more substance at the molecular level to the protein conformational change hypothesis in coupled phosphorylation I recently elaborated upon this point including information about nucleotide binding regions of proteins and known effects of adenine nucleotides on the conformational changes of coupling factor ATPases (Baltscheffsky, 1977). Here I can only refer to some recent publications on the current hypotheses for the coupling mechanism in oxidative phosphorylation and photophosphorylation (Boyer et al., 1977; 1978) which may serve as a broader background to a new concept for the coupling mechanism in electron transport phosphorylation. As this new concept is based on specific functional changes in polypeptide secondary structures of the proteins involved in the redox and phosphorylation reactions, we may for the sake of clarity tentatively call it "the secondary structural change hypothesis".

Before formulating its essential aspects in detail, it will be necessary to discuss briefly some of its elements.

Two alternatives exist with respect to the energy balance when a concerted hydrogen bond change occurs in a protein.

If the breakage of a hydrogen bond is linked to a release of about 1.5 kcal/mol, a concerted change in the hydrogen bond pattern of a secondary structure may lead to appreciable liberation, or consumption, of energy. Two extreme situations exist in connection with the concerted formation or breakage of such secondary structures. One alternative is that no or practically no other molecular change will occur which would compensate for the energetic effect of the hydrogen bond pattern change. In this situation a concerted formation of, say, 10 H-bonds would result in the conservation of about 15 kcal/mol. The other alternative is energy compensation by other changes in the protein molecule, for example by changes in the hydrophobic force pattern. Such energy compensation may be partial or complete, and there could even be either overcompensation, or reinforcement of the energetic effect obtained by the concerted H-bond pattern change.

The former alternative, with energetically uncompensated H-bond pattern change, may lead to energy conservation in coupled phosphorylation. The requirement for this is, as is shown in Figure 7, that energy liberated during oxidation-reduction induces concerted formation of H-bonds. This may occur as a change from a random coil into an α- or β-structure, for example as a prolongation of the hydrogen bond pattern between two strands of a β-pleated sheet. Subsequent breaking of this structure with concomitant formation of a corresponding secondary structure in a coupling factor polypeptide would set the stage for the energy-conserving phosphorylation reaction. In Figure 7 this possibility has been described, with the contact between oxidation-reduction reaction and phosphorylation reaction visualized as either direct or involving a proton pump mechanism. The energy transfer as schematized

14

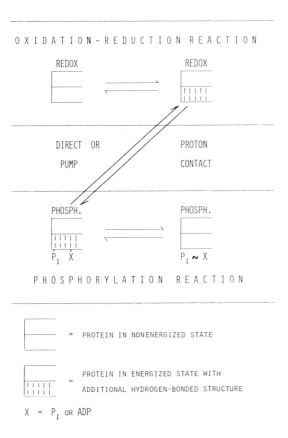

OXIDATION-REDUCTION REACTION

REDOX   REDOX

DIRECT OR   PROTON

PUMP   CONTACT

PHOSPH.   PHOSPH.

P_I X $P_I \sim X$

PHOSPHORYLATION REACTION

⊢ = PROTEIN IN NONENERGIZED STATE

⊢ = PROTEIN IN ENERGIZED STATE WITH
     ADDITIONAL HYDROGEN-BONDED STRUCTURE

X = P_I OR ADP

Fig. 7. Scheme for energy coupling according to the secondary structural change hypothesis. The contact between oxidation-reduction and phosphorylation may be either direct or over a proton pump

for the phosphorylation reaction, when a pyrophosphate bond and free ATP (or PPi) are formed at the expense of energy liberated during concerted breakage of a hydrogen bond pattern, does not differentiate between whether the energy requirement is for pyrophosphate bond formation or for liberation of the phosphorylated product from the coupling factor protein.

In this connection it may be noted that (1) there are well-known similarities in protein secondary structure between nucleotide-binding dehydrogenases and kinases, specially in the nucleotide-binding region (Rossmann et al., 1975); (2) the coenzyme fragments AMP and ADP can induce a conformational change in lactate dehydrogenase in the same way as do NAD^+ and NADH, and the ADP difference electron density map shows two partially occupied sites for the second phosphate, when ADP is bound to the enzyme (Chandrasekhar et al., 1973); and (3) beef heart mitochondrial ATPase has recently been shown to bind Pi to a specific site, apparently located in or near an adenine nucleotide site and equivalent to the γ-phosphate position of ATP with data supporting the conclusion that the Pi binding site is the site to which Pi binds for ATP synthesis in oxidative phosphorylation (Kasahara and Penefsky, 1977).

The secondary structural change hypothesis, as applied on the mechanism of electron transport phosphorylation in photosynthesis and respiration, may now be presented:

1) Energy is captured for conservation in electron transport chains by reversible formation of structures containing several hydrogen bonds (α- or β-structures: formation of X H-bonds requires approximately X x 1.5 kcal/mol = Y kcal/mol).

2) Coupling to energy-rich phosphate formation occurs directly or over an intermediary intra- or inter-membrane charge separation (possibly mediated by concerted changes in hydrogen bond patterns).

3) Energy is released for the formation of the free energy-rich phosphate compound by a reversible breaking of several hydrogen bonds on the coupling factor enzyme.

An attempt to visualize the generally accepted link between proton pump and energy coupling in electron transport phosphorylation systems, on the basis of the secondary structural change hypothesis, is shown in Figure 8. Here the cell membrane of *R. rubrum*, with formation of both ATP and PPi at the expense of light energy, is taken as an example. The proton pump mechanism may, in part, or in essence (Nagle and Morowitz, 1978), be based on a fundamental structural element of continuous chains of hydrogen bonds. Such a hydrogen bond pattern is indicated in the proton channels in Figure 8. The direct contact alternative is given for the coupling between redox and phosphorylation proteins, which are schematized as in Figure 8. As in Mitchell's chemiosmotic hypothesis, a recent version of which should be referred to (Mitchell, 1978), a protonmotive mechanism through the membrane, or as in Williams' membrane charge separation model a localized charge separation, may well be essential for coupling to occur by the secondary structural change variety of the general conformational change mechanism. But one may as well consider these phenomena as such to be only intimately related to, and *in part the same* as the hydrogen bond pattern changes of proteins directly participating in, the energy coupling, as is also indicated in Figure 8. Thus, energy coupling by a specific secondary structural change mechanism may or may not need the energy-linked chemiosmotic or membrane charge separation phenomena, at least with respect to their full extent.

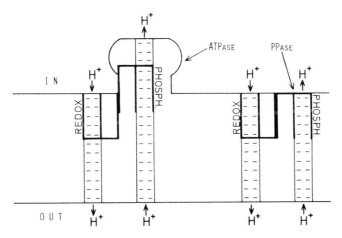

Fig. 8. Possible connection between proton pump and energy coupling according to the secondary structural change mechanism. The *Rhodospirillum rubrum* cell membrane with its differently located ATPase and PPase coupling factor systems is used as model. The structures of the proteins involved in a direct coupling reaction between the redox and phosphorylation reactions are given as in Figure 7

Possible Fundamental Similarity Between Biological Information Transfer and Energy Transfer

Low molecular weight oligopeptides, such as some of those involved in ion transport, notably valinomycin, or the enkephalins, seem to exert a remarkable ability to exist in various three-dimensional structures characterized by different hydrogen bond patterns. Recent evidence, as has been discussed above, indicates that concerted hydrogen bond pattern changes in proteins of high molecular weight occur and may be of great functional importance. Based on such data, a secondary structural change hypothesis has been proposed for various aspects of protein function.

Application of the secondary structural change hypothesis on the problem of energy coupling and conservation has led to a simple and detailed tentative formulation of the mechanisms involved. The hypothesis is presented in the hope of getting some additional light through what Mitchell (1978) calls "the conformational 'black box'".

Furthermore, the secondary structural change hypothesis points toward a general similarity, at the fundamental level of molecular function, between essential aspects of biological information transfer exerted by the polynucleotides, and biological energy transfer exerted by the proteins involved in photosynthetic and respiratory oxidation-reduction and phosphorylation reactions: a reversible concerted hydrogen bond pattern formation.

Finally, not only in the information transfer systems, but also in the energy-transfer systems, such functional hydrogen bond pattern changes in the three-dimensional structures under consideration would accordingly *not just be linked* to the fundamental mechanisms but *rather constitute the very mechanisms themselves*.

Acknowledgment. This work was supported by Grant no K2292-035 from the Swedish Natural Science Research Council.

Note added in proof: A new automatic identification of secondary structure in globular proteins, which generally gives extra residues that are missed from the earlier reported assignments, in particular with respect to β-sheet residues, assigns to *Peptococcus aerogenes* ferredoxin 4 short β-strands and 2 α-helical regions (Levitt, M., Greer, J.: J. Mol. Biol. 114, 184-239 (1977)). This may be compared with our predicted β-structure pattern in bacterial ferredoxin apoproteins and may be taken to strengthen our evolutionary argument to the extent that it signals a pathway in ferredoxin apoprotein evolution from all β- to mixed α- and β- secondary structures. Furthermore, the result exemplifies that the sharper analytical resolution obtained may also contribute to a closer understanding of both evolutionary and functional hydrogen bond patterns in proteins.

Very recent evidence indicates that the "Archaebacteria" encompass several distinct subgroups including methanogens, extreme halophiles, and various thermoacidophiles (Woese, C.R., Magrum, L.J., Fox, G.E.: J. Mol. Evol. 11, 245-252 (1978)).

References

Adman, E.G., Sieker, L.C., Jensen, L.H.: The structure of a bacterial ferredoxin. J. Biol. Chem. 248, 3987-3996 (1973)

Baltscheffsky, H.: Inorganic pyrophosphate and the origin and evolution of biological energy transformation. In: Molecular evolution, Vol. 1. Buvet, R., Ponnamperuma, C. (eds.) pp. 466-474. Amsterdam-London: North-Holland 1972

Baltscheffsky, H.: A new hypothesis for the evolution of biological electron transport. Origins of Life 5, 387-395 (1974a)

Baltscheffsky, H.: On the evolution of electron transport in photosynthesis and respiration. In: BBA Library 13, Dynamics of energy-transducing membranes. Ernster, L., Estabrook, R.W., Slater, E.C. (eds.) pp. 21-27. Amsterdam-London-New York: Elsevier 1974b

Baltscheffsky, H.: Protein β-structure and the molecular evolution of biological energy conversion. In: Living systems as energy converters. Buvet, R., Allen, M.J., Massué, J.-P. (eds.) pp. 81-88. Amsterdam: Elsevier/North Holland 1977

Baltscheffsky, H., von Heijne, G., Blomberg, C.: Protein secondary structures and the molecular evolution of biological electron transport. In: Evolution of protein molecules. Matsubara, H., Yamanaka, T. (eds.). Tokyo: Japan University Press, in press 1978

Baltscheffsky, H., von Stedingk, L.-V.: Bacterial photophosphorylation in the absence of added nucleotide. A second intermediate stage of energy transfer in light-induced formation of ATP. Biochem. Biophys. Res. Commun. 22, 722-728 (1966)

Baltscheffsky, H., von Stedingk, L.-V., Heldt, H.-W., Klingenberg, M.: Inorganic pyrophosphate: formation in bacterial photophosphorylation. Science 153, 1120-1121 (1966)

Baltscheffsky, M.: Conversion of solar energy into energy-rich phosphate compounds. In: Living Systems as Energy Converters. Buvet, R., Allen, M.J., Massué, J.-P. (eds.), pp. 199-207. Amsterdam: Elsevier/North Holland 1977

Baltscheffsky, M., Randahl, H.: (1978) manuscript in preparation

Boyer, P.D., Chance, B., Ernster, L., Mitchell, P., Racker, E., Slater, E.G.: Oxidative phosphorylation and photophosphorylation. Ann. Rev. Biochem. 46, 955-1026 (1977)

Boyer, P.D., Hackney, D.D., Choate, G.L., Janson, C.: Relations of protein conformation to enzyme catalysis and to proton and calcium transport. In: The proton and calcium pumps. Azzone, G.F., Avron, M., Metcalfe, J.C., Quagliariello, E., Siliprandi, N. (eds.), pp. 17-28. Amsterdam: Elsevier/North-Holland 1978

Brack, A., Orgel, L.E.: β-structures of alternating polypeptides and their possible prebiotic significance. Nature (London) 256, 383-387 (1975)

Brändén, C.-I., Eklund, H.: Coenzyme induced conformational changes and substrate binding in liver alcohol dehydrogenase. CIBA Foundation monograph No. 60 in press (1978)

Bui, P.T., Mortenson, L.E.: Mechanism of the enzymic reduction of N_2: The binding of adenosine-5'-triphosphate and cyanide to the N_2-reducint system. Proc. Natl. Acad. Sci. U.S.A. 61, 1021-1027 (1968)

Carter, Jr., C.W., Kraut, J.: A proposed model for interaction of polypeptides with RNA. Proc. Natl. Acad. Sci. U.S.A. 71, 283-287 (1974)

Chandrasekhar, K., McPherson, Jr., A, Adams, M.J., Rossmann, M.G.: Conformation of coenzyme fragments when bound to lactate dehydrogenase. J. Mol. Biol. 76, 503-518 (1973)

Demerec, M.: In: Enzymes: units of biological structure and function. Gaebler, O.H. (ed.), pp. 131-134. New York: Academic Press 1956

Dickerson, R.E., Timkovich, R.: Cytochromes c. In: The enzymes. Boyer, P. (ed.), Vol. XII, pp. 397-547. New York-San Francisco-London: Academic Press 1975

Eilen, E., Krakow, J.S.: Cyclic AMP-mediated intersubunit disulfide crosslinking of the cyclic AMP receptor protein of *Escherichia coli*. J. Mol. Biol. 114, 47-60 (1977)

Fox, G.E., Magrum, L.J., Balch, W.E., Wolfe, R.S., Woese, C.R.: Classification of methanogenic bacteria by 16S ribosomal RNA characterization. Proc. Natl. Acad. Sci. U.S.A. 74, 4537-4541 (1977)

Hall, D.O., Cammack, R., Rao, K.K.: The iron-sulphur proteins: evolution of a ubiq-
uitous protein from model systems to higher organisms. In: Cosmochemical evolution
and the origins of life. Oró, J., Miller, S.L., Ponnamperuma, C., Young, R.S. (eds.)
Vol. I, pp. 363-386. Dordrecht-Boston: Reidel 1974

Heijne, G.von, Blomberg, C., Baltscheffsky, H.: Early evolution of cellular electron transport: molecular models for the ferredoxin-rubredoxin-flavodoxin region.
Origins of Life 9(1978)

Horowitz, N.H.: On the evolution of biochemical syntheses. Proc. Natl. Acad. Sci.
U.S.A. 31, 153-157 (1945)

Horowitz, N.H.: The evolution of biochemical syntheses - retrospect and prospect.
In: Evolving genes and proteins. Bryson, V., Vogel, H.J. (eds.), pp. 15-23.
New York-London: Academic Press 1965

Kasahara, M., Penefsky, H.S.: Specific binding of Pi by beef heart mitochondrial
ATPase. In: BBA Library 14, structure and function of energy-transducing membranes.
van Dam, K., van Gelder, B.F. (eds.), pp. 295-305. Amsterdam-London-New York:
Elsevier 1977

Lumry, R.: Structure-function relationships in proteins and their possible bearing
on the photosynthetic process. In: Photosynthetic mechanisms of green plants, pp.
625-634. NAS-NRC Publication 1145, 1963

Mitchell, P.: Protonmotive chemiosmotic mechanisms in oxidative and photosynthetic
phosphorylation. TIBS 3, N58-N61 (1978)

Nagle, J.F., Morowitz, H.J.: Molecular mechanisms for proton transport in membranes.
Proc. Natl. Acad. Sci. U.S.A. 75, 298-302 (1978)

Orgel, L.E.: Evolution of the genetic apparatus. J. Mol. Biol. 38, 381-393 (1968)

Orgel, L.E.: A possible step in the origin of the genetic code. Isr. J. Chem. 10,
287-292 (1972)

Pette, D., Luh, W., Bücher, Th.: A constant-proportion group in the enzyme activity
pattern of the Embden-Meyerhof chain. Biochem. Biophys. Res. Commun. 7, 419-424
(1962)

Rossmann, M.G., Liljas, A., Brändén, C.-I., Banaszak, L.J.: Evolutionary and structural relationships among dehydrogenases. In: The enzymes. Boyer, P. (ed.) Vol. XI.
pp. 61-102. New York-San Francisco-London: Academic Press 1975

Ryrie, I.J., Jagendorf, A.T.: Correlation between a conformational change in the
coupling factor protein and the high energy state in chloroplasts. J. Biol. Chem.
247, 4453-4459 (1972)

Tanaka, M., Haniu, M., Yasunobu, K.T., Mortenson, L.E.: The amino acid sequence of
Clostridium pasteurianum iron protein, a component of nitrogenase. J. Biol. Chem.
252, 7093-7100 (1977)

Venyaminov, S.Y., Rodikova, L.P., Metlina, A.L., Poglazov, B.F.: Secondary structure
change of Bacteriophage T4 sheath protein during sheath contraction. J. Mol. Biol.
98, 657-664 (1975)

Vogel, H., Bruschi, M., Le Gall, J.: Phylogenetic studies of two subredoxins from
sulfate reducing bacteria. J. Mol. Evol. 9, 111-119 (1977)

Warrant, R.W., Kim, S.-H.: α-Helix-double helix interaction shown in the structure
of a protamine-transfer RNA complex and a nucleoprotamine model. Nature (London)
271, 130-135 (1978)

Woese, C.R., Fox, G.E.: Phylogenetic structure of the prokaryotic domain: the primary kingdoms. Proc. Natl. Acad. Sci. U.S.A. 74, 5088-5090 (1977)

Zumft, W.G., Mortenson, L.E., Palmer, G.: Electron-paramagnetic-resonance studies
on nitrogenase. Eur. J. Biochem. 46, 525-535 (1974)

Organization of the Mitochondrial Respiratory Chain[*]

Y. HATEFI and Y. M. GALANTE

In mitochondria, the machinery for oxidative phosphorylation is contained in the inner membrane, which is composed of approximately 70%
protein and 30% lipid. Systematic fractionation of inner membrane preparations from bovine heart mitochondria showed in 1961 that the mitochondrial electron transport system could be fragmented into four
protein-lipid enzyme systems with the following catalytic properties:

    I. NADH:ubiquinone oxidoreductase
    II. Succinate:ubiquinone oxidoreductase
    III. Reduced ubiquinone:ferricytochrome *c* oxidoreductase
    IV. Ferrocytochrome *c*:oxygen oxidoreductase

The relative stability of these fragments and the constant ratio of
their components, which was found to be essential for enzymic activity, suggested that the above preparations are structural units of
function, and were designated enzyme complexes I, II, III, and IV,
respectively (1-4). It was also shown in these early studies that

Table 1. Components of complexes I, II, III, and IV

Complex	Component	Concentration (per mg protein)
I. NADH-Q reductase	FMN	1.4-1.5 nmol
	nonheme iron	23-26 ng atom
	labile sulfide	23-26 nmol
	ubiquinone	4.2-4.5 nmol
	lipids	0.22 mg
II. Succinate-Q reductase	FAD	4.6-5.0 nmol
	nonheme iron	36-38 ng atom
	labile sulfide	32-38 nmol
	cytochrome *b*	4.5-4.8 nmol
	lipids	0.2 mg
III. QH$_2$-cytochrome *c* reductase	cytochrome *b*	8.0-8.5 nmol
	cytochrome *c*$_1$	4.0-4.2 nmol
	nonheme iron	10-12 ng atom
	labile sulfide	6-8 nmol
	ubiquinone	> 2 nmol
	lipids	0.4 nmol
IV. Cytochrome *c* oxidase	cytochromes *aa*$_3$	10.5-10.9[a] nmol
	copper	10-12[a] ng atom
	lipids	0.3-0.4 mg

[a] Corrected for biuret protein overestimate of 25%.
From: Hatefi, Y., Galante, Y.M., Stiggall, D.L., Djavadi-Ohaniance,L.:
In: The Structural Basis of Membrane Function: Hatefi, Y., Djavadi-
Ohaniance, L. (eds.). New York: Academic Press, 1976, pp. 169-188

[*] The work of this laboratory was supported by grants USPHS AM 08126 and NSF PCM76-
01378 A01 to Y.H.

Table 2. Molecular weights, composition, and relative ratios of complexes I, II, III, IV, and V

Complex	Mol. Wt. x 10^6 (Min., protein only)	Polypeptides	Constituent proteins purified	Ratio in inner membrane
I	≤ 0.7	16-18	NAD(P)H dehydrogenase, Center 2 FeS protein	1
II	0.2	4	Succinate dehydrogenase	2
III	0.25	8-9	Cyt. c_1, FeS protein	3
IV	0.15-0.2	7	--------	7
V	≥ 0.5	11	F_1, OSCP, DCCD-binding protein, F_6, Factor B	3

complexes I, II, III, and IV can be recombined (where necessary, in the presence of cytochrome c, which separates during fractionation) to reconstitute the entire electron transport system, or segments thereof, with the expected overall activities and inhibitor-response properties (1-4). Data on the composition of complexes I-IV, their molecular weights, and ratio in the mitochondrial inner membrane, as well as on reconstituted electron transport systems are compiled in Tables 1, 2 and 3. More recently, a fifth enzyme complex (complex V) was isolated whose function in mitochondria is ATP synthesis and hydrolysis. In the isolated state, complex V catalyzes oligomycin-sensitive ATP hydrolysis and ATP-^{33}Pi exchange (5). The exchange reaction, which is also sensitive to uncouplers, is indicative of the energy-conserving property of complex V. As shown in Figure 1, complexes I-V can be isolated from the same batch of mitochondria or inner membrane preparations.

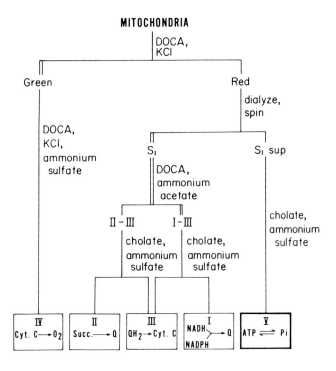

Fig. 1. Scheme for resolution of the mitochondrial inner membrane into complexes I, II, III, IV, and V. *DOCA*, deoxycholate; Q and QH_2, oxidized and reduced ubiquinone. From: Hatefi, Y., Hanstein, W.G., Galante, Y., Stiggall, D.L.: Fed. Proc. 34, 1699-1706 (1975)

Table 3. Reconstituted electron transport systems and their activities

System	Reaction	Activity[a]	Activity per mole flavin (FMN or FAD)[b]	Specific inhibitors
I + III	NADH → cyt c	28.7[c]	22,700	Rotenone, piericidin A, antimycin A
II + III + Q$_2$	Succ → cyt c	24.2[d]	6,000	TTFA[h], antimycin A
I + II + III	NADH → cyt c	21.5[c]	17,000	Rotenone, piericidin A, antimycin A
I + II + III	Succ → cyt c	28.3[d]	7,000	TTFA, antimycin A
(I-III) + IV + cyt c	NADH → O$_2$	30.6[d]	7,600	Rotenone, piericidin A, antimycin A, cyanide, azide
I + II + III + IV + cyt c	NADH → O$_2$	7.3[e]	5,200	Rotenone, piericidin A, antimycin A, cyanide, azide
I + II + III + IV + cyt c	Succ → O$_2$	14.4[e]	7,000	TTFA, antimycin A, cyanide, azide
ETP + cyt c	NADH → O$_2$	2.0[f]	20,000[g]	Rotenone, piericidin A, antimycin A, cyanide, azide
ETP + cyt c	Succ → O$_2$	1.5[f]	7,500[g]	TTFA, antimycin A, cyanide, azide

[a] Micromoles NADH or succinate oxidized per min per mg of complex I(c), complex II(d), or complex IV(e) at 38°. Data calculated from: Hatefi, Y., Haavik, A.G., Fowler, L.R., Griffiths, D.E.: J. Biol. Chem. 237, 2661-2669 (1962).

[b] Moles NADH or succinate oxidized per min per mole of FMN (acid-extractable) or FAD (acid-nonextractable), respectively.

c-e see Footnote a.

[f] Micromoles NADH or succinate oxidized per min per mg submitochondrial particle (ETP) protein at 30°C.

[g] Values used for NADH dehydrogenase and succinate dehydrogenase flavins were, respectively, 0.1 and 0.2 nmol/mg of ETP protein.

[h] 2-Thenoyltrifluoroacetone.

In mitochondria, the segments corresponding to complexes I, III, IV, and V are capable of energy transduction. In the isolated state also these enzyme complexes have been shown to be energy-transducing systems by virtue of their ability, when embedded in phospholipid vesicles, to catalyze transmembrane proton translocation coupled to substrate oxidation (in the case of complexes I, III, and IV) or ATP hydrolysis (in the case of complex V or similar preparations from other sources) (6-11). In addition, as expected from their proton translocation property, there is evidence that the four energy-transducing complexes span the mitochondrial inner membrane (12-14). It is now clear from the fractionation, genetic, and biosynthetic studies of others than mitochondria from various species and organs also contain units of function analogous to the five enzyme complexes of bovine heart mitochondria. Therefore, it has become generally accepted now that for electron transport and oxidative phosphorylation in eukaryotic systems the structural unit of energy transduction is the protein-lipid enzyme complex. The well-characterized electron transfer components of complexes I, II, III and IV are shown in Figure 2. These components are essentially of two types: quinones (flavoproteins and ubiquinone) and transition metal complexes (heme proteins, iron-sulfur proteins, copper proteins). Thiol groups are involved in the enzymic activities of the respiratory chain, but whether they and other protein moieties participate in the act of electron and proton transfer is not known.

Complex I. As seen in Figure 2, complex I contains FMN and 5 iron-sulfur centers. It catalyzes electron transfer from NADH to ubiquinone, and this reaction is inhibited by mercurials, rotenone, piericidin A, barbiturates, and Demerol. Complex I has been resolved with chaotropic ions, and the primary NADH dehydrogenase and an iron-sulfur protein fraction with electron paramagnetic resonance (epr) characteristics of iron-sulfur center 2 of complex I have been isolated. The NADH dehydrogenase has been purified recently in our laboratory. It is an

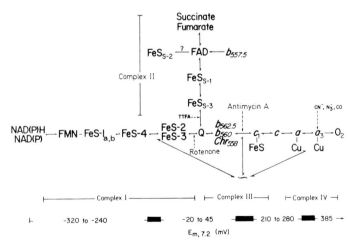

Fig. 2. Kinetic profile of the well-characterized components of the mitochondrial electron transport system. *FeS*, iron-sulfur center; *Chr*$_{558}$, chromophore-558; a, b, c, cytochromes a, b, and c; Q, ubiquinone; *TTFA*, 2-thenoyltrifluoroacetone. The *bottom line* shows the isopotential regions of the NADH oxidase pathway, and the regions of large E_m drop (*shaded*) which correspond to the coupling sites in complexes I, III, and IV shown by *long arrows* leading to $\sim$ (energized membrane)

iron-sulfur flavoprotein with a molecular weight of 75,000 ± 6%, three polypeptides of M_r = 51,000, 24,000, and ≤ 10,000, and a FMN:Fe:S^2- ratio of 1:5-6:5-6 per mol. The isolated dehydrogenase is water soluble and catalyzes the oxidation of NADH by quinones, ferric compounds, and NAD. Earlier studies (15) showed that the reduced dehydrogenase can be oxidized by the isolated iron-sulfur protein fraction, which in turn was oxidized by ubiquinone. Subsequent epr studies have confirmed and extended these early observations concerning the position of iron-sulfur center 2 relative to FMN and ubiquinone (16-18). Contrary to what was generally assumed, we have also shown recently that the mitochondrial respiratory chain *is* capable of oxidizing NADPH directly and without the intervention of the NADPH → NAD transhydrogenase reaction, and that direct NADPH dehydrogenation is catalyzed also by NADH dehydrogenase (19, 20).

Complex II. This enzyme complex catalyzes the oxidation of succinate by ubiquinone. This reaction is inhibited by thiol inhibitors, 2-thenoyl-trifluoroacetone, and the competitive inhibitors of succinate oxidation such as oxaloacetate and malonate (21). As isolated, complex II contains about 10% complex III impurity (22). Otherwise, it contains only four polypeptides that are present in roughly equimolar amounts (23, 24). Two of these polypeptides with M_r = 70,000 ± 7% and 27,000 ± 5% belong to succinate dehydrogenase. The other two polypeptides have molecular weights of about 13,000 and 15,000. One or both belong to a low potential cytochrome *b* which fractionates into complex II. This cytochrome, whose absorption spectrum is shown in Figure 3, does not appear to be reduced by succinate, but its chemically reduced form in complex II is rapidly oxidized by fumarate and less rapidly by ubiquinone-2 (22). We feel that this cytochrome, designated cytochrome $b_{557.5}$, might be an entry point for an electron tributary of the respiratory chain. Succinate dehydrogenase, which had been extensively studied for nearly three decades, was purified in our laboratory in 1971 (25, 26). It has a molecular weight of 97,000 ± 4%, and is composed of two subunits as stated above. The enzyme is an iron-sulfur flavoprotein containing one mole of FAD covalently bound to the larger subunit, 8 g-atoms of Fe, and 8 moles of acid-labile sulfide per mol. The two subunits have been separated in the presence of potent

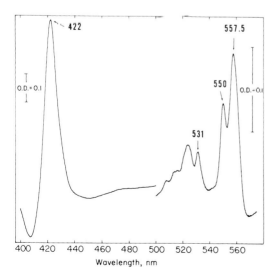

Fig. 3. Reduced-*minus*-oxidized difference spectrum of cytochrome $b_{557.5}$ of complex II at 77 K. From: Davis, K.A., Hatefi, Y., Poff, K.L., Butler, W.L.: Biochim. Biophys. Acta 325, 341-356 (1973)

chaotropic ions with retention of tertiary structure and their iron-sulfur chromophores. These studies have shown that iron and labile sulfide distribute equally between the two subunits. Thus, the large subunit has a flavin:Fe:S^{2-} ratio of 1:4:4, and the small subunit appears to be a soluble iron-sulfur protein containing 3-4 g-atoms of Fe and 3-4 mol of S^{2-} per mol. According to the epr studies of Ohnishi et al. (27), the iron and labile sulfide of the subunits of succinate dehydrogenase form three iron-sulfur centers, a 4Fe-4S center in the small subunit, and two 2Fe-2S centers in the large subunit. Recently, Yu et al. (28) isolated and partially purified a mitochondrial protein with a molecular weight of 15,000. They have shown that succinate-ubiquinone reductase activity can be reconstituted by combining this preparation with a partially purified ($\sim$ 30%) preparation of succinate dehydrogenase. They believe that the above protein is a specific ubiquinone-binding protein required for linking succinate dehydrogenase to cytochromes b and c_1 of complex III. For detailed reviews on complexes I and II and on NADH and succinate dehydrogenases the reader is referred to references 21, 29-31.

Complex III. Complex III catalyzes the oxidation of reduced ubiquinone by cytochrome c, and this reaction is specifically inhibited by antimycin A or 2-alkyl-4-hydroxyquinoline-N-oxide (32). These inhibitors interrupt electron flow from the b cytochromes to cytochromes c_1 and c (Fig. 2). Complex III is composed of eight (possibly nine) polypeptides, and has a minimum molecular weight of 230-250 x 10^3 (32-35). The known electron carriers of complex III are two b cytochromes, b_{566} and b_{562} (at 77 K the major α peaks are at 562.5 and 560 nm, respectively), cytochrome c_1 (absorption spectra shown in Fig. 4), an iron-

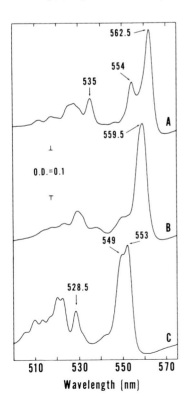

Fig. 4. Reduced-*minus*-oxidized difference spectra of cytochromes b_{566} *(trace A)*, b_{562} *(trace B)*, and c_1 *(trace C)* of complex III at 77 K. From: Davis, K.A., Hatefi, Y., Poff, K.L., Butler, W.L.: Biochim. Biophys. Acta 325, 341-356 (1973)

sulfur protein with a molecular weight of 26,000 and a 2Fe-2S core
(35), and a cytochrome b-like component designated chromophore-558
because of its reduced-*minus*-oxidized absorption peak at liquid nitrogen temperature at 558 nm (22). Among the components of complex III,
the iron-sulfur protein and cytochrome c_1 have been isolated from
beef heart (35,36), and Weiss, (37) has purified a cytochrome b of
Neurospora complex III. It is known that cytochrome c_1 is the electron
donor to cytochrome c, but the precise sequence of electron transfer
in complex III and the roles of the two b cytochromes, the iron-sulfur
protein, and chromophore-558 are not clear.

Interesting control devices are built into complex III. For example,
in the absence of energy or antimycin A, cytochrome b_{566} is not reducible by substrates, and it has been shown that this cytochrome
undergoes an apparent change in reduction potential (from -30 mV to
+245 mV) upon membrane energization (38)[1]. A similar effect is observed
upon treatment of particles with antimycin A. Our studies have shown
that even in the presence of antimycin A, reduced b_{566} is not detectable
when electron flux from the substrates to the ubiquinone-cytochrome b
region is slow. In addition, it has been shown that in antimycin A-
treated preparations, prereduction of cytochrome c_1 (or a hypothetical
component X with E_m = +140 mV) (39) inhibits the reduction of b_{566}
(40) and slows down the reduction of b_{562} (41). Chromophore-558 in
antimycin-treated preparations also responds similarly to b_{566} both
with respect to the rate of electron flux from substrates and the reduced state of cytochrome c_1. However, unlike b_{566} and similar to
b_{562}, chromophore-558 is reducible by substrates in the absence of
antimycin A (22). Thus, it appears that in mitochondria the b cytochromes and chromophore-558 are capable of modulating electron flow
from substrates to oxygen by responding to the rates of electron flow to
and from complex III (or to the E_h of their reductants and oxidants).
This is an interesting possibility because the b cytochromes are located in the respiratory chain at an important electron transfer junc-
ture. On the substrate side, the ubiquinone pool is the recipient of
electrons from several oxidative pathways, namely, those for the oxidation of NADH, NADPH, succinate, α-glycerol-3-phosphate, fatty acids,
and choline, while on the oxygen side, electron flow from cytochrome
c_1 to c, aa_3, and oxygen is extremely rapid. The next two speakers in
the program will discuss the properties of complex III and its components much more fully than time and space will allow for this presentation, and detailed reviews of complex III are available (35, 42).

Complex IV. For similar reasons, our discussion of complex IV (cytochrome c oxidase) will be limited, because the subject matter will no
doubt receive authoritative treatment by the last speaker of this session, Professor Chance. Cytochrome c oxidase has been isolated and
studied from various organisms. As seen in Table 4, the preparations
from beef heart, yeast, *Neurospora*, and locust are very similar in
polypeptide composition and in the molecular weights of the subunits.
However, whether the heart enzyme is composed of six or seven subunits
is not certain. In yeast, the three largest subunits are translated on
mitochondrial ribosomes, while the four smaller subunits are products
of the cytoribosomal protein synthesis. Eytan et al. (13) have shown
that in beef heart mitochondria the subunits with M_r values of 22.5,
9.8, and 7.3 X 10^3 are located on the outer side of the inner membrane,

---

[1] A discussion of the apparent redox midpotential changes of cytochrome b_{566} (b_T)
and other respiratory chain components as possible energy-transducing devices is
outside the scope of this article. For this purpose, the reader is referred to (38)
and (61-63)

Table 4. Polypeptides of cytochrome c oxidase from different sources[a]

Poly-peptide	Source Ref.	Beef Heart						Yeast	*Neuro-spora*	Locust
		(b)	(c)	(d)	(e)	(f)	(g)	(h)	(i)	(j)
1		33-38	40	40	40	37	47.5	37-41	36	38
2		19-25	22.5	25	21		20	28-34	28	24
3		19-25		19		19	14.5	22-24	20	19
4		14-16	15	14	15	14	14.5	14-16	14	14.5
5		9-12.5	11	10	13	10	13	12-14	13	12.5
6		6-8.5	10	8	12		11	10-13	11	10
7		3-5	7					4.5-5	8	8

[a]Numbers under columns 2 to 10 indicate M_r values in thousands rounded to the nearest 0.5 thousand.
[b]Capaldi, R.A., Bell, R.L., Branchek, T.: Biochem. Biophys. Res. Commun. 74, 425-433 (1977).
[c]Rubin, M.S., Tzagoloff, A.: J. Biol. Chem. 248, 4269-4274 (1973).
[d]Yamamoto, T., Orii, Y.: J. Biochem. (Tokyo) 75, 1081-1089 (1974).
[e]Kuboyama, M., Yong, F.C., King, T.E.: J. Biol. Chem. 247, 6375-6383 (1972).
[f]Kierns, J.J., Yang, C.S., Gilmour, M.V.: Biochem. Biophys. Res. Commun. 45, 835-841 (1971).
[g]Pahn, S.H., Mahler, H.R.: J. Biol. Chem. 251, 257-269 (1976).
[h]Poyton, R.O., Schatz, G.: J. Biol. Chem. 250, 752-761 (1975).
[i]Sebald, W., Weiss, H,. Jackl, G.: Eur. J. Biochem. 30, 413-417 (1972).
[j]Weiss, H., Lorenz, B., Kleinow, W.: FEBS Lett. 25, 49-51 (1972).

the subunit with an M_r value of 15 X 10^3 is situated on the matrix side, while those with M_r values of 40 and 11.2 X 10^3 are located between the others in the interior of the membrane. Birchmeier et al. (43) showed that yeast cytochrome c activated at its single sulfhydryl group with 5,5'-dithiobis(2-nitrobenzoate) binds to cytochrome oxidase and forms a mixed disulfide bridge with subunit 3 (M_r = 24,000). According to Briggs and Capaldi (44), when a 1:1 complex of cytochrome c and beef heart cytochrome oxidase is cross-linked with the 11-Å bridging bifunctional reagent dithiobissuccinimidyl propionate, cytochrome c binds to subunit 2 of the oxidase. The cross-linked cytochrome c-cytochrome oxidase subunit of Birchmeier et al. was reported to have a M_r of 38,000, and that of Briggs and Capaldi a M_r of 35,000-37,000. Birchmeier et al. cleaved their cross-linked complex and showed that the participating cytochrome oxidase component is indeed subunit 3. Therefore, the difference between the conclusions of the above workers might be (a) because subunit 3 of the yeast enzyme would correspond to subunit 2 of beef heart cytochrome oxidase if the latter has only 6 subunits, (b) due to proximity of subunits 2 and 3 and the thiol specificity of the cross-linking reagent of Birchmeier et al., or (c) real, and characteristic of the yeast and the beef heart enzymes. Kinetic studies have revealed the presence of two cytochrome c-binding sites on cytochrome oxidase (45, 46). The first, high-affinity binding site is considered to involve the hydrophobic surface of cytochrome c provided by residues 81 through 83 with adjacent cationic charges on lysines 13 and 72 (47). This area is on one side of the protein cleft and includes the upper half of the exposed heme edge through which electron transfer is considered to occur. The high-affinity binding site is considered to be the principal pathway for respiratory chain electron transfer, while the second site with lower affinity is thought to be involved mainly in intermembrane (and interchain) electron shuttle (47). A third binding site at very high cytochrome c concentration has also been detected, which, however, is considered to be of no physiologic significance (47). A copper-containing protein with a molecular weight of 22-25 X 10^3 has been isolated by various investigators from cytochrome oxidase (48, 49), and Yu et al. (50) reported the isolation of a heme a containing polypeptide with molecular weight of 11,600. A recent review on cytochrome c oxidase is available (51).

Complex V. As stated above this enzyme complex is involved in mitochondria in ATP synthesis, and in membrane energization by ATP hydrolysis. The activities of the isolated complex are oligomycin-sensitive ATP hydrolysis, and oligomycin- and uncoupler-sensitive ATP-Pi exchange (Table 5). Several years ago, we showed, with the use of the radioactive

Table 5. Enzymatic properties of complex V[a]

Activity	Rate
	$nmol \cdot min^{-1} \cdot mg \ protein^{-1}$
1. Exchange	
ATP-Pi + phospholipids	270-310 (350-470)[b]
- phospholipids	100-130 (150-195)
ITP-Pi	$\sim$ 20
GTP-Pi	$\sim$ 12
UTP-Pi	0.0
2. Hydrolysis[c]	
ATP (K_m = 0.14-0.18 mM)	8-10 x 10^3
GTP (K_m = 14-16 mM)	6 x 10^3
ITP (K_m = 6.5-7.5 mM)	4 x 10^3
UTP	0.0

[a] From: Stiggall, D.L., Galante, Y.M., Hatefi, Y.: J. Biol. Chem. 253, 956-964 (1978).

[b] Values in parentheses are corrected rates for ATP hydrolysis during exchange.

[c] The activities given are at V_{max}.

Table 6. Specific uncoupler binding capacity of complexes I, III, IV, and V

Preparation	Capacity (nmol/mg protein)
Mitochondria	0.35
Complex I	$\leq$ 0.05
Complex III	< 0.01
Complex IV	< 0.01
Complex V	0.81

From: Stiggall, D.L., Galante, Y.M., Hatefi, Y.: J. Biol. Chem. 253, 956-964 (1978)

photoaffinity labeling uncoupler, 2-azido-4-nitrophenol, that mitochondria contain a specific uncoupler-binding site at a concentration comparable to F_1-ATPase or respiratory chain components (52).

Using the above compound, the existence of a specific uncoupler-binding site has been confirmed by others (53, 54), and again recently by Katre and Wilson (55) using 2-nitro-4-azidocarbonylcyanide phenylhydrazone as another photoaffinity labeling uncoupler. As seen in Table 6, the specific uncoupler-binding site of mitochondria fractionates into complex V at a concentration 2-3 times that of mitochondria, but is absent from the other energy-transducing complexes. Recently, we obtained a highly purified preparation of complex V containing 11 polypeptides and have identified ten of them. They are: the five subunits of F_1-ATPase, the uncoupler-binding protein ($M_r \simeq 31,000$) identified by Hanstein (56) with the use of 2-azido-4-nitrophenol, the oligomycin-sensitivity-conferring protein, the dicyclohexylcarbodiimide (DCCD)-binding protein, coupling factor F_6, and provisionally a dithiol protein with $M_r = 11$-12×10^3, which was purified by You and Hatefi (57). The dithiol protein appears to be the active component in different preparations designated Factor(s) B by Lam et al. and Shankaran et al. (58, 59). Three final points regarding the components of complex V might be added: (a) Similar to mitochondria and submitochondrial particles, when complex V is treated with 2-azido-4-nitrophenol, two polypeptides are labeled. One is the polypeptide designated uncoupler-binding protein by Hanstein; the other is the largest (α) subunit of F_1-ATPase. Hanstein (56) believes that the two polypeptides form a pocket for uncoupler binding, but whether one or both are involved in the act of uncoupling is not clear. (b) The dithiol protein (Factor B) is required for all the reactions that lead to and from ATP, namely, ATP synthesis by oxidative phosphorylation, ATP-Pi exchange, ATP-energized reverse electron transfer from succinate to NAD, and ATP-energized transhydrogenation. It does not appear to be required, however, for transhydrogenation energized by succinate oxidation. Therefore, Factor B appears to participate in complex V in the process of energy transfer to and from the site of ATP synthesis and hydrolysis. (c) The eleventh, unidentified polypeptide of complex V is comparable in molecular weight to the polypeptide of yeast oligomycin-sensitive ATPase shown recently by Griffiths, Criddle and their colleagues (60) to contain pantothenic acid. Whether complex V or oligomycin-sensitive ATPase preparations from beef heart also contain pantothenic acid remains to be seen.

Acknowledgments. The authors thank Dr. R. Kiehl for the synthesis of radioactive dicyclohexylcarbodiimide and his help in identifying the DCCD-binding polypeptide of complex V. The expert technical assistance of Mrs. S.-Y. Wong is also gratefully acknowledged.

References

1. Hatefi, Y., Haavik, A.G., Griffiths, D.E.: Biochem. Biophys. Res. Commun. 4, 441-446 and 447-453 (1961)
2. Fowler, L.R., Hatefi, Y.: Biochem. Biophys. Res. Commun. 5, 203-208 (1961)
3. Hatefi, Y., Haavik, A.G., Fowler, L.R., Griffiths, D.E.: J. Biol. Chem. 237, 2661-2669 (1962)
4. Hatefi, Y.: Compr. Biochem. 14, 199-231 (1966)
5. Stiggall, D.L., Galante, Y.M., Hatefi, Y.: J. Biol. Chem. 253, 956-964 (1978)
6. Ragan, C.I., Hinkle, P.C.: J. Biol. Chem. 250, 8472-8476 (1975)
7. Leung, K.H., Hinkle, P.C.: J. Biol. Chem. 250, 8467-8471 (1975)
8. Wikström, M.K.F.: Nature (London) 266, 271-273 (1977)
9. Serrano, R., Kanner, B.I., Racker, E.: J. Biol. Chem. 251, 2453-2461 (1976)
10. Ryrie, I.J.: Arch. Biochem. Biophys. 184, 464-475 (1977)

11. Kagawa, Y., Sone, N., Yoshida, M., Hirata, H., Okamoto, H.: J. Biochem. (Tokyo) 80, 141-151 (1976)
12. Harmon, H.J., Hall, J.D., Crane, F.L.: Biochim. Biophys. Acta 344, 119-155 (1974)
13. Eytan, G., Carroll, R.C., Schatz, G., Racker, E.: J. Biol. Chem. 250, 8598-8603 (1975)
14. Mairouch, H., Godinot, C.: Proc. Natl. Acad. Sci. U.S.A. 74, 4185-4189 (1977)
15. Hatefi, Y., Stempel, K.E.: Biochem. Biophys. Res. Commun. 26, 301-308 (1967)
16. Orme-Johnson, N.R., Orme-Johnson, W.H., Hansen, R.E., Beinert, H., Hatefi, Y.: Biochem. Biophys. Res. Commun. 44, 446-452 (1971)
17. Orme-Johnson, N.R., Hansen, R.E., Beinert, H.: J. Biol. Chem. 249, 1922-1927 (1974)
18. Hatefi, Y., Bearden, A.J.: Biochim. Biophys. Res. Commun. 69, 1032-1037 (1976)
19. Djavadi-Ohaniance, L., Hatefi, Y.: J. Biol. Chem. 250, 9397-9403 (1975)
20. Hatefi, Y., Galante, Y.M.: Proc. Natl. Acad. Sci. U.S.A. 74, 846-850 (1977)
21. Hatefi, Y., Stiggall, D.L.: In: The Enzymes. Boyer, P.D. (ed.) New York: Academic Press 1976, Vol. XIII, pp. 175-297
22. Davis, K.A., Hatefi, Y., Poff, K.L., Butler, W.L.: Biochim. Biophys. Acta 325, 341-356 (1973)
23. Hatefi, Y., Galante, Y.M., Stiggall, D.L., Ragan, C.I.: In: Methods in Enzymology. Fleischer, S., Packer, L. (eds.) New York: Academic Press, in press
24. Capaldi, R.A., Sweetland, J., Merli, A.: Biochemistry 16, 5707-5710 (1977)
25. Davis, K.A., Hatefi, Y.: Biochemistry 10, 2509-2516 (1971)
26. Hanstein, W.G., Davis, K.A., Ghalambor, M.A., Hatefi, Y.: Biochemistry 10, 2517-2524 (1971)
27. Ohnishi, T., Lim, J., Winter, D.B., King, T.E.: J. Biol. Chem. 251, 2105-2109 (1976)
28. Yu, C.A., Yu, L., King, T.E.: Biochem. Biophys. Res. Commun. 78, 259-265 (1977)
29. Ohnishi, T.: Biochim. Biophys. Acta 301, 105-128 (1973)
30. Ragan, C.I.: Biochim. Biophys. Acta 456, 249-290 (1976)
31. Singer, T.P., Gutman, M., Massey, V.: In: Iron-Sulfur Proteins. Lovenberg, W. (ed.) New York: Academic Press, 1973, Vol. I, pp. 225-301
32. Hatefi, Y., Haavik, A.G., Fowler, L.R., Griffiths, D.E.: J. Biol. Chem. 237, 1681-1685 (1962)
33. Bell, R.L., Capaldi, R.A.: Biochemistry 15, 996-1001 (1976)
34. Marres, C.A.M., Slater, E.C.: Biochim. Biophys. Acta 462, 531-548 (1977)
35. Rieske, J.S.: Biochim. Biophys. Acta 456, 195-247 (1976)
36. Yu, C.A., Yu, L., King, T.E.: J. Biol. Chem. 247, 1012-1019 (1972)
37. Weiss, H.: Biochim. Biophys. Acta 456, 291-313 (1976)
38. Wilson, D.F., Brocklehurst, E.S.: Arch. Biochem. Biophys. 158, 200-212 (1973)
39. Rieske, J.S.: Arch. Biochem. Biophys. 145, 179-193 (1971)
40. Erecińska, M., Chance, B., Wilson, D.F., Dutton, P.L.: Proc. Natl. Acad. Sci. U.S.A. 69, 50-54 (1972)
41. Hatefi, Y.: In: Dynamics of Energy Transducing Membranes. Ernster, L., Estabrook, R., Slater, E.C. (eds.) Amsterdam: Elsevier Scientific Publishing Co., 1974, pp. 125-141
42. Wikström, M.K.F.: Biochim. Biophys. Acta 301, 155-193 (1973)
43. Birchmeier, W., Kohler, C.E., Schatz, G.: Proc. Natl. Acad. Sci. U.S.A. 73, 4334-4338 (1976)
44. Briggs, M.M., Capaldi, R.A.: Biochem. Biophys. Res. Commun. 80, 553-559 (1973)
45. Errede, B., Haight, G.P., Jr., Kamen, M.D.: Proc. Natl. Acad. Sci. U.S.A. 73, 113-117 (1976)
46. Ferguson-Miller, S., Brautigan, D.L., Margoliash, E.: J. Biol. Chem. 251, 1104-1115 (1976)
47. Ferguson-Miller, S., Brautigan, D.L., Margoliash, E.: J. Biol. Chem. 253, 149-159 (1978)
48. MacLennan, D.H., Tzagoloff, A.: Biochim. Biophys. Acta 96, 166-168 (1965)
49. Tanaka, M., Haniu, M., Zeitlin, S., Yasunobu, K.T., Yu, C.A., Yu, L., King, T.E.: Biochem. Biophys. Res. Commun. 66, 357-367 (1975)
50. Yu, C.A., Yu, L., King, T.E.: Biochem. Biophys. Res. Commun. 74, 670-676 (1977)
51. Caughey, W.S., Wallace, W.J., Volpe, J.A., Yoshikawa, S.: In: The Enzymes. Boyer, P.D. (ed.) New York: Academic Press, 1976, Vol. XIII, pp. 299-344

52. Hanstein, W.G., Hatefi, Y.: J. Biol. Chem. <u>249</u>, 1356-1362 (1974)
53. Cyboron, G.W., Dryer, R.L.: Arch. Biochem. Biophys. <u>179</u>, 141-146 (1977)
54. Ramakrishna Kurup, C.K., Sanadi, D.R.: J. Bioenerg. Biomembr. <u>9</u>, 1-15 (1977)
55. Katre, N., Wilson, D.F.: Arch. Biochem. Biophys. <u>184</u>, 578-585 (1977)
56. Hanstein, W.G.: Biochim. Biophys. Acta <u>456</u>, 129-148 (1976)
57. You, K.-S., Hatefi, Y.: Biochim. Biophys. Acta <u>423</u>, 398-412 (1976)
58. Lam, K.W., Warshaw, J.B., Sanadi, D.R.: Arch. Biochem. Biophys. <u>119</u>, 477-484 (1967)
59. Shankaran, R., Sani, B.P., Sanadi, D.R.: Arch. Biochem. Biophys. <u>168</u>, 394-402 (1975)
60. Criddle, R.S., Edwards, T.L., Partis, M., Griffiths, D.E.: FEBS Lett. <u>84</u>, 278-282 (1977)
61. Chance, B., Wilson, D.F., Dutton, P.L., Erecińska, M.: Proc. Natl. Acad. Sci. U.S.A. <u>66</u>, 1175-1182 (1970)
62. Ohnishi, T., Pring, M.: In: Dynamics of Energy Transducing Membranes. Ernster, L., Estabrook, R.W., Slater, E.C. (eds.) Amsterdam: Elsevier Scientific Publishing Co., 1974, pp. 161-180
63. Lambowitz, A.M., Bonner, W.D., Jr., Wikström, M.K.F.: Proc. Natl. Acad. Sci. U.S.A. <u>71</u>, 1183-1187 (1974)

Contribution of the Mitochondrial Genetic System to the Biogenesis of Ubiquinone: Cytochrome c Oxidoreductase

H. WEISS

Introduction

The main energy-transducing enzymes of the mitochondrial inner membrane are organized as distinct multiprotein complexes (Fig. 1), oxidoreductases (complexes I, III, and IV), which transduce the energy generated by multiple oxidation-reduction reactions into an intermediate energy form that can be used by the ATP-synthetase (complex V) for the formation of ATP (for review see Hatefi, 1978).

The biogenesis of three of these complexes, namely, the ubiquinone: cytochrome c oxidoreductase, the cytochrome c:O_2 oxidoreductase, and ATP-synthetase requires genes present in mitochondrial DNA as well as genes present in nuclear DNA and two separate systems of protein synthesis, the mitochondrial and the cytoplasmic (for review see Saccone and Kroon, 1976; Bücher et al., 1976; Bandlow et al., 1977).

A brief review of the biogenesis of the ubiquinone:cytochrome c oxidoreductase complex with special reference to the contribution of the mitochondrial genetic system is attempted in this article. This multiprotein complex is known to consist of two b cytochromes, one cytochrome c_1, one iron-sulfur protein, and five subunits without known prosthetic groups (for review see Hatefi, 1978; Rieske, 1976). The cytochrome constituents can be easily detected by their characteristic light absorbance bands, and the enzymatic activity can be assayed spectrophotometrically by following the reduction of ferricytochrome c by reduced ubiquinone derivatives. A further characteristic of this complex is that the antifungal antibiotics antimycin (Keilin and Hartree, 1955), diuron (Colson et al., 1977), mucidin (Subik et al., 1974), and funiculosin (Moser and Walter, 1975), specifically inhibit electron transport from the b cytochromes to cytochrome c_1.

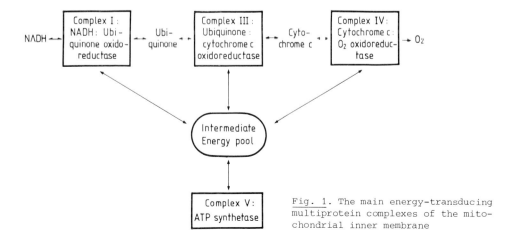

Fig. 1. The main energy-transducing multiprotein complexes of the mitochondrial inner membrane

Mitochondrial Genes Required for the Biogenesis of Ubiquinone: Cytochrome *c* Oxidoreductase

The best-studied mitochondrial DNA is that of *Saccharomyces cerevisiae*. It is a circular duplex molecule with an estimated length of 25 μ corresponding to a molecular weight of 5×10^7 daltons (for review see Borst, 1972; Borst et al., 1977). Two broad classes of genes are present on this DNA (Tzagoloff et al., 1976):

1. Genes that are essential for the development of respiratory-competent mitochondria. These genes might include both regulatory and structural genes that code for subunits of the above-mentioned three complexes (Tzagoloff et al., 1975, 1976; Slonimski and Tzagoloff, 1976).

2. Genes that code for the RNAs of the small and large subunits of mitochondrial ribosomes and all the transfer RNAs (Reijnders et al., 1972; Reijnders and Borst, 1972; Casey et al., 1974; Fukuhara et al., 1976).

The following nomenclature for the different classes of mutants of *Saccharomyces cerevisiae* affected in mitochondrial DNA is generally used (Slonimski and Tzagoloff, 1976):

ρ^+ designates a yeast cell that has a functional system of mitochondrial protein synthesis. ρ^- is a mitochondrial mutant that lacks mitochondrial protein synthesis as a result of a large deletion in mitochondrial DNA. ρ^0 is a mutant that is deleted of all mitochondrial DNA.

Mit$^-$ is a mitochondrial mutant that has retained a functional mitochondrial protein-synthesizing machinery, but, as a result of small deletions or point mutations in mitochondrial DNA, has become respiratory deficient - this deficiency might be restricted to a single enzyme.

Ant^R is a mitochondrial mutant capable of growth in the presence of antibiotics or drugs that normally inhibit growth of wild-type yeast.

Selection of Mitochondrial Mutants of *Saccharomyces cerevisiae* Which Lack Ubiquinone: Cytochrome *c* Oxidoreductase

The following procedure is now generally used for the selection of mit$^-$ mutants with a specific lesion in ubiquinone:cytochrome *c* oxidoreductase (or cytochrome c:O_2 oxidoreductase or ATP-synthetase) (Tzagoloff et al., 1976a): Cells are mutagenized with manganese since it induces point mutations and deletions primarily in mitochondrial DNA (Putrament et al., 1973; Tzagoloff et al., 1975b). Then they are spread on a medium containing a low concentration of glucose but a high concentration of glycerol. Strains that have become respiratory deficient are incapable of growing on the nonfermentable glycerol, and thus stop growing when the fermentable glucose is exhausted. Consequently, they form small colonies. These small colonies consist mainly of ρ^- and mit$^-$ mutants. To resolve the two broad classes of mutants, the small colonies are assayed for mitochondrial protein synthesis (Tzagoloff et al., 1975a). Strains in which mitochondrial protein synthesis is still functional are assayed for ubiquinone: cytochrome c oxidoreductase (or cytochrome c:O_2 oxidoreductase or ATP-synthetase).

These mutants have to be shown to be affected in mitochondrial DNA and not in nuclear DNA. The following criteria have to be fulfilled:

1. Respiratory deficiency is not restored by mating the mit⁻ mutants with ρ⁰ tester strains, which contain a normal complement of nuclear genes for mitochondrial respiratory functions but which are deleted of all mitochondrial DNA. Respiratory deficiency is, however, restored by mating with ρ⁺ strains that have intact mitochondrial DNA.

2. Respiratory deficiency of the mit⁻ mutants segregates in a non-Mendelian fashion during meiosis.

3. Respiratory deficiency of the mit⁻ mutants segregates during mitosis when the mutant strains are mated to respiratory-competent ρ⁺ strains with normal mitochondrial DNA. Different mitochondrial genotypes segregate out during mitosis, and thus restore the genetic homogeneity of mitochondria within one cell (for review see Williamson et al., 1977).

Selection of Mitochondrial Mutants of *Saccharomyces cerevisiae* with Resistance to Inhibitors of Ubiquinone:Cytochrome *c* Oxidoreductase

The antifungal antibiotics, antimycin, diuron, mucidin, and funiculosin, normally stop growth of wild-type yeast strains on nonfermentable substrates, since they inhibit the electron transport from the *b* cytochromes to cytochrome c_1. Mitochondrial mutants with resistance to these inhibitors, namely, to antimycin (Pratje and Michaelis, 1977; Burger et al., 1976; Obbink et al., 1977), diuron (Colson et al., 1977), mucidin (Subik et al., 1977), and funiculosin (Pratje and Michaelis, 1977), were produced from spontaneous mutations of wild-type strains grown on media that contain these inhibitors. In these mutants, the in vivo electron transport of isolated mitochondria or submitochondrial has become resistant, or, less sensitive to these drugs. Consequently, as a result of mutations in mitochondrial DNA (part of) ubiquinonene:cytochrome *c* oxidoreductase appears to be modified.

Localization in Mitochondrial DNA of the Mutations Expressed in Deficiency or Inhibitor Resistance of Ubiquinone:Cytochrome *c* Oxidoreductase

An average map of the mitochondrial DNA of *Saccharomyces cerevisiae* has been proposed (Dujon et al., 1977) using sectors delimited by boundaries (Fig. 2). Five boundaries have been defined. They have been termed B1 to B5 and they represent the genetic locations of the most often studied, classical drug-resistant mutations, namely, mutations conferring resistance to chloramphenicol (B1), erythromycin (B2), oligomycin (B3) and (B4), and paronomycin (B5). The regions of the mitochondrial DNA in which mutations are expressed by a deficiency of ubiquinone:cytochrome *c* oxidoreductase or by resistance to the inhibitors of this complex were localized as follows: ubiquinone:cytochrome *c* oxidoreductase-deficient mutants (mit⁻), or the inhibitor-resistant mutant (Ant^R), were mated with (1) respiratory-competent mit⁺ strains carrying the alleles that confer resistance to the above-mentioned classical drugs or (2) ρ-strains in which the retained segments of mitochondrial DNA contained some of the classical drug-resistant markers. From the frequency of recombinants produced from these crosses, the relative genetic distances were evaluated. By this means both types of mutations, i.e., those expressed in ubiquinone:cytochrome *c* oxidoreductase deficiency and those expressed in inhibitor

34

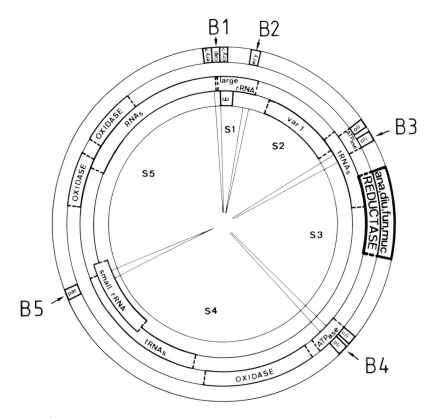

Fig. 2. Average genetic map of mitochondrial DNA. B1-B5 correspond to boundaries and S1-S5 to sectors of the average map (see text). The main types of mutations and genes have been placed on concentric circles. *First circle* (external): Mutations and genes conferring drug resistance. *Second circle*: Mutations and genes affecting respiratory or phosphorylating functions. *Third circle*: Genes of rRNAs and tRNAs. *Fourth circle*: Other mutations and genes. *REDUCTASE* represents ubiquinone: cytochrome *c* oxidoreductase; *ana* represents antimycin; *diu*, diuron; *fun*, funiculosin; and *muc*, mucidin. (From Dujon et al., 1977)

resistance, were localized in the same region of mitochondrial DNA, namely, in sector 4, which is delimited by the two oligomycin resistance-conferring loci, the boundaries 3 and 4 (Fig. 2) (Slonimski and Tzagoloff, 1976; Tzagoloff et al., 1976b).

The availability of a large number of different ubiquinone:cytochrome *c* oxidoreductase-deficient mit⁻ mutants and Ant^R mutants with resistance to specific inhibitors made possible the construction of a detailed map of this region of mitochondrial DNA (Tzagoloff et al., 1976b; Kotylak and Slonimski, 1977; Colson and Slonimski, 1977).

The following criteria have been used to establish that two phenotypically similar mutations are closely linked and located in the same locus:

1. When strains with mit⁻ and/or Ant^R mutations in the same locus are mated, diploid respiratory-competent (mit⁺) and/or inhibitor-sensitive (Ant^S) strains are produced at a low frequency (below 1%). The

frequency, however, is high (up to 15%) in crosses of mutants belonging to different loci.

2. Mit⁻ and/or Ant^R mutants are restored with respect to functional respiration and inhibitor sensitivity by the same set of ρ^- strains.

Based on these criteria, initially two (Tzagaloff et al., 1976b; Pratje and Michaelis, 1977; Colson et al., 1977; Subik et al., 1977) and subsequently three (Colson et al., 1977) different genetic loci have been distinguished. At the first locus (located toward B4) diuron, antimycin, funiculosin-resistant, and ubiquinone-cytochrome c oxido-reductase-deficient alleles are found. At the second locus (located in the middle of this DNA region) diuron, antimycin, mucidin-resistant, and ubiquinone:cytochrome c oxidoreductase-deficient alleles are found, and at the third locus (located toward B3) mucidin-resistant or ubiquinone:cytochrome c oxidoreductase-deficient alleles are found (for review see Dujon et al., 1977). From the high recombination frequency observed between these three loci, the existence of three genes coding for distinct gene products has been suggested. The allelism relationship between respiratory-competent inhibitor-resistant loci and ubiquinone:cytochrome c oxidoreductase-deficient loci suggested that each of the loci codes for a structural gene product that is part of the ubiquinone:cytochrome c oxidoreductase complex (Colson and Slonimski, 1977).

Translation Site of the Subunits of Ubiquinone: Cytochrome c Oxidoreductase

Selective Labeling of Proteins According to Their Translation Site Mitochondrial or Cytoplasmic Ribosomes

The proteins of mitochondria can be labeled selectively according to their site of translation, namely, the mitochondrial or the cytoplasmic ribosomes, by in vivo incorporation of radioactive amino acids in the presence of either cycloheximide, a specific inhibitor of cytoplasmic protein synthesis (Beattie, 1968; Sebald et al., 1969; Schatz and Saltzgaber, 1969) or chloramphenicol, a specific inhibitor of mitochondrial protein synthesis (Mager, 1960; Clark-Walker and Linnane, 1966; Sebald et al., 1972; Weiss et al., 1973). For these labeling studies, the fungi *Neurospora crassa* or *Saccharomyces cerevisiae* have been widely used, due to the fact that the catabolism of radioactive amino acids added to cultures of these microorganisms is small compared to the incorporation into proteins. The flight muscle of the insect *Locusta migratoria* also proved to be an appropriate source of mitochondria for these studies. After the imaginal moult it rapidly develops from a precursor muscle, showing a 50-fold increase of mitochondrial mass within a few days (Brosemer et al., 1963). During this time injected radioactive leucine is incorporated into the muscle protein at a sufficient rate.

In the above three organisms, the cytochrome b subunits of ubiquinone: cytochrome c oxidoreductase have been identified as mitochondrial translation products (Weiss and Ziganke, 1974; Lorenz et al., 1974; Katan et al., 1976). The procedure followed with *Neurospora crassa* will be described: The proteins were double labeled, firstly by the incorporation of [^{14}C] leucine in the absence of inhibitors and then secondly by the incorporation of [^{3}H] leucine in the presence of either cycloheximide or chloramphenicol, the ubiquinone:cytochrome c oxidoreductase was isolated and analyzed for ^{14}C and ^{3}H-radioactivity.

Purification and Stepwise Resolution of Ubiquinone:Cytochrome *c* Oxidoreductase from *Neurospora crassa*

Ubiquinone:cytochrome *c* oxidoreductase was purified from *Neurospora crassa* by means of an affinity chromatographic technique, in the presence of the non-ionic detergent Triton X-100. The method is based upon the affinity binding of the solubilized complex to immobilized ferricytochrome *c* (which can be regarded as a substrate of the complex) and its subsequent affinity release by converting the immobilized ferricytochrome *c* into ferrocytochrome *c* (which can be regarded as a product of the complex) using ascorbate as reductant (Weiss et al., 1978; Weiss and Juchs, 1978).

This affinity chromatographic procedure takes advantage of functional specificity and can be performed at low ionic strength. Low ionic strength during the isolation procedure was found to be of critical importance since the Triton-solubilized complex readily dissociates as the ionic strength increases.

The isolated complex was found to be dimeric (see also von Jagow et al., 1977). It has a molecular weight of about 500,000, and consists of four *b* cytochromes (M_r each ~ 30,000), two *c* cytochromes (M_r ~ 30,000), two iron-sulfur proteins (M_r ~ 25,000), and five additional subunits without known prosthetic groups (M_r ~ 50,000, 45,000, 14,000, 10,000, and 8000) probably present in a 2:4:2:2:2 stoichiometry (Fig. 3). The complex binds about 0.2 g Triton X-100 per g protein, which corresponds to one Triton X-100 micelle.

By isopycnic centrifugation on a sucrose gradient containing lecithin liposomes, the Triton X-100 micelle is replaced by a phospholipid bilayer and a vesicular preparation is formed. This leads to a restoration of enzymatic activity (Weiss and Wingfield, in preparation).

A tetrameric cytochrome *b* probably forms the membranous part of the complex and spans the lipid bilayer. Cytochrome c_1 appears to be orientated toward the outer surface of the mitochondrial inner membrane, and the large subunits (M_r 50,000 and 45,000) without known prosthetic groups, to the inner surface of the membrane (Weiss and Wingfield, in preparation) (Fig. 4).

The Triton X-100-solubilized complex from *Neurospora crassa* is very fragile and can be stepwise dissociated under relatively gentle conditions (Weiss et al., 1978) (Fig. 4).

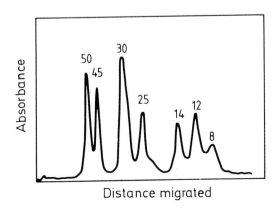

Fig. 3. Dodecyl sulfate gel electrophoresis of ubiquinone:cytochrome *c* oxidoreductase from *Neurospora crassa*. The protein bands have been stained with Coomassie brilliant blue. The *numbers* represent the apparent molecular weights of the subunits

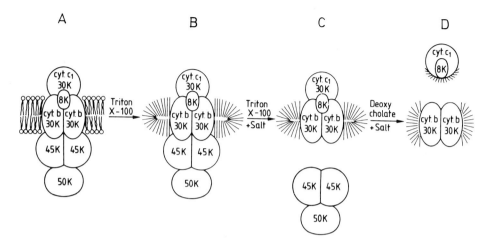

Fig. 4A-D. Preliminary and incomplete model of the arrangement of subunits of
ubiquinone:cytochrome c oxidoreductase and a schematic representation of the resolution of the complex. (A) Complex in the phospholipid bilayer of the membrane. The
complex in situ is probably dimeric, i.e., each of the schematically drawn subunits
is present twice. The 25,000 iron-sulfur protein and the 14,000 and 12,000 subunits
are not included in the scheme, because so far no information exists regarding their
arrangement. (B) Complex solubilized by a Triton X-100 micelle. Also the solubilized
complex is probably dimeric. (C and D) Partially resolved complex (see text)

Increasing ionic strength first leads to monomerization of the complex
and then to the loss of the iron-sulfur protein. Further increase in
the ionic strength causes the dissociation of the large subunits
(M_r 50,000 and 45,000), which then are rendered water soluble. Finally,
treatment with deoxycholate causes the dissociation of cytochrome c_1
and the smaller subunits without known prosthetic group. A dimeric
cytochrome b remains (Weiss, 1976; Weiss et al., 1978).

Identification of the Cytochrome b Subunits as Mitochondrial Trans-lation Product

The complex was isolated from $Neurospora$ $crassa$ cells that had incorporated [^{14}C] leucine in the absence of inhibitors and [^{3}H] leucine
in the presence of cycloheximide or chloramphenicol. Then the subunits were separated according to their molecular weights by dodecyl
sulfate gel electrophoresis (Fig. 3) and were analyzed for ^{3}H and
^{14}C radioactivity. This led to the identification of the translation
site of the M_r subunits 50,000, 45,000, 25,000, 14,000, 10,000, and
8000. The protein bands attributed to these subunits were found to
be ^{3}H-labeled when the complex was isolated from chloramphenicol-
treated cells but were not ^{3}H-labeled when the complex was isolated
from cycloheximide-treated cells (Weiss and Ziganke, 1977). This indicated that these subunits were translated on cytoplasmic ribosomes.
The 30,000-M_r cytochrome b and cytochrome c_1 subunits, however, were
not separated from each other by gel electrophoresis in the presence
of dodecyl sulfate (Fig. 3). They migrated as one band, which was
found to be ^{3}H-labeled in both the cycloheximide and the chloramphenicol
experiments. Thus, an unambiguous identification of the translation
site of the cytochrome b and c_1 subunits was not possible by this
means. Furthermore, dodecyl sulfate treatment leads to the removal of
the noncovalently bound heme groups of cytochrome b, so that the

38

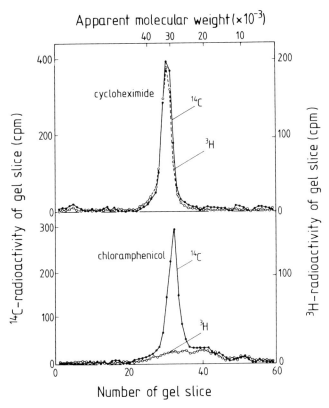

Fig. 5. Dodecyl sulfate gel electrophoresis of the dimeric cytochrome b from cells having incorporated [^{14}C]leucine in the absence of inhibitors and [^{3}H] leucine in the presence of cycloheximide or chloramphenicol

identification of the cytochrome b subunits among the protein bands obtained by dodecyl sulfate gel electrophoresis of the whole complex was also not possible (Weiss and Ziganke, 1977).

As previously mentioned, the purified complex from *Neurospora crassa* can be stepwise dissociated, and since under these conditions the prosthetic groups remain attached to the apoproteins, cytochrome b and cytochrome c_1 subunits can be separated and identified (Fig. 4). Measurement of radioactivity in the various preparations obtained by this means demonstrated that cytochrome b was a mitochondrial translation product (Fig. 5) and that cytochrome c_1 was a cytoplasmic translation product (Weiss and Ziganke, 1977).

Identity or Nonidentity of the Cytochrome b Subunits?

The following observations suggested the nonidentity of the cytochrome b subunits, i.e., suggested that the dimeric cytochrome b preparation is an α,β-type heme protein and, thus, that the tetrameric cytochrome b in ubiquinone:cytochrome c oxidoreductase is a $2\alpha,2\beta$-type heme protein.

1. In the intact mitochondria and in isolated ubiquinone:cytochrome *c* oxidoreductase, two functionally different cytochrome *b* species can be distinguished with respect to light absorbance spectra, redox potential, and redox kinetics (for review see Wikström, 1973).

2. More than one structural gene has been suggested to code for the subunits of ubiquinone:cytochrome *c* oxidoreductase in yeast (see above) but only cytochrome *b* has (so far) been identified as a mitochondrial translation product (Colson and Slonimski, 1977; Claisse et al., 1977).

3. The protein moiety of the dimeric cytochrome *b* preparation from *Neurospora crassa* can be separated into two protein peaks of about equal amounts by hydroxylapatite chromatography in the presence of dodecyl sulfate (Weiss, 1976).

To date no differences in the protein chemical properties of the two protein peaks obtained by dodecyl sulfate hydroxylapatite of the cytochrome *b* apoprotein can be detected. The amino acid compositions of the two protein peaks are very similar. Differences are found essentially only in the lysine contents. However, whether this difference really exists or whether it is caused by lysine-rich contaminations still present in the preparation cannot yet be decided. Furthermore, fingerprinting by two-dimensional thin-layer chromatography showed no significant differences between the two protein peaks. Thus, the question of whether or not cytochrome *b* consists of subunits of identical or nonidentical primary structure cannot yet be answered on a structural basis (Weiss, 1976; Weiss and Ziganke, 1977).

Is (are) the Cytochrome *b* Subunit(s) Coded for by a Gene(s) in Mitochondrial DNA?

The total set of mitochondrial translation products can be visualized by autoradiography of an (SDS) electropherogram from mitochondria, in which the mitochondrially translated proteins are radioactively labeled as described above. A distinct number of polypeptide bands are observed. Comparison of the band pattern obtained from wild-type yeast (or *Neurospora crassa* or *Locusta migratoria*) with those from ubiquinone:cytochrome *c* oxidoreductase-deficient mit⁻ mutants in most cases revealed the absence of the 30,000-M_r polypeptide, which is now attributed to the cytochrome *b* subunit. With some of the mit⁻ mutants, however, new mitochondrially synthesized polypeptides are visible with apparent molecular weights smaller than 30,000 (Tzagoloff et al., 1976b; Claisse et al., 1977). Interestingly, the order of genetic loci in which mutations seem to modify the cytochrome *b* polypeptide(s) follows the order of the molecular weight of the putative cytochrome *b* fragments. It is approximately 13,000 when the mutation in the ubiquinone:cytochrome *c* oxidoreductase region of the mitochondrial DNA is located toward oligomycin locus II (boundary 4), about 18,000 when the mutation is located in the middle of this DNA region, and about 26,000 when it is located toward oligomycin locus I (boundary 3). A possible interpretation is that a single structural gene for cytochrome *b* covers this entire mitochondrial DNA region, that this gene is read in the direction from boundary 4 to boundary 3, and that mutations might generate chain-terminating codons (Claisse et al., 1977).

Concluding Remarks

The experimental data obtained by a number of groups make it fairly
certain that the cytochrome b apoprotein of the ubiquinone:cytochrome
c oxidoreductase complex is coded for by mitochondrial DNA and translated on mitochondrial ribosomes. The other proteins of the complex
are coded for by nuclear DNA and are synthesized in the cytoplasm. Furthermore, the cytochrome b protein appears to be the site of action of
the antifungal antibiotics, which inhibit the electron transport activity of the complex. What remains to be clarified is the number of genes
in mitochondrial DNA-determining cytochrome b and which of these genes
are regulatory or structural. Also the question as to whether or not
cytochrome b consists of two polypeptide chains of identical primary
structure has not yet been satisfactorily answered. Finally, future
studies have to elucidate how the nuclear genetic system and the mitochondrial genetic system interact with each other, i.e., how the one
system influences the expression of the other and vice-versa.

References

Bandlow, W., Schweyen, R.J., Wolf, K., Kaudewitz, F.: Mitochondria 1977, Genetics
  and Biogenesis of Mitochondria. Berlin, New York: Walter de Gruyter, 1977
Beattie, D.S.: Studies on the Biogenesis of Mitochondrial Protein Components in Rat
  Liver Slices. J. Biol. Chem. 243, 4027-4033 (1968)
Borst, P.: Mitochondrial Nucleic Acids. Ann. Rev. Biochem. 41, 333-376 (1972)
Borst, P., Bos, J.L., Grivell, L.A., Groot, G.S.P., Heyting, C., Moorman, A.F.M.,
  Sanders, J.P.M., Talen, J.L., van Kreijl, C.F., van Ommen, G.J.B.: The Physical
  Map of Yeast Mitochondrial DNA Anno 1977 (A Review). In: Mitochondria 1977, Genet-
  ics and Biogenesis of Mitochondria. Bandlow, W., Schweyen, R.J., Wolf, K.,
  Kaudewitz, F. (eds.) Berlin, New York: Walter de Gruyter, 1977, pp. 213-254
Brosemer, R.W., Vogell, W., Bücher, Th.: Morphologische und enzymatische Muster bei
  der Entwicklung indirekter Flugmuskeln von Locusta migratoria. Biochem. Z. 338,
  854-910 (1963)
Bücher, Th., Neupert, W., Sebald, W., Werner, S.: Genetics and Biogenesis of Chloro-
  plasts and Mitochondria. Amsterdam-New York-Oxford: North Holland, 1976
Burger, G., Lang, B., Bandlow, W., Schweyen, R.J., Backhaus, B., Kaudewitz, F.:
  A New Mutation on the mit DNA Conferring Antimycin Resistance on the Mitochondrial
  Respiratory Chain. Biochem. Biophys. Res. Commun. 72, 1201-1208 (1976)
Casey, J.W., Hsu, H.J., Getz, G.S., Rabinowitz, M., Fukuhara, H.: Transfer RNA
  Genes in Mitochondrial DNA of Grande (Wild-type) Yeast. J. Mol. Biol. 88, 735-
  747 (1974)
Claisse, M., Spyridakis, A., Slonimski, P.P.: Mutations at any one of the Three Un-
  linked Mitochondrial Genetic Loci Box 1, Box 4 and Box 6 Modify the Structure of
  Cytochrome b Polypeptide(s). In: Mitochondria 1977, Genetics and Biogenesis of
  Mitochondria. Bandlow, W., Schweyen, R.J., Wolf, K., Kaudewitz, F. (eds.) Berlin,
  New York: Walter de Gruyter, 1977, pp. 337-344
Clark-Walker, G.D., Linnane, A.W.: In Vivo Differentiation of Yeast Cytoplasmic and
  Mitochondrial Protein Synthesis with Antibiotics. Biochem. Biophys. Res. Commun.
  25, 8-13 (1966)
Colson, A.M., Slonimski, P.S.: Mapping of Drug-Resistant Loci in the Coenzyme QH_2-
  Cytochrome C Reductase Region of the Mitochondrial DNA Map in Saccharomyces cere-
  visiae. In: Mitochondria 1977, Genetics and Biogenesis of Mitochondria. Bandlow,
  W., Schweyen, R.J., Wolf, K., Kaudewitz, F. (eds.) Berlin, New York: Walter de
  Gruyter, 1977, pp. 185-196
Colson, A.M., The Van, L., Convent, B., Briquet, M., Goffeau, A.: Mitochondrial
  Heredity of Resistance to 3, (3,4-Dichlorophenyl)-1,1-dimethylurea, an Inhibitor
  of Cytochrome b Oxidation in Saccharomyces cerevisiae. Eur. J. Biochem. 74, 521-
  526 (1977)

Dujon, B., Colson, A.M., Slonimski, P.O.: The Mitochondrial Genetic Map of Saccharomyces cerevisiae: Compilations of Mutations, Genes, Genetic and Physical Maps. In:
Mitochondria 1977, Genetics and Biogenesis of Mitochondria. Bandlow, W., Schweyer,
R.J., Wolf, K., Kaudewitz, F. (eds.) Berlin, New York: Walter de Gruyter, 1977,
pp. 577-669

Fukuhara, H., Martin, N., Rabinowitz, M.: Isoaccepting Mitochondrial Glutamyl-tRNA
Species Transcribed from Different Regions of the Mitochondrial Genome of Saccharomyces cerevisiae. J. Mol. Biol. 101, 285-296 (1976)

Hatefi, Y., Galante, J.M.: Organization of the Mitochondrial Respiratory Chain.
This volume, pp.

Jagow, G., von, Schägger, H., Riccio, P., Klingenberg, M., Kolb, H.J.: Bc$_1$ Complex
from Beef Heart: Hydrodynamic Properties of the complex Prepared by a Refined
Hydroxyapatite Chromatography in Triton X-100. Biochim. Biophys. Acta 462, 549-
558 (1977)

Katan, M.B., Van-Harten-Loosbroek, N., Groot, G.S.P.: The Cytochrome bc$_1$ Complex
of Yeast Mitochondria. Site of Translation of the Polypeptides in vivo. Eur. J.
Biochem. 70, 409-417 (1976)

Keilin, D., Hartree, E.F.: Relationship between Certain Components of the Cytochrome
System. Nature (London) 176, 200-206 (1955)

Kotylak, Z., Slonimski, P.P.: Fine Structure Genetic Map of the Mitochondrial DNA
Region Controlling Coenzyme QH$_2$-Cytochrome c Reductase. In: Mitochondria 1977,
Genetics and Biogenesis of Mitochondria. Bandlow, W., Schweyen, R.J., Wolf, K.,
Kaudewitz, F. (eds.) Berlin, New York: Walter de Gruyter, 1977, pp. 161-172

Lorenz, B., Kleinow, W., Weiss, H.: Mitochondrial Translation of Cytochrome b in
Neurospora crassa and Locusta migratoria. Z. Physiol. Chem. 355, 300-304 (1974)

Mager, J.: Chloramphenicol and Chlortetracycline Inhibition of Amino Acid Incorporation into Protein Cell Free System from Tetrahymina Pyriformis. Biochim. Biophys.
Acta 38, 10-15 (1960)

Moser, U.K., Walter, P.: A New Specific Inhibitor of the Respiratory Chain in Rat
Liver Mitochondria. FEBS Lett. 50, 279-282 (1975)

Obbink, D.J.G., Spithill, T.W., Maxwell, R.J., Linnane, A.W.: A Mitochondrial Mutation conferring Resistance to an Antimycin A-Like Contaminant in Mikamycin. Mol.
Gen. Genet. 151, 127-136 (1977)

Pratje, E., Michaelis, G.: Allelism Studies of Mitochondrial Mutants Resistant to
Antimycin A or Funiculosin in Saccharomyces cerevisiae. Mol. Gen. Genet. 152,
167-174 (1977)

Putrament, A., Baranowska, H., Pazmo, W.: Induction by Manganese of Mitochondrial
Antibiotic Resistance Mutations in Yeast. Mol. Gen. Genet. 126, 357-366 (1973)

Reijnders, L., Borst, P.: The Number of 4-S RNA Genes on Yeast Mitochondrial DNA.
Biochem. Biophys. Res. Commun. 47, 126-133 (1972)

Reijnders, L., Kleisen, C.M., Grivell, L.S., Borst, P.: Hybridization Studies with
Yeast Mitochondrial RNAs. Biochem. Biophys. Acta 272, 396-407 (1972)

Rieske, J.S.: Composition, Structure and Function of Complex III of the Respiratory
Chain. Biochim. Biophys. Acta 456, 195-247 (1976)

Saccone, C., Kroon, A.M.: The Genetic Function of Mitochondrial DNA. Amsterdam-
Oxford-New York: North-Holland, 1976

Schatz, G., Saltzgaber, J.: Protein Synthesis by Yeast Promitochondria in vivo.
Biochem. Biophys. Res. Commun. 37, 996-1001 (1969)

Sebald, W., Schwab, A.J., Bücher, Th.: Cycloheximide Resistant Amino Acid Incorporation into Mitochondrial Protein from Neurospora crassa. FEBS Lett. 4, 243-246
(1969)

Sebald, W., Weiss, H., Jackl, G.: Inhibition of the Assembly of Cytochrome Oxidase
in Neurospora crassa by Chloramphenicol. Eur. J. Biochem. 30, 413-417 (1972)

Slonimski, P.P., Tzagoloff, A.: Localization in Yeast Mitochondrial DNA of Mutations Expressed in a Deficiency of Cytochrome Oxidase and/or Coenzyme QH$_2$-Cyto-
chrome c Reductase. Eur. J. Biochem. 61, 27-41 (1976)

Subik, J., Behun, M., Musilek, V.: Antibiotic Mucidin, a New Antimycin A-Like Inhibitor of Electron Transport in Rat Liver Mitochondria. Biochem. Biophys. Res.
Commun. 57, 17-22 (1974)

Subik, J., Kovacova, V., Takacsova, G.: Mucidin Resistance in Yeast. Isolation
Characterization and Genetic Analysis of Nuclear and Mitochondrial Analysis of

Nuclear and Mitochondrial Mucidin Resistant Mutants of Saccharomyces cerevisiae. Europ. J. Biochem. 73, 275-286 (1977)

Tzagoloff, A., Akai, A., Needleman, R.B.: Assembly of the Mitochondrial Membrane System: Isolation of Nuclear and Cytoplasmic Mutants of Saccharomyces cerevisiae with Specific Defects in Mitochondrial Functions. J. Bacteriol. 122, 826-831 (1975)

Tzagoloff, A., Akai, A., Needleman, R.B., Zulch, G.: Assembly of the Mitochondrial Membrane System. Cytoplasmic Mutants of Saccharomyces cerevisiae with Lesions in Enzymes of the Respiratory Chain and in the Mitochondrial ATPase. J. Biol. Chem. 250, 8236-8242 (1975)

Tzagoloff, A., Foury, F., Akai, A.: Resolution of the Mitochondrial Genome. In: The Genetic Function of Mitochondrial DNA. Saccone, C., Kroon, A.M. (eds.) Amsterdam-Oxford-New York: North-Holland, 1976a, pp. 155-161

Tzagoloff, A., Foury, F., Akai, A.: Assembly of the Mitochondrial Membrane System XVIII. Genetic Loci on Mitochondrial DNA Involved in Cytochrome b Biosynthesis. Mol. Gen. Genet. 149, 33-42 (1976b)

Weiss, H.: Subunit Composition and Biogenesis of Mitochondrial Cytochrome b. Biochim. Biophys. Acta 456, 291-313 (1976)

Weiss, H., Juchs, B.: Isolation of a Multiprotein Complex Containing Cytochrome b and c_1 from Neurospora crassa Mitochondria by Affinity Chromatography on Immobilized Cytochrome c. Difference in the Binding between Ferri- and Ferro-cytochrome c to the Multiprotein Complex. Eur. J. Biochem., in press

Weiss, H., Juchs, B., Ziganke, B.: Complex III from Mitochondria of Neurospora crassa: Purification, Characterization and Resolution. In: Methods in Enzymology. Biomembranes. Fleischer, S., Packer, L. (eds.) New York: Academic Press, in press

Weiss, H., Sebald, W., Schwab, A.J., Kleinow, W., Lorenz, B.: Contribution of Mitochondrial and Cytoplasmic Protein Synthesis to the Formation of Cytochrome b and Cytochrome aa_3. Biochimie 55, 815-821 (1973)

Weiss, H., Ziganke, B.: Cytochrome b in Neurospora crassa Mitochondria. Site of Translation of the Heme Protein. Eur. J. Biochem. 41, 63-71 (1974)

Weiss, H., Ziganke, B.: Partial Identification, Stoicheometry and Site of Translation of the Subunits of Ubiquinone: Cytochrome c Oxidoreductase. In: Mitochondria 1977, Genetics and Biogenesis of Mitochondria. Bandlow, W., Schweyen, R.J., Wolf, K., Kaudewitz, F. (eds.) Berlin, New York: Walter de Gruyter, 1977, pp. 463-472

Wikström, M.R.F.: The Different Cytochrome b Components in the Respiratory Chain of Animal Mitochondria and their Role in Electron Transport and Energy Conservation. Biochim. Biophys. Acta 301, 155-193 (1973)

Williamson, D.H., Johnston, L.H., Richmond, K.M.V., Game, J.C.: Mitochondria DNA and the Heritable Unit of the Yeast Mitochondrial Genome: a Review. In: Mitochondria 1977, Genetics and Biogenesis of Mitochondria. Bandlow, W., Schweyen, R.J., Wolf, K., Kaudewitz, F. (eds.) Berlin, New York: Walter de Gruyter, 1977, pp. 1-24

Beef Heart Complex III: Isolation and Characterization of the Heme *b*-Carrying Subunits

G. v. JAGOW, H. SCHÄGGER, W. D. ENGEL, H. HACKENBERG, and H. J. KOLB

Introduction

This short comment concentrates on only one functional aspect of the complex, namely, on which of its polypeptide chains may function as the electrogenic proton carrier.

Papa et al. (1), Lawford and Garland (2), Hinkle and Leung (3), and Nelson et al. (4) have shown on submitochondrial particles and on a complex III reincorporated in phospholipid vesicles, that electron flow from ubiquinone to cytochrome c is accompanied by synchronous proton translocation across the mitochondrial membrane, or across the phospholipid bilayer, respectively.

Figure 1 demonstrates one of the recently published schemes (5), which describes the coupling of electron flow from ubiquinone (QH_2 in the scheme) to oxygen to the transmembrane proton transport in the region of complex III. In model (a) Papa discusses, in addition to ubiquinone, an unknown polypeptide ZH_2 as a mobile hydrogen ion carrier and in model (b), the redox centers of the metalloprotein Z as an effective hydrogen carrier. He has stated (6) that: "No classical hydrogen carrier is known to exist in complex III. However, metal electron carriers might act as effective hydrogen carriers and might be responsible for transmembrane proton translocation, if oxidoreduction of the metal were linked to proton equilibria (pK shifts) of acidic groups in the apoprotein. In this case the redox potential of the electron carrier has to be pH dependent."

As will be shown in the following, we have searched for a pH dependence of the midpoint potential of isolated cytochrome b (7).

A new isolation procedure for cytochrome b had to be developed, since, with the exception of the *Neurospora* preparation, all preparations were

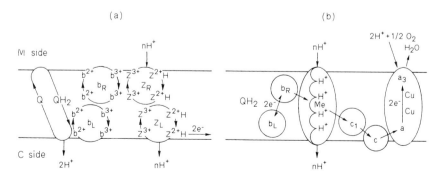

Fig. 1a and b. Scheme for the coupling of electron flow to proton gradient in the region of complex III as reported by Papa in 1976 (5)

Table 1. Properties of the published cytochrome *b* preparations and characteristics of the applied isolation techniques

Author	Goldberger et al. (8) (1961)	Ohnishi (9) (1966)	Yu et al. (10) (1975)	Weiss (11) (1974)
Species	Beef	Beef	Beef	*Neurospora crassa*
Detergent used	Dodecylsulfate	Cholate plus proteinase	Dodecylsulfate	Cholate, desoxycholate
Isolation technique	Precipitation, solubilization	Precipitation, solubilization	Precipitation, solubilization	Chromatography
Minimum mol. wt.	28,000	21,000	37,000; 17,000	32,000
State of aggregation	Polymerized	Polymerized	Monodisperse	Monodisperse
Midpoint potential (mV)	-340 (12)	-21	not dt.	-70

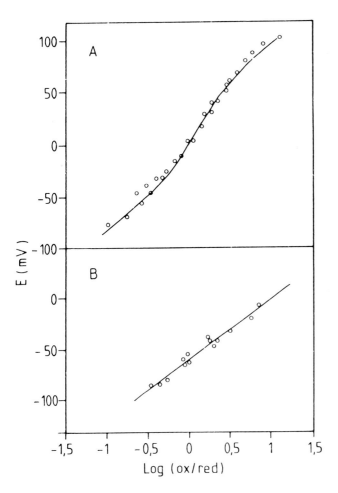

Fig. 2. Redox titration curves of (A) the *b*-type cytochromes in *Neurospora crassa* mitochondria and (B) purified *Neurospora crassa* cytochrome *b* as published by Weiss (14)

more or less altered or partially denatured. Table 1 contains a list
of the characteristics of the applied procedures and some of the properties of the published preparations. All of the isolation procedures
include the resolubilization of the highly hydrophobic and precipitated
protein by a denaturing detergent, e.g., sodium dodecylsulfate(SDS).
Ohnishi (9) treated the precipitated cytochrome b with a protease. The
result was denatured proteins with either an extremely negative midpoint potential (12) or partially degraded proteins (9). Only Weiss
and Ziganke (13) have isolated a rather native and monodisperse cytochrome b, since they applied for the first time a chromatographic procedure for isolation and used the relatively mild detergent system
cholate-deoxycholate.

However, as is evident from the lower redox titration curve in Figure
2 (14), this preparation has only one midpoint potential of about -70 mV.
The upper curve reflects the redox properties of cytochrome b, when
it is still integrated in the mitochondrial membrane. Under this condition it possesses two different midpoint potentials. They are about
-40 mV and +60 mV at pH 7.0.

In an attempt to obtain a protein in a more functionally active state,
a change was made from the detergent system cholate-deoxycholate to
the nonionic detergent Triton X-100. The chromatographic technique
consisted of a series of preserving hydroxyapatite chromatographies.

Results

Figure 3 is a summary of the procedure for isolating cytochrome b from
beef heart (7). An antimycin-loaded complex III, which can be prepared
in Triton X-100 by simple hydroxyapatite chromatography (15), is first
bound to a hydroxyapatite column. In step 2), it is dissociated into
its individual subunits by 1.5 M guanidine. After guanidine cleavage,
cytochrome c_1 remains bound on the column, whereas the other polypeptides can be found in the eluted fractions. Without going into further
procedural details, cytochrome b is finally isolated by two subsequent
hydroxyapatite chromatographies, which have to be interrupted by two
desalting steps adjusting the appropriate ionic and detergent conditions.

Figure 4 shows a densitogram from polyacrylamide gel electrophoresis.
The preparation reveals in the presence of SDS, only one band with a
molecular weight of 31,000 daltons. The molecular weight is in fair
agreement with the minimum molecular weight of 33,000 daltons, which
has been calculated from the heme b contents. Some of the preparations
contain traces of the core protein even after the third hydroxyapatite
column. The homogeneity and the state of aggregation of the isolated
cytochrome b were tested by high-speed ultracentrifugal sedimentation
velocity and equilibrium runs. The runs were performed in the presence
of 0.5% Triton X-100. The heme b was monitored at 416 nm. Figure 5
shows a plot of the cytochrome b concentration versus the square of
the distance to the center of rotation of a sedimentation equilibrium
run. It is a straight line over almost the whole sector of the cell.
From this result it can be concluded that the preparation is homogeneous. With knowledge of the specific detergent binding, the molecular
weight may be calculated from such runs.

As listed in Table 2, Triton binding was determined with tritium-
labeled Triton X-100. In chromatographic runs on Sephadex G-200, it
amounts to 1.5 g Triton per gram of protein. After correction for the

46

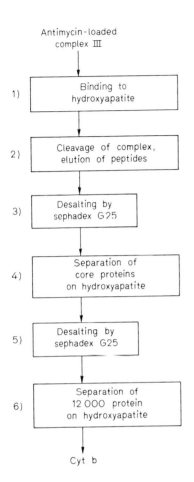

Fig. 3. Flow scheme for the isolation of cytochrome *b* from beef heart

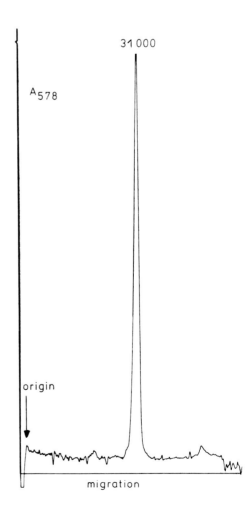

Fig. 4. Densitogram of dode-
cylsulfate-polyacrylamide
gel electrophoresis of isolated cytochrome *b* from beef
heart

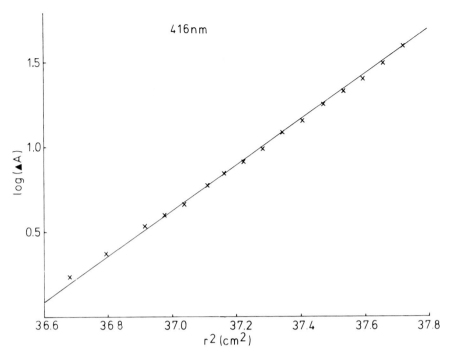

Fig. 5. Sedimentation equilibrium run of beef heart cytochrome b in Triton X-100.
The logarithm of recorder deflection in mm (proportional to the absorption) is
plotted versus the square of the distance to the center of rotation, r^2 (cm^2)

Table 2. Physicochemical data of the isolated cytochrome
b of beef heart

Triton X-100 binding (tritium-labeled Triton X-100)	1.5 g/g protein
Antimycin binding (tritium-labeled antimycin)	nil
Minimum mol. wt. (pyridine hemochromogen)	33,000 daltons
(dodecylsulfate electrophoresis)	31,000 daltons
Corrected mol. wt. (sedimentation equilibrium runs)	62,000 daltons

bound Triton (16), the molecular weight was calculated to be 62,000
± 3000 daltons (7). This shows convincingly that cytochrome b from
beef heart, when isolated in Triton, is present as a dimer in similar
manner as in the case of *Neurospora crassa* , although in the latter case,
cholate, deoxycholate were used as detergents (14).

Using the cytochrome b dimer, the redox potential characteristics were
studied. For comparison, the properties of the b-type cytochromes of
the isolated complex III were investigated. The redox titrations were
performed with a combined potentiometric spectrophotometric technique
under anaerobic conditions with electrodes in the presence of redox

48

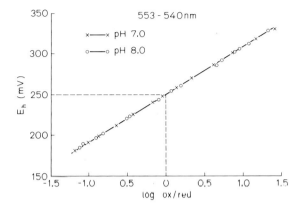

Fig. 6. Nernst plot of a redox titration of isolated cytochrome c_1 from beef heart. The titration was performed as described in the text

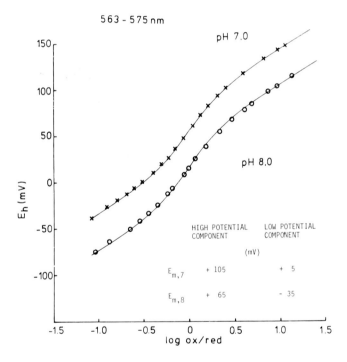

Fig. 7. Nernst plot of a redox titration of b-type cytochromes of complex III from beef heart, isolated according to (7)

mediators according to Dutton et al. (17). Titration was performed in a negative potential direction by addition of dithionite, in a positive direction by addition of ferricyanide.

Figure 6 shows a control titration with isolated beef heart cytochrome c_1. As in all subsequent titrations, it was performed in the presence of 0.05% Triton. The crosses indicate the titration at pH 7.0, the circles at pH 8.0 It is obvious that, as expected, the redox behavior of cytochrome c_1 is pH independent. The curve is a straight line with an angle of 60 mV per unit fraction (ox) over fraction (red). The midpoint potential of the redox component can be taken directly from the curve and is +250 mV.

Figure 7 shows two titration curves of the cytochromes b, when they are integrated in the isolated complex. This complex, isolated according to the hydroxyapatite method in Triton X-100 (15), consists of six polypeptides. It is present in a dimeric aggregation state, does not contain phospholipids or ubiquinone, and has bound antimycin (unpublished). The complex is held in a dispersed state by interacting with one Triton micelle per dimer. It was estimated that the Triton molecules can occupy only part of the surface of the multienzyme complex (15).

A membrane potential cannot influence the measured midpoint potentials under these conditions. Some objections made in the past (13, 19) regarding interaction between the membrane potential and the redox potential, and the existence of accessibility barriers do not seem to be valid for the subsequent titrations.

An analysis of the demonstrated sigmoidal titration curves of complex III gives an equal contribution of two components to the measured absorbance changes, provided it is assumed that there is no heme-heme interaction involving the redox potentials. The midpoint potentials of the two components - as indicated in the insert - are, at pH 7.0, +105 mV and +5 mV; and at pH 8.0, +65 mV and -35 mV, respectively. At

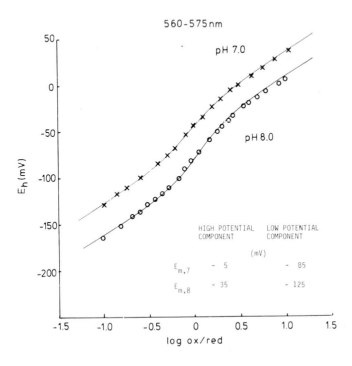

Fig. 8. Nernst plot of a redox titration of the isolated cytochrome b dimer from beef heart. Purification and redox titration were performed as described in the text

both pH values the midpoint potential difference of the two components is 100 mV. Accordingly, the pH dependence of the two midpoint potentials is 40 mV per unit pH.

The curves of Figure 8 depict the redox behavior of the isolated cytochrome b dimer. Consider, that in its isolated form, cytochrome b has lost the tight protein-protein interaction with its neighboring polypeptide subunits. The titration curves are shifted to a more negative potential region of about 100 mV, but the two different midpoint potentials, as well as the pH dependence of both midpoint potentials, are maintained. The midpoint potential difference of the two components has become a little smaller: instead of 100 mV, now about 80 mV. The pH dependence remains about 40 mV per unit pH.

Discussion

Analysis of the titration curves - as first performed by Wilson and Dutton (20) for cytochrome b of rat liver mitochondria - does assume the existence of two different b-type cytochromes, which do not possess a cooperativity of their redox centers involving the midpoint potentials. It is not yet known whether the titrated cytochrome b dimer consists of an $\alpha\beta$-dimer or of an $\alpha\alpha$-dimer. Hopefully protein chemistry will give a definite answer in the near future. However, under the assumption that the isolated cytochrome b is an $\alpha\alpha$-dimer, the cytochrome b titration curves have to be reanalyzed by a model implying a negative cooperativity of the midpoint potentials of the two redox centers (21).

As was mentioned in the Introduction, in order for a redox component to function as a hydrogen carrier, it must show a pH dependence of its midpoint potential. That seems to be the case for the isolated cytochrome b dimer.

If the acidic groups of the apoproteins of the two monomers would change from a completely protonated to a completely unprotonated state, depending on the redox state of the heme centers, the pH dependence should be 60 mV per unit pH for both subunits.

In the case of the isolated cytochrome b dimer and the isolated dimeric complex III, it amounts to only 40 mV per unit pH. This may rely on a small change of the pK of the respective groups and/or on a change of the surrounding pH, compared to the condition in the mitochondrion. However, the cytochrome b dimer seems to be the likely candidate for hydrogen ion transport in the region of the second phosphorylation site.

How can the mechanism of the vectorial Bohr effect of the isolated cytochrome b be visualized? Figure 9 combines part of a scheme given by Boyer (22) and part of a scheme given by Klingenberg (23). These schemes were created to elucidate the conformational coupling to the synthesis of ATP of a membrane potential and a proton gradient. The presented scheme offers a possible explanation of the conformational coupling of electron transfer between three redox components (ubiquinone $\rightarrow$ cytochrome b $\rightarrow$ cytochrome c_1) to a proton translocation. It implies that cytochrome b, or the cytochrome b dimer, span the mitochondrial membrane (this has not been shown so far experimentally). Fe^{3+} and Fe^{2+} may indicate the iron of the heme centers in its ferric and ferrous form, respectively. B^- and BH reflect the unprotonated and the protonated form of a hypothetical migratory group of a

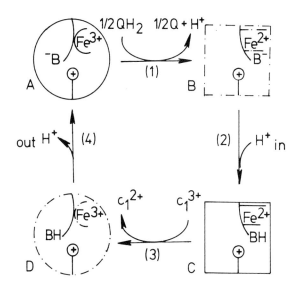

Fig. 9. Hypothetic scheme for the conformational coupling of electron transfer to protein gradient for cytochrome b. The different conformational states of cytochrome b are depicted by *circles* and *rectangles* with *full* and *dotted lines*, respectively. Fe^{3+} and Fe^{2+} may indicate the heme center in its ferric and ferrous forms. They are located more to the matrix side of cytochrome b. B^- represents the unprotonated, BH the protonated migratory group. The positive charge at the proton channel is supposed to prevent free diffusion of protons (22). Transfer of one electron from ubiquinone to cytochrome c_1 serves to drive one proton from the matrix to the cytosolic side

hypothetical proton channel. The positive charge of the channel is sup-posed to prevent free diffusion of protons [cf. Boyer (22)].

The key event in the conformational coupling may be the redox changes of the heme centers of the b cytochrome. The order of the reaction sequence may be explained by following through one cycle of the scheme: While one electron is transferred from ubiquinone to cytochrome b due to a conformational change, the group B^- may turn from a cytosolic to a matrix position (step 1). In this position group B^- may become pro-tonated to BH (step 2). The degree of protonation may depend on its pK in this position and on the present pH of the environment. In the third step the electron of ferrocytochrome b is transferred to cyto-chrome c_1. Synchronously with a second conformational change group BH may turn from a matrix to a cytosolic position. Finally group BH may become deprotonated due to its cytosolic position in dependence on the prevailing pK and pH (step 4).

One cycle may give a net translocation of one proton per electron transferred, provided that the migratory group B becomes completely protonated, respectively deprotonated, in its two positions.

Now the question arises, whether only the cytochrome b dimer, or each monomer spans the mitochondrial membrane. In the first case one might expect that one proton per electron may be transferred, whereas in the second case a stoichiometry of two protons per electron may be en-visaged. For both of these ratios conflicting experimental data are available in the literature (2, 24, 25).

To sum up, the primary event of the electrogenic proton translocation by cytochrome b may be the conformational changes of the protein, due to the redox changes of the heme centers. These may involve the movement of a basic group in the domain of a proton channel. Protonation and deprotonation of this group seem to be secondary events. The proposed mechanism requires no protein rotation for net translocation of protons.

The presented data have shown the so-called "Bohr effect" (26) for the isolated cytochrome b dimer. Functional studies with this preparation, reincorporated into phospholipid vesicles, may reveal whether cytochrome b is capable of creating a proton gradient, i.e. exerting a "vectorial Bohr effect" across a membrane.

Acknowledgments. The authors are grateful to Professor M. Klingenberg for his helpful suggestions and criticism. They thank Mrs. I. Werkmeister and Mrs. J. Drechsel for skillful technical assistance. This work was supported by a grant from the Deutsche Forschungsgemeinschaft.

References

1. Papa, S., Lorusso, M., Guerrieri, F.: Biochim. Biophys. Acta 387, 425 (1975)
2. Lawford, H.G., Garland, P.B.: Biochem. J. 130, 1029 (1973)
3. Hinkle, P.C., Leung, K.H.: In: Membrane Proteins in Transport and Phosphorylation. Azzone, G.F. et al. (eds.) Amsterdam: North-Holland Publ. Co., 1975, p. 73
4. Nelson, B.D., Mendel-Hartwig, J., Guerrieri, F., Gellerfors, P.: In: Genetics, Biogenesis and Bioenergetics of Mitochondria. Bandlow, W. et al. (eds.) Berlin: de Gruyter, 1976, p. 315
5. Papa, S.: Biochim. Biophys. Acta 456, 39 (1976)
6. Papa, S., Guerrieri, F., Lorusso, M.: Biophys. J. 15, 963 (1975)
7. v. Jagow, G., Schägger, H., Engel, W.D., Machleidt, W., Machleidt, J., Kolb, H.J.: FEBS Lett. 91, 121 (1978)
8. Goldberger, R., Smith, A.L., Tisdale, H., Bomstein, R.: J. Biol. Chem. 236, 2788 (1961)
9. Ohnishi, K.: J. Biochem. 59, 1 (1966)
10. Yu, C., Yu, L., King, E.: Biochem. Biophys. Res. Commun. 66, 1194 (1975)
11. Weiss, H.: Eur. J. Biochem. 30, 469 (1972)
12. Goldberger, R., Pumphrey, A., Smith, A.: Biochim. Biophys. Acta 58, 307 (1962)
13. Weiss, H., Ziganke, B.: Eur. J. Biochem. 41, 63 (1974)
14. Weiss, H.: Biochim. Biophys. Acta 456, 291 (1976)
15. v. Jagow, G., Engel, W.D., Riccio, P., Klingenberg, M., Kolb, H.J.: Biochim. Biophys. Acta 462, 459 (1977)
16. Tanford, C., Nozaki, Y., Reynolds, J.A., Makino, S.: Biochemistry 13, 2369 (1974)
17. Dutton, P.L., Wilson, D.F., Lee, C.P.: Biochemistry 9, 5077 (1970)
18. Caswell, A.H.: Arch. Biochem. Biophys. 144, 445 (1971)
19. Wikström, M.K.F.: Biochim. Biophys. Acta 301, 155 (1973)
20. Wilson, D.F., Dutton, P.L.: Biochem. Biophys. Res. Commun. 39, 59 (1970)
21. Wikström, M.K.F., Harmon, H.J., Ingledew, W.J., Chance, B.: FEBS Lett. 65, 259 (1976)
22. Boyer, P.D.: FEBS Lett. 58, 1 (1975)
23. Klingenberg, M.: In: Energy Transformation in Biological Systems. (Ciba Foundation Symposium 31) Amsterdam: Elsevier, Excerpta Medica, North Holland, 1975, p. 23
24. Papa, S., Lorusso, M., Guerrieri, F.: Biochim. Biophys. Acta 387, 425 (1975)
25. Brand, M.D., Reynafarje, B., Lehninger, L.: Proc. Natl. Acad. Sci. U.S.A. 73, 437 (1976)
26. Wyman, J.: Q. Rev. Biophys. 1, 35 (1968)

Sequence Homology of Cytochrome Oxidase Subunits to Electron Carriers of Photophosphorylation

G. BUSE

I would like to add to Dr. Hatefi's report, that our chemical studies on beef heart cytochrome c oxidase indicate that up to 12 components can be isolated from complex IV of the respiratory chain. These are listed in Table 1 according to their fractionation on BioGel P 60 columns, together with their molecular weights, site of biosynthesis, and N-terminal end groups (4).

Table 1. Polypeptide composition of beef heart cytochrome c oxidase

No. in prep. separation	Apparent mol. wt.	Synthesis	N-terminus
I	36,000 D	Mitochondrial	Blocked, F-Met
II	24,000 D	Mitochondrial	Blocked, F-Met
III	19,000 D	(Disputed)	Thr
IV	16,000 D	Cytoplasmic	Ala
V ·	11,600 D	Cytoplasmic	Ser
VI a,b,c	12,500-10,000 D	Probably cytoplasmic	All Ala
	(substoichiometric extrinsic parts?)		
VII	9 500 D	Cytoplasmic	Blocked
VIII a,b,c	6 000 D	Probably cytoplasmic	Ser, Ile, Phe

We also obtained this same pattern from enzyme preparations from other laboratories. The main constituents I, II, III, IV, V, and VII are apparently present in a one-to-one ratio.

Fraction VIII contains three small peptides, which, from their primary structures (1), may have, among others, the function of cytoplasmically synthesized "precursors" specifically cleaved off at the complex formation in the inner mitochondrial membrane.

Recent results of our sequence work on several of the main components show that polypeptide II belongs to the azurin/plastocyanin protein family (see, e.g., Table 2a) and thus is one of the copper subunits of the oxidase, containing the invariant two histidine, cysteine, and methionine residues, which have now been demonstrated to bind copper in plastocyanin (2).

Following the first invariant histidine, we find most of the tyrosine and tryptophan residues of this chain concentrated in a short segment, which may be involved in charge transfer in this protein.

Again as concluded from sequence similarity around the invariant heme-Fe complexing methionine of cytochrome f and C 555, polypeptide VII (VI in the numbering of some other investigators) should belong to the cytochrome c protein family and may be the heme a moiety (Table 2b).

Table 2. Alignment of characteristic cytochrome *c* oxidase sequences to sequences of copper and heme proteins (residues possibly involved in copper or iron coordination in italics)

(a)		
Cyt. ox. pp II	Met-Gly-*His*-Gln-Trp-Tyr-Trp-Ser-	
Azurin (e.g., *Pseudomonas fluorescens*)	Met-Gly-*His*-Asn-Trp-Val-Leu-Ser-	
Plastocyanin (e.g., french bean)	- - *His*-Asn-Val-Val-Phe -	
Cyt. ox. pp II	Tyr-Tyr-Gly-Gln-*Cys*-Ser-Glu-Ile-Cys-Gly-Ser-Asn-*His*-Ser - - Phe-*Met*-	
Azurin	Tyr-Met-Phe-Phe-*Cys*-Ser - Phe-Pro-Gly - *His*-Ser - - Ala-*Met*-	
Plastocyanin	Tyr-Ser-Phe-Tyr-*Cys*-Ser - Pro - *His*-Gln-Gly-Ala-Gly-*Met*-	
(b)		
Cyt. ox. pp VII	Ala-*Met*-Thr-Ala-Lys-Gly-Asp-Val-Ser-Val-Cys-Glu-Glu-Tyr-Arg-	
Cyt. *c*555 (e.g., *Chlorobium limicola*)	Phe-*Met*-Pro-Ala-Lys-Gly-Gly-Asn-Pro-Asp-Leu-Thr-Asp-Lys-Gln-Val-	
Cyt. *f* (e.g., *Porphyra tenera*)	Ala-*Met*-Pro-Ala-Phe-Gly-Gly - Arg-Leu-Val-Asp-Glu-Asp-Ile -	

As Dr. Chance just reported, both polypeptides have been shown to face the outer side of the inner mitochondrial membrane and have even been cross-linked with cytochrome c (3).

Looking upon some 3×10^9 years of evolution of the electron transfer chains of mitochondria, chloroplasts, and (photo-) phosphorylating bacteria, the common origin of their terminal electron carrier segments is thus clearly indicated.

References

1. Buse, G., Steffens, G.: Z. Physiol. Chemie 359, 1005-1009 (1978)
2. Colman, P.M., Freeman, H.C., Guss, J.M., Murata, M., Noris, V.A., Ramshaw, J.A.M., Venkatappa, M.P.: Nature (London) 272, 319-324 (1978)
3. Erecińska, M., Wilson, A.F., Blasie, J.K.: 11th FEBS-Congress, Copenhagen, abstr. no. A4-1122-4, 1977
4. Steffens, G., Buse, G.: Z. Physiol. Chem. 357, 1125-1137 (1976)

Low Temperature Electron Transport in Cytochrome c in the Cytochrome c-Cytochrome Oxidase Reaction: Evidence for Electron Tunneling

B. CHANCE, A. WARING, and C. SARONIO

Introduction

Chairman Decker's charge to provide an overview of the field and at
the same time to incorporate new information is a difficult one to
which I shall respond as well as possible in terms of current research
on bioenergetics. Recent reviews on the mechanism of energy coupling
(7) make special reference to the multiple hypotheses of energy coupling by chemical intermediates, conformation change, and charge separation as the most important contributors to the conservation of
energy in mitochondrial membranes. The reason for the multiplicity of
the hypotheses in oxidative phosphorylation is clear: There is too
little information on fundamentals of electron transfer reactions to
decide among them. In fact, thus far, we do not have a clear insight
into how electrons are transferred in the macromolecules of the respiratory chain and find it essential to afford an explanation of this
basic phenomenon before building hypothetical houses of ephemeral cards.
So firstly, I will present what I consider to be "useful knowledge" in
terms of the American Philosophical Society's motto (1) before I can
embark upon hypotheses of energy coupling mechanisms.

The problem of electron transfer between pairs of hemoproteins seemed
simple in the context of heme-discs protruding to the periphery of the
protein with edge-to-edge accessibility of the two proteins and electron transfer simply by contact. Contact electron transfer mechanisms
were supplemented by electron hopping over amino acid chains (23, 26)
and by electron transfers between aromatic amino acids (23) to the
point that mechanisms appeared to take into account electron transfers
between hemoproteins operating on a larger scale of dimensions. Furthermore, all of these reactions were Arrhenius "over the barrier"
electron transfers. However, the concept of the exposed heme edge seems
appropriate for the 13,000 molecular weight cytochrome c (23), but does
not seem currently appropriate for the electron donor and acceptor complexes bc_1 and aa_3 respectively for cytochrome c whose equivalent molecular weight lies in the vicinity of 10^5 daltons. In fact, a variety of
methods of increasing precision indicate that the distances from
the periphery of cytochrome oxidase to heme a and a_3 are greater than 15 Å
and furthermore that the distance between the heme of cytochrome c and that
of cytochrome $a + a_3$ is in the vicinity of 25-30 Å - a value obtained independently from two laboratories (24,40). Although one may again invoke the possibilities of electron hopping along peptide chains and aromatic to
aromatic transfers, the fact remains that the first process involves
a barrier approaching the ionization potential, and the second is
unlikely because the reactivity of cytochrome c seems inpedendent
of the presence of aromatic residues essential to these mechanisms (23). In fact, even the recent formulation of Hopfield (32)
for a simplified tunneling calculation depends upon the presence of
peripheral aromatics on both donor and acceptor. The fact that electron tunneling from cytochrome c does occur is evidenced by what I now
term the "Chromatium experiment" in which the temperature-independent

portion of the kinetics from 4 to 125 K gives clear-cut testimony to
the possibility that cytochrome c itself can be an electron donor in
the tunneling process over distances estimated in a rudimentary calculation to be approximately 30 Å (21). A number of more sophisticated
calculations have been based upon both electron and nuclear tunneling
possibilities (34). In this paper we shall consider the specific case
of electron transfer from cytochrome c to cytochrome oxidase where
the capability of cytochrome c to participate in tunneling processes
in respiration as well as in photosynthesis (14, 16) and the recent
data on the high probability of large distances from the heme-cyto-
chrome c to the electron acceptor heme in cytochrome oxidase (24, 40)
suggest tunneling as the preferred process.

The impact of this concept upon the organization of the respiratory
chain is significant since it affords a possibility to have large macromolecular membrane-bound complexes, each relatively inaccessible
to each other, as electron transfer acceptor-donors capable of participating by a tunneling process as a functional role in electron transport
and in energy coupling as well. However, basic questions of regulation
of electron flow and maintenance of high specificity need to be considered in order to be consistent with the observed characteristics of
high degree of control and specificity in the respiratory chain (16).

In order to estimate precisely the tunneling parameters for the cytochrome c-cytochrome oxidase electron transfer reaction, it seems necessary to have detailed knowledge of the chemical and physical properties of both molecules. While many such data are available for
cytochrome c, our knowledge of these properties for cytochrome oxidase
is rudimentary. Thus, a complete description of the electron transfer
reaction according to a tunneling mechanism is not yet possible. On
the other hand, significant increases in our knowledge of cytochrome
oxidase have occurred from the standpoint of electron transfer mechanisms and structural relationship of the components of the oxidase.
Thus, the first portion of this talk is a summary of results obtained
in a variety of laboratories on the structural features of cytochrome
oxidase and is followed by a discussion of recent work on the intermediate compounds of cytochrome oxidase and oxygen (12, 13) together
with some recent data on the nature and role of the iron and copper
atoms of the oxidase, especially the copper atom structurally associated
with heme a_3, which we here identify as Cu_{a_3}; i.e., $a_3 \cdot Cu_{a_3}$ and that
kinetically associated with heme a, as Cu_a; i.e., $a \cdot Cu_a$.

Physical and Chemical Methods

In view of the ambiguities of the nature of the "solubilized oxidase"
(29), - the possibility of heterogeneity, the low activity of some
preparations, and the real possibility of structural alterations in
the relationships of cytochrome c to cytochrome oxidase - our initial
experiments have been conducted with suspensions of intact mitochondria
in which the cytochrome c is maintained at its physiologic concentration or has been removed by salt washing.

Preparations from pigeon hearts exhibit their full phosphorylating
activity as do,to a smaller extent,those from beef heart. In addition, frozen-thawed beef heart mitochondria, less capable of respiratory control, have also been used. Preparations retain their energy
coupling properties to $\sim$ -15°C (11), and apparently energized states
may be stabilized at low temperatures by rapid freeze-trapping processes.

It appears therefore that the electron transfer reactions described in this paper are from configurations that are characteristic of physiologic functions in energy coupling of the cytochrome oxidase (44,45), rather than from configurations that are capable of electron transfer only.

Generally, optical methods have proved most satisfactory in studying cytochrome, primarily because of selectivity and speed of read-out of the data. One important point in this contribution is the extent to which the heme and copper portions of cytochrome oxidase can be identified by physical methods. Optical spectroscopy appears to provide a read-out of the oxidized copper components in near infrared regions of the spectrum (30,15,43). In reduced mitochondria there is negligible absorption or change of absorption with liganding of the heme in the infrared region (43,8), which strongly indicates that there is no measurable heme absorption. The identification of the copper associated with heme a_3 as a Type I copper atom suggests that a portion of the oxidized spectrum at 609 nm may be due to oxidized copper (collaborative studies with Bell Telephone Laboratories and Stanford Synchrotron Radiation Laboratories, L. Powers et al. 38a).

A more penetrating optical method is afforded by magnetocircular dichroism, which is especially appropriate for the study of low-spin hemes (2). The possibility of applying this to the intermediate compounds of cytochrome oxidase and oxygen has not yet been achieved but will be an important diagnostic method for understanding the nature of these and other compounds. Measurement of infrared stretching frequencies of cytochrome oxidase-CO, has already been achieved and the similar extension of this method to the oxygen compounds also seems possible (3). The "invisibility" of the copper atom associated with heme a_3 and the lack of diagnostic hyperfine structure in the signal from the copper associated with heme a make electron spin resonance (ESR) a less desirable technique for studying the copper atoms of cytochrome oxidase than the optical methods. Furthermore, the assignment of high-spin signals to heme a and a_3 appears complicated (5).

One of the few methods that appear capable of obtaining characteristically different signals from both the oxidized and the reduced forms of the two copper atoms of cytochrome oxidase is specific X-ray photon absorption at the K edge, particularly when the X-rays are obtained at very high fluxes, as in the 20-100 MA 3.8 GeV synchrotron radiation at the Stanford Linear Accelerator. A recent work in this case allows the deconvolution of the four possible species of the two copper atoms and makes appropriate assignments in the oxidized and reduced forms of these two metal atoms (collaborative studies with Bell Telephone Laboratories and Stanford Synchrotron Radiation Laboratories, L. Powers, et al. 38a). Preliminary results indicate that $a_3 \cdot Cu_{a_3}$ is a Type I, blue "ionic" copper protein such as stellacyanin, while Cu_a is a nonblue "covalent" copper such as copper imidazole. The eventual application of this method to the heme iron atoms is feasible, as has been the case with hemoglobin and iron-sulfur proteins (39). Further studies of extended edge absorption fine structure (EXAFS) show second shell reflections at 3.4 Å from the iron atom of oxidized, but not reduced, cytochrome a_3, suggesting an important conformation change in the iron-copper relationship in the oxidation reduction cycle (L. Powers et al., 38a).

In summary, optical methods seem essential for exploratory studies of the nature of cytochrome oxidase compounds and of their reaction kinetics. Since the optical transitions are inherently complicated, supplementary methods are invaluable in describing the chemical nature of the optically identified intermediates and electron transfer reactions.

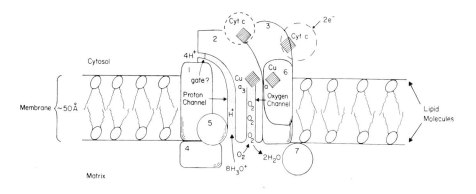

<u>Fig. 1</u>. A schematic diagram of the location of the subunits, heme groups (a, a_3), copper atoms, oxygen pocket, proton channel, and other features currently proposed for cytochrome oxidase structure and function. (Courtesy Elsevier, Ref. 14)

Experimental and Theoretical Background

Current Knowledge of Cytochrome Oxidase Structure

An important experimental observation that has resulted in much of the structural information obtained from low-angle X-ray and low-dose electron microscopy was that of Wakabayashi et al. (42), in which they found that two-dimensional crystalline arrays of cytochrome oxidase could be obtained by centrifuging the vesicular preparations containing the oxidase. Similarly, the oriented specimens can be obtained by centrifuging submitochondrial particles and even mitochondria (6). Thus, the variety of structural studies restricted to crystalline arrays can now be applied to the cytochrome oxidase molecule. In fact, considerable progress has already been made with the unstained image state by Henderson et al. (31), which suggests the existence of four 30 Å columns of peptide chains perpendicular to the plane of the membrane as clearly visualized on a smaller scale in the much smaller purple membrane protein.

A summary of current data on the cytochrome oxidase structure and function is afforded by Figure 1 (14, 24). In addition, the unusual properties of cytochrome oxidase as a membrane protein are emphasized. Low-angle X-ray diffraction together with image processed "low-dose" electron microscopy (31) suggests that as much of the 120 K-dalton molecule extends beyond the membrane as is located within it. Thus, much of the molecule consists of hydrophilic protein extending into the water phase, external to the membrane, and only one of the subunits seems completely within the hydrophobic phase of the membrane. Cytochrome c interaction exhibits two binding sites, one to subunit 3 and the other to both subunits 2 and 3. The oxidase molecule is large and the dimeric form is equivalent to 240,000 daltons. Thus, the diagram illustrates only one of the two molecules and no information is available concerning the function of cytochrome oxidase as a dimer. The existence and the final identification of the seven subunits that bear the hemes are not yet available. However, the possibility that hemes a and a_3, with their associated coppers, are in separate subunits is indicated by the diagram together with the approximate size of the subunits.

The diagram indicates the hemes to be considerably displaced from the periphery of the subunits so that paramagnetic ions such as nickel (9) and dysprosium (Salerno, J., Leigh, J.S. and Ohnishi, T., personal communication) cause little or no line broadening of the ESR signals of hemes a and a_3; thus, location of these hemes from points on the periphery of the protein accessible to the paramagnetic ions is at distances greater than approximately 15 Å. Somewhat larger distances are indicated by resonance energy transfer from zinc cytochrome c to heme a and a_3 of cytochrome oxidase, because these distances include a contribution of 5-10 Å from cytochrome c itself (40). Thus, the two methods seem to give concordant results and place hemes a and a_3 at such a distance from cytochrome c that tunneling mechanisms seem essential. Distance measurements, however, do not distinguish the four components of cytochrome oxidase, currently available resonance energy transfer data represent resonance transfers between the cytochrome c heme and that of a or a_3 or both. Similar studies with derivatives of cytochrome c fluorescing in the infrared region and showing resonances with absorption of the copper component of hemes a and a_3 appear to be essential for further resolution of the geometry of the electron donor-acceptor (41 and Vanderkooi, J.M., personal communication). The proximity of the copper associated with heme a_3 and that of heme a_3 itself is deduced from the lack of a distinguishable g = 2 electron paramagnetic resonance (EPR) signal from this copper atom (4), which is considered by various authors to be antiferromagnetically coupled with that of heme a_3 and thus only a few angstroms distant. Interaction of heme a_3 with the copper of a is suggested by the considerable increase of the g = 6 EPR signal of heme a caused by photolysis of the a_3CO compound at very low temperatures (but cf. (5)).

The EPR undetectability of Cua_3^{2+} suggests antiferromagnetic coupling with heme a_3 and indicates a small distance between the two atoms. Their proximity is confirmed by the preliminary results obtained by extended K edge absorption line structure (EXAFS) that indicates the presence of a heavy metal atom near the iron in the oxidized state and its absence in the reduced state, (L. Powers et al. 38a). K edge absorption studies of copper also tell us that Cua_3 is a stellacyanin -like protein, (38a,) the peptide sequence of which is found in subunit 2 (Buse et al., this colloquium). The above-mentioned short distance between copper and iron places heme a_3 and Cua_3 in the same subunit, and the other iron and Cu near enough for electron transfer by a tunneling process, i.e., 20-30 Å, although the subunit location remains unknown. In addition the rapid redoxequilibration of heme a and Cua requires them to be within tunneling distance also.

The eventually more precise determination of these distances by physical methods is of course of great interest, and many latent possibilities exist: for example, the determination of the iron-iron distance by neutron diffraction using ^{57}Fe (Blasie, J.K., personal communication) and the further studies with synchrotron radiation of the nearest and next-to-nearest neighbors of the iron and copper atoms; finally, the X-ray diffraction studies of anomalous diffraction in highly ordered multilayers of cytochrome oxidase seem possible (Blasie, J.K. and Leigh, J.S., personal communication).

Two structural features of the oxidase are based upon functional rather than structural approaches and are illustrated in the diagram in Figure 1 as an oxygen pocket (14) and a proton channel (45). The evidence for an oxygen pocket arises from low-temperature kinetic experiments on the freeze-trapped oxidase in which a bimolecular reaction of oxygen with heme a_3 is observed over a significant variation of oxygen concentrations and temperatures (12, 13). In the vicinity of -105°C, the rate of formation of oxy-cytochrome oxidase (compound A_1)

is proportional to the oxygen concentration in solution at the time of
freeze trapping the sample. Thus a representative sample of oxygen
concentration present at the time of freeze trapping is contained in
the vicinity of the active site and is capable of free diffusion at
-105°C. This feature is confirmed by the studies of Beinert (this colloquium) that identify ESR signal changes of the oxidized enzyme in response to the occupancy of the active site by various nonreacting gases.
The second functional feature is based on observations of the stoichiometry of H^+: e^- in excess of 1:1 as observed in the studies of
Wikström (45). Presumably, the amplification of the stoichiometry from
one hydrogen ion per electron to higher values requires couplings between the electron transfer reaction and structural features involving
proteolytic groups on both sides of the membrane and within the channel altering their pKs (10, 38). Thus, depending upon the redox state
of the cytochrome oxidase molecule, protons may be exchanged by a
variety of groups and a net change of proton concentration in the
required stoichiometry and on the required side may be obtained. The
process of coupling of the pKs of peripheral groups on the protein to
the state of liganding the iron is similar to the "Bohr" effect in
hemoglobin and as an analogous process, is termed "membrane Bohr effect"
(10), and appears in purple membranes following light activation of the
chromophore as described by Oesterhelt in this colloquium (pp. 140-151).
The channel shown in Figure 1 may involve the columns of amino acids
orthogonal to the plane of the membrane identified in the purple membrane protein and as larger structures in cytochrome oxidase itself
by optically processed electron microscopy (31).

The rotational mobility of the cytochrome oxidase protein in the biological membrane can be precisely evaluated by photo-induced dichroism
(35) as indicated by the diagram in Figure 2. Flash illumination of
the CO compound of cytochrome a_3 establishes a dichroic component of
the light absorption which randomizes as the molecules execute their
motions in the membrane. However, no change of dichroism is observed
and the molecules can be regarded as essentially stationary in the
time-frame of electron transport, which occurs at rates of 300 per
second at room temperature (25).

The complex containing cytochromes b and c_1 is of similar size and
may well be immobilized (25) (Fig. 3). Thus, cytochrome c may function to communicate electrons between two macromolecular immobilized
membrane proteins see Fig. 3.

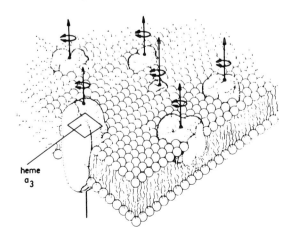

heme
a_3

Fig. 2. Junge's diagram of cytochrome oxidase in phospholipid
bilayer membranes indicating
possibilities for rotational
motion. The position of the heme
plane has now been determined
to be perpendicular to the plane
of the membrane. [Courtesy of
W. Junge (35)]

The process by which cytochrome c accomplishes its function involves rotational and translational diffusion, causing an approach of cytochrome-c heme close enough to the active center of the donor or acceptor for tunneling to occur. The positions from which the tunneling occurs are presumably the identifiable binding sites for cytochrome c. This electron transport model differs from the "solid state model", in which cytochrome c carries out its electron transfer function as an immobilized component (27).

Tunneling Processes

In electron transfer reactions controlled by the Arrhenius law, the electron must achieve an activation equal to or greater than the energy barrier in the reaction coordinate system. This leads to the simple exponential dependence upon temperature that is characteristic of the Arrhenius law. The quantum mechanical probability that the electron may, instead of going over the barrier, tunnel through the barrier, leads to a different formulation where the temperature dependence may be zero. Since, however, the probability that the electron may be capable of tunneling through the barrier may depend on a variety of thermodynamic and configurational properties, temperature dependence may also be observed in electron tunneling reactions. Thus, temperature independence itself is not the essential criterion for establishing whether or not electron tunneling occurs (32). When temperature independence is observed, as for example in the chromatium experiment (Fig. 4), tunneling seems to be the only explanation (21, 22,28,32,34). However, when a temperature-dependent electron transfer reaction occurs over a distance greater than that for which contact electron transfer mechanisms are feasible, then tunneling seems to be the only possible electron transfer mechanism. Since we already have a unique example of temperature-independent electron transfer in the Chromatium experiment, the extension of the tunneling idea to cases where the reaction is temperature dependent,but in which the distances are large,seems highly appropriate, and is indeed the guiding hypothesis of this contribution.

It is not possible within this contribution to provide more than a superficial introduction to the processes of electron tunneling, nor to deal with a number of the unresolved issues in tunneling mechanisms,

ELECTRON TRANSPORT MODEL

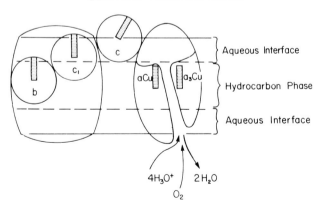

Fig. 3. The function of cytochrome c in transferring electrons from the relatively immobilized donor complex to the acceptor complex by lateral and rotational diffusion

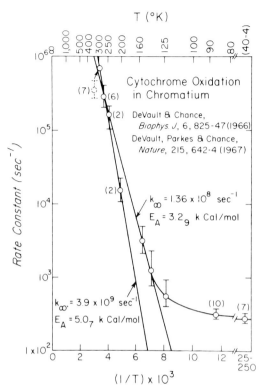

Fig. 4. DeVault's graph of experimental data on the kinetics of cytochrome *c* by the light-oxidized reaction center of the purple sulfur bacteria *Chromatium* at temperatures from 4 to 300 K. The half-time at very low temperatures is 2 ms and at room temperature is 1 μs. The range of limiting slopes of the temperature-dependent portion is 3.9×10^{-9} to 1.36×10^{8} s^{-1}. [Courtesy D. DeVault (22)]

for example, the role of nuclear and electron tunneling. However, in biological systems, the feature of a tunneling mechanism that distinguishes tunneling from other electron transfer mechanisms is the possibility of transferring electrons over longer distances than would be possible by contact electron transfer (the electron transfer is over much longer distances than the nuclear tunneling). Thus, the "feasibility criterion" insofar as biological reactions are concerned is the maximum distance between electron donor and acceptor. Two conditions of rate control are encountered. First, an alteration of nuclear coordinates by tunneling to a new configuration may be accompanied by an instantaneous tunneling of the electron. This is explicit in the Born-Oppenheimer approximation under which calculations of nuclear tunneling are made. Secondly, the electron tunneling may be slower than nuclear tunneling and thus a rate-controlling process. For regulation of electron flow, nuclear tunneling appears more appropriate as a controllable rate-controlling process. Obviously, tunneling over long distances is highly desirable from the standpoint of possibilities of organization of the respiratory carriers.

The physicist's result is indicative of the biological possibility. A gallium-selenide barrier between an aluminum and a gold electrode is indicated in Figure 5 (37). The abscissa gives the experimentally determined thickness of the barrier and the ordinate, the observed time constant. More useful experimentally is the half-time for the electron transfer reaction in seconds. For biologically interesting electron transfer times, particularly (10^{-2}-10^{-3} s), it is clear that this model would afford electron transfer over a more than adequate distance (approximately 90 Å); for rates consistent with the photosynthetic systems (10^{-6} s), allowable distances would be approximate-

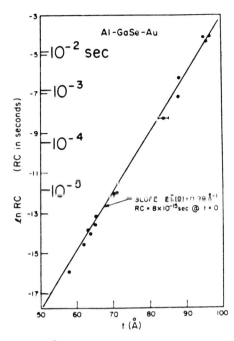

Fig. 5. The relationship between the tunneling time (abscissa) and the tunneling
barrier width (Å) for a gallium selenide
barrier of various width flanked by aluminum and gold electrodes [Courtesy
of Phys. Rev. (37)]

ly 60 Å. This simple diagram therefore demonstrates the feasibility
of electron tunneling over somewhat longer distances than those adequate to meet the requirements of the biological system.

Experimental Results

Role of Iron and Copper in the Reaction of Cytochrome Oxidase and Oxygen

In considering the electron pathway of cytochrome oxidase, it is necessary to identify the nature of the electron acceptor from cytochrome
c. Conventionally, this is considered to be heme a and its associated
copper ($a \cdot Cu_a$), and indeed, numerous experimental data have been put
forward that justify this viewpoint. However, at low temperature, we
observe that $a \cdot Cu_a$ is not oxidized to a measurable extent, and thus any
oxidation of cytochrome c in the low temperature range involves a novel
electron transfer pathway. The intermediate, which remains in a steady

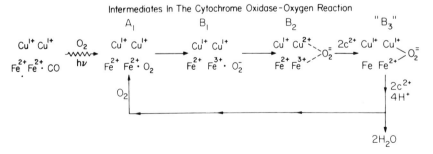

Intermediates In The Cytochrome Oxidase-Oxygen Reaction

Fig. 6. A mechanism for the reduction of compounds B_2 and B_3 of cytochrome oxidase
and oxygen with cytochrome c in a two-step reaction at temperatures of $-80°$ to $-20°C$

state during the oxidation of cytochrome c at low temperatures, is compound B_2, the second in a series of intermediates formed when the completely reduced CO-inhibited oxidase is flash photolyzed in the presence of oxygen and the exchange of oxygen for CO as ligand is accomplished (Fig. 6), leading to the oxy-compound A_1 and to compound B_2 (12).

Compound A_1 is stable at -120°C and at-100°C is slowly converted into a second intermediate compound B_2, proposed to be a bridge peroxide. The possibility that electron delocalization in the oxygen compound makes a superoxide anion intermediate (B_1) is included only for completeness; no evidence of this intermediate is observed when the iron and the copper are initially in the fully reduced state. Compound B_2 is in a steady state to a temperature of -40°C, above which the oxidation of the reduced $a \cdot Cu_a$ occurs, leading to the fully oxidized or resting state of the enzyme as indicated in the diagram.

Reaction of compound B_2 with reduced cytochrome c could reduce either the iron and copper or the peroxide. Thus, cytochrome oxidase could again react with oxygen to recycle through the compound B_2 state without, from the formal point of view, having produced the fully oxidized or resting state. This would appear to be different from the oxidation reaction that has hitherto been considered for the room temperature role of cytochrome oxidase.

Role of Cytochrome a and a_3 in the Formation of Compound B_2

Experiments with cytochrome oxidase prepared from mammalian tissues give obscure results on the role of cytochromes a and a_3 due to the overlap of the absorption bands of these cytochromes in the α and γ band regions. On this basis it is difficult to state unequivocally that compound B is formed without participation of cytochrome a. *Candida utilis* cells grown in a copper-deficient medium (19) show little if any synthesis of cytochrome oxidase. If, however, the cells are incubated with 200 mg/liter of copper, cytochrome oxidase is synthesized in profusion (36), the low temperature absorption spectrum of the reduced cytochrome oxidase reveals two bands, one at 599 and the other at 609 nm. The 9-nm spread between the two bands is adequate at low temperature for readily discerning them. The longer wave band is responsive to CO and is also diminished on reaction with oxygen and formation of compound B_2 (as shown in Fig. 7), leaving the 599 nm band near its original intensity. Compound B_2 forms after 20 s at -74°C. Thus, in the case of mitochondria prepared from *Candida utilis*, cytochrome a_3 is identified with the longer wavelength absorption and cytochrome a with the shorter.

Subsequent steps in the reaction can be observed by dual-wavelength spectrophotometry, and summaries of experimental data presented elsewhere (14,18) show that a steady-state level of compound B_2 is maintained during the oxidation of cytochrome c at -65°, -58°, and -49°C (Fig. 8). In those data, it is clear that compound B_2 can be formed at -100°C with a half-time of 2 min and that it is only at -70°C that cytochrome c oxidation kinetics are observed to have a half-time of 2 min. At the same time the appearance of ferricytochrome a_3 occurs. Only at -50°C does cytochrome a itself become oxidized appreciably and its oxidation is cytochrome c dependent.

Figure 9 affords a summary of kinetic data on compound B at 608-630 nm, cytochrome a_3^{3+} at 630-650 nm, and cytochrome c^{2+} at 550-540 nm at three temperatures (-71°, -61°, and -51°C). At -71°C, flash photolysis produces the ligand exchange, and compound A_1, followed by a nearly constant

66

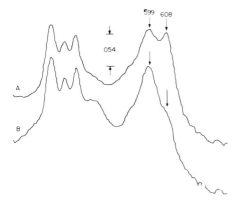

Fig. 7. The reduced-minus oxidized difference spectrum of the cytochromes of
yeast mitochondria prepared from *C.
utilis* grown in 200 µg/ml of added copper sulfate. The *top trace (A)* represents the fully reduced state of the
CO liganded oxidase with reference to
the oxidized state. The *lower trace
(B)* represents the formation of compound B_2 at 20 s after flash illumination of the reduced CO compound,
with the corresponding loss of absorption at 608 nm. Candida Utilis 0.3 M
Mannitol 20% Ethylene Glycol 10 mM
Phosphate Buffer 1.2 mM CO 100 µM O_2
pH 7.4

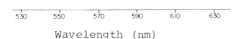

Wavelength (nm)

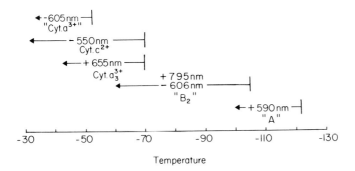

Fig. 8. A summary of the temperatures at which there is half completion in 2 min of
the response attributable to the various spectroscopically identifiable components
of the reaction of cytochrome oxidase and oxygen

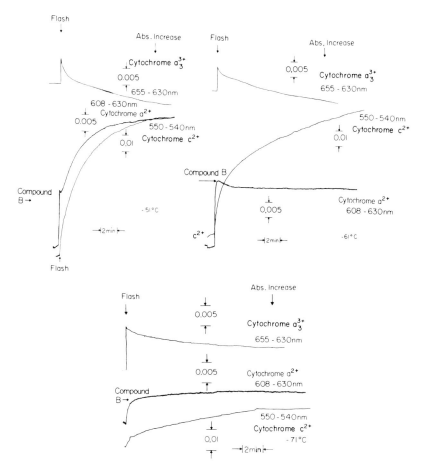

Fig. 9. A composite of recordings of compound B (608-630 nm), cytochrome c (550 540 nm), and a_3^+ (655-630 nm) at -71°, 61°, and -51°C. The reaction is initiated by flash photolysis of the CO compound of cytochrome oxidase in the presence of excess oxygen concentration using beef heart mitochondria under contitions similar to those of Fig. 7

concentration of compound B, is formed. At the same time that the cytochrome c is oxidized, parallel kinetics for the formation of cytochrome a_3^+ appear (top trace). Throughout this interval, compound B appears to remain in its constant steady-state level. This phenomenon is even clearer at -61°C, where a much more abrupt formation of compound B_2 occurs together with a slight overshoot of its kinetics. Cytochrome c oxidation is now very rapid and biphasic as well. The formation of the ferric-cytochrome a_3 has an initial fast phase like that of cytochrome c and then continues in a roughly parallel way. At the highest temperature, -51°C, the kinetics of cytochrome c oxidation now find for the first time a parallel response in the kinetics of compound B, which is here interpreted to be the oxidation of cytochrome a, registered in this case at the same wavelength as the kinetics of cytochrome a_3 since beef heart mitochondria were used. Under these conditions there is a rapid initial formation of the a_3^+ followed by a much slower turnover, again paralleling the cytochrome c oxidation kinetics as if the two were linked, both in the kinetics and in the stoichiometry. Of considerable interest is the total absorbancy measured at 608 nm at -51°C, being 2 1/2 times greater than that which corresponds to compound B formation. Thus the involvement of heme a_3 in compound B formation causes only 40% of the a band of cytochrome oxidase to disappear, again consistent with the nearly exclusive involvement of cytochrome a_3 in compound B_2 formation. The kinetics of the slower phase at 608 nm and of corresponding cytochrome c changes exhibit almost a 1:1 correspondence.

Effect of Cytochrome c on Cytochrome Oxidase Kinetics

It is essential here to reiterate a key experiment concerning the effects of cytochrome c on the kinetics of the slow phase of the 608 nm changes illustrated in Figure 9 (see Fig. 10). In this experiment, the reaction at 608-630 nm and at 550-540 nm is shown, full details having been published elsewhere (18,20). With the normal cytochrome c content, the correlation between the slow phase of the cytochrome oxidase trace and the cytochrome traces is apparent. In the lower set of traces the cytochrome c content has been depleted to approximately one-fifth of the normal value by KCl extraction (33). While the intermediates form at the previous rates, the oxidation of cytochrome a as well as that of cytochrome c is slowed, suggesting a central role for cytochrome c in the oxidation of cytochrome a.

A diagram representing the various pathways for reaction of cytochrome c with cytochrome oxidase is given in Figure 11. In this diagram, the low temperature pathway of cytochrome c to compound B_2 is detectable at -80°C and occurs with a 2-min half-time at -50°C, as indicated in Figure 6 above. In accordance with the kinetic data, it is unnecessary to postulate any other pathway for cytochrome c oxidation up to a temperature of about -20°C. Nevertheless, we must account for the oxidation of cytochrome a (slow phase at -51°C, Fig. 9). In order to explain the relatively slow oxidation of cytochrome a, which is not observed until a temperature below -51°C and which depends upon the presence of cytochrome cf, two possibilities may be cited, both of which are based upon the well-known reversibility of the reaction of cytochrome c with cytochrome a and the reaction of cytochrome a with cytochrome a_3. Cousidering the first case, cytochrome c oxidized by compound B could in turn oxidize cytochrome a and Cu_a and thereby be reduced. This could account for the very close parallelism of the kinetics of cytochromes c and a observed at -51°C. Alternatively, the oxidation of cytochrome

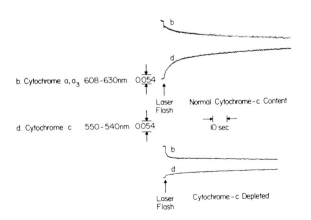

b. Cytochrome a, a_3 608-630nm 0.054

Laser Flash   Normal Cytochrome-c Content

d. Cytochrome c   550-540nm 0.054

10 sec

Laser Flash   Cytochrome-c Depleted

Fig. 10. Effect of cytochrome c depletion (*trace* d) upon the response of cytochrome a (*trace* b) as measured at 608-630 nm in beef heart mitochondria at -38°C (See Ref. 18)

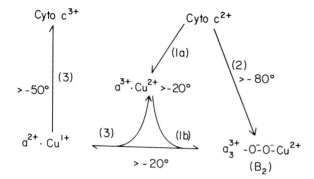

Fig. 11. Chemical equations for the cytochrome c-cytochrome oxidase reaction for four temperature ranges

a by compound B_2 at low temperature could be so slow that each molecule of cytochrome a that was oxidized would be reduced immediately to give a constant absorbance level at 608 nm. When the reaction of cytochrome a with compound B becomes more rapid than the reaction of cytochrome c with cytochrome a, a significant concentration of cytochrome a can be obtained. This latter possibility is, however, rendered extremely unlikely, as indicated by the diagrams of Figures 10 and 11. *Note added in proof:* The resting state may contain $O^=$, according to W. Blumberg and J. Peisack (personal communication). Thus the last equation would be $a_3{}^{2+}-O^-.O^-.Cu^{1+} \longrightarrow a_3{}^{3+}-O^{2-}.O^{2-}.Cu^{2+} \xrightarrow{+2H^+} a_3{}^{3+}.O^{2-} + Cu^{2+} + H_2O$.

Discussion

The Reaction Mechanism

The reaction mechanism is summarized in the diagram in Figure 12. The first step below -120°C, in agreement with Figure 6 above, corresponds to the formation of compound A_1 and at -100°C to its conversion to compound B_2.

Above -120° $a_3^{2+}.Cu^{1+} + O_2 \longrightarrow a_3^{2+}.O_2\,Cu^{1+} \longrightarrow a_3^{3+}-O-O-Cu^{2+}$
                    (A_1) (B_2)

Above -80° $2c^{2+} + a_3^{3+} - O^-.O^- Cu^{2+} \longrightarrow 2c^{3+} + a_3^{2+}-O^-O^- Cu^{1+}$
                 $a_3^{2+}-O-O\ Cu^{1+} + 4H^+ \longrightarrow a_3^{3+}.Cu^{2+} + 2H_2O$
                        (B_3) (MVS)

Above -50° $2c^{3+} + a^{2+}.Cu^{1+} \longrightarrow 2c^{2+} + a^{3+}.Cu^{2+}$

Above -20° $2c^{2+} + a^{3+}.Cu^{2+} \longrightarrow 2c^{3+} + a^{2+}.Cu^{1+}$
                 $a^{2+}.Cu^{1+} + a_3^{3+}-O^-.O^- Cu^{2+} \longrightarrow a^{3+}.Cu^{2+} + a_3^{2+}-O^-O^- Cu^{1+}$
                 $a_3^{2+}-O-O-Cu^{1+} + 4H^+ \longrightarrow a_3^{3+}.Cu^{2+} + 2H_2O$
                        (Resting/Oxidized)

Fig. 12. A schematic diagram of the low temperature electron transfer pathway in the cytochrome c cytochrome oxidase reactions. (*1a,1b*), conventional room temperature pahtway; (*2*) low temperature pathway, (*3*) the oxidation of cytochrome $a^{2+}.Cu^1-$ by a^{3+} at -50°C

Above -80°C, the oxidation of cytochrome c by compound B_2 is depicted as forming oxidized cytochrome c and reduced compound B_2. In this case we illustrate that the electrons are accepted by the iron and copper atoms giving an as yet unidentified compound which we term B_3; i.e., the ferrous-cuprous bridge peroxide compound. This compound may break down in manifold ways, which are outside the scope of this discussion, to give a mixed state in which the $a_3.Cu_{a_3}$ are oxidized and $a.Cu_a$ are reduced, a mixed valence compound (MVS).[a3]This compound is then responsible for the increased absorption at 655-nm changes due to a_3^{3+} and at 790 nm due to Cu_a^{2+}. In order to explain the slow cytochrome c-dependent oxidation of $a.Cu_a$ between -50° and 20°C, we propose a reversal of the usual pathway, the oxidation of $a.Cu_a$ by cytochrome c. Finally, the overall reaction of cytochrome oxidase that is usually observed is depicted as occurring by the reaction of cytochrome c with $a^{3+}.Cu_a^{2+}$, which causes its reduction and the reaction of $a^{2+}.Cu_a^{1+}$ with compound B_2 to give the same ferrous cuprous peroxide compound as that postulated above (compound B_3). The compound can dismute in this case to give the fully oxidized resting oxidase.

Role of Tunneling in the Reaction Mechanism

While the elucidation of these mechanisms may be in preliminary stages of development, the possible impact on electron transport mechanisms of the tunneling theory is obvious. The diagram in Figure 12 is consistent with electron transfer among components at distances as great as 30 Å, at which the tunneling mechanisms are feasible. It is further apparent that a primary tunneling process may be from cytochrome c to compound B_2 rather than from cytochrome c to cytochrome a; i.e., the electron equilibration between cytochromes a and a_3 appears to occur at a higher temperature than the primary reaction between cytochrome c and compound B_2. Also, cytochrome c can transfer electrons to cytochrome a_3 at lower temperatures than those at which cytochrome a transfers electrons to cytochrome a_3. Assuming that reaction rates are governed by tunneling distances, the shortest distance would be from cytochrome c to compound B_2 and the longest distance from cytochrome a to cytochrome a_3. This model differs from that of Figure 1 and suggests a triangular arrangement that has roughly equal distances from cytochrome c and to compound B_2. With more precise distance determinations, the scale of the model can be more precisely determined.

Summary

A reaction mechanism for cytochrome oxidase based upon the optical measurements of intermediate compounds of heme a_3 and its associated copper atom (Cu_{a_3}) and of electron transfer reactions to oxygen and from cytochrome c over wide ranges of temperature indicates novel electron transfer pathways in cytochrome oxidase. The first is from cytochrome c directly to a ferric cupric bridge-peroxide compound above -80°C. The usual reaction of cytochrome c with cytochrome a and its associated copper ($a.Cu_a$) is observed at temperatures above 20°C. A second novel pathway is the reversal of the usual one from cytochrome c to cytochrome a; between 0° and -20°C the oxidation of a Cu_a depends upon the presence of cytochrome c. Nuclear and electron tunneling appear to afford explanations of appropriate control of electron transfer in the different pathways and over the long distances involved (~ 25 Å). On the basis that the rate of electron transfer is determined by the distance between donor and acceptor in the tunneling process, the distances increase in the order cytochrome c to cyto-chrome a_3 (as compound B_2), cytochrome c to cytochrome a (for the particular condition of electron equilibration at $-50^\circ C$), and finally cytochrome a to a_3. However fine control of the electron flow by struc-tural parameters of nuclear tunneling is also possible.

References

1. American Philosophical Society Yearbook. 1977: p. 11.Philadelphia, PA, George H. Buchanan, Co. 1977.
2. Babcock, G.T., Vickery, L.E., Palmer, G.: The electronic state of heme in cyto-chrome oxidase II. Oxidation-reduction potential interactions and heme iron spin state behavior observed in reductive titrations. J. Biol. Chem. 253, 2400-2411 (1978)
3. Barlow, C.H., Maxwell, J.C., Wallace, W.J., Caughey, W.S.: Elucidation of the mode of oxygen binding to hemoglobin by infrared spectroscopy. Biochim. Biophys. Res. Comm. 55, 91-95 (1973)
4. Beinert, H., Griffiths, D.E., Warton, D.C., Sands, R.H.: Properties of the copper associated with cytochrome oxidase as studied by paramagnetic resonance spectros-copy. J. Biol. Chem. 237, 2337-2346 (1962)
5. Beinert, H., Shaw, R.W.: On the identity of the high spin heme components of cytochrome c oxidase. Biochim. Biophys. Acta 462, 121-130 (1977)
6. Blasie, J.K., Erecinska, M., Samuels, S., Leigh, Jr., J.S.: The structure of a cytochrome oxidase-lipid model membrane. Biochim. Biophys. Acta 501, 33-52 (1978)
7. Boyer, P.D., Chance, B., Ernster, L., Mitchell, P., Racker, E., Slater, E.C. (eds.) In: Oxidative phosphorylation and photophosphorylation. Ann. Rev. Bio-chem. 46, 955-1026 (1977)
8. Camerino, P.W., King, T.E.: Studies on cytochrome oxidase II. A reaction of cyanide with cytochrome oxidase in soluble and particulate forms. J. Biol. Chem. 241, 976 (1966)
9. Case, G.D., Leigh, Jr., J.S.: Intramitochondrial positions of cytochrome haem groups determined by dipolar interactions with paramagnetic cations. Biochem. J. 160, 769-783 (1976)
10. Chance, B., Crofts, A.R., Nishimura, M., Price, B.: Fast membrane H^+ binding in the light-activated state of *Chromatium* chromatophores. Eur. J. Biochem. 13, 364-374 (1970)
11. Chance, B., Saronio, C., Leigh, Jr., J.S.: Mobility and heterogeneity of elec-tron transport reactions in mitochondria at low temperatures. Fed. Proc. Abs. 33, 374, p. 1289 (1974)
12. Chance, B., Saronio, C., Leigh, Jr., J.S.: Functional intermediates in the reac-tion of membrane-bound cytochrome oxidase with oxygen. J. Biol. Chem. 250, 9226-9237 (1975)
13. Chance, B., Saronio, C., Leigh, Jr., J.S.: Functional intermediates in reaction of cytochrome oxidase with oxygen. Proc. Natl. Acad. Sci. U.S.A. 72, 1635-1640 (1975)
14. Chance, B., Leigh, Jr., J.S., Waring, A.: Structure and function of cytochrome oxidase and its intermediate compounds with oxygen reduction products. In: Struc-ture and function of energy transducing membranes. K. van Dam, B.F. van Gelder (eds.) pp. 1-10. Amsterdam: Elsevier-North Holland 1977
15. Chance, B., Leigh, Jr., J.S.: Oxygen intermediates and mixed valence states of cytochrome oxidase: Infrared absorption difference spectra of Compounds A, B, and C of cytochrome oxidase and oxygen. Proc. Natl. Acad. Sci. U.S.A. 74, 4777-4780 (1977)
16. Chance, B.: Electron transfer reactions in cytochrome oxidase. In: Tunneling in biological reactions. Chance, B., DeVault, D., Frauenfelder, H., Marcus, R., Schrieffer, R., Sutin, N. (eds.) New York: Academic Press 1978 (in press)
17. Chance, B., Smith, J., Bashford, L., Powers, L.: Intrinsic and extrinsic probes of localized and delocalized charges in and across biological membranes. Int. Symp. on Membrane Structure and Function, New Zealand, September 12-15, 1978 (in press)

18. Chance, B., Saronio, C., Waring, A., Leigh, Jr., J.S.: Cytochrome c-cytochrome oxidase interaction at subzero temperatures. Biochim. Biophys. Acta (1978) In press
19. Chance, B., Keyhani, E., Saronio, C.: Oxygen metabolism in yeast cells. In: Biochemistry and genetics of yeasts. Horecker, B., Bacila, M. (eds.) New York: Academic Press 1978 In press
20. Chance, B., Waring, A.: Electron tunneling in cytochrome c cytochrome oxidase reaction. Biophys. Soc. Mtg., Washington, DC, March 27-30, 1978. In press
21. DeVault, D., Chance, B.: Studies of Photosynthesis using a pulsed laser I. Temperature dependence of cytochrome oxidation rate in *Chromatium*. Evidence of tunneling. Biophys. J. 6, 826-847 (1966)
22. DeVault, D., Chance, B.: The effects of high hydrostatic pressure on light-induced electron transfer and proton binding in Chromatium. In: Tunneling in biological reactions. Chance, B., DeVault, D., Frauenfelder, H., Marcus, R., Schrieffer, R., Sutin, N, (eds.) New York: Academic Press 1978 in press
23. Dickerson, R.E., Timkovich, R.: Cytochrome c. In: The enzymes, Vol XI, Part A. p. 483. New York: Academic Press 1975
24. Dockter, M.E., Steinemann, A., Schatz, G.: Topographical mapping of yeast cytochrome c oxidase by fluorescence resonance energy transfer. In: Biochim. Biophys. Acta Library 14, Structure and function of energy transducing membranes. van Dam, K., van Gelder, B.F. (eds.) pp. 169-176. Amsterdam: Elsevier-North Holland 1977
25. Erecinska, M.: Mitochondrial electron transfer at phosphorylation sites 2 and 3. In: Tunneling in biological reactions. Chance, B., DeVault, D., Frauenfelder, H., Marcus, R., Schrieffer, R., Sutin, N. (eds.) New York: Academic Press 1978 in press
26. Evans, M.G., Gergely, J.: A discussion of the possibility of bands of energy levels in proteins: Electronic interaction in non-bonded systems. Biochim. Biophys. Acta 3, 188-197 (1949)
27. Ferguson-Miller, S., Brautigan, D.L., Margoliash, E.: Definition of cytochrome c binding domains by chemical modification, III. Kinetics of reaction of carboxydinitriphenyl cytochromes c with cytochrome oxidase. J. Biol. Chem. 253, 149-159 (1978)
28. Goldanskii, V.I.: Quantum chemical reactivity new absolute zero: biological, chemical and astrophysical aspects. In: Tunneling in biological reactions, Chance, B., DeVault, D., Frauenfelder, H., Marcus, R., Schrieffer, R., Sutin, N. (eds.) New York: Academic Press 1978 in press
29. Greenwood, C., Gibson, Q.H.: The reaction of reduced cytochrome c oxidase with oxygen. J. Biol. Chem. 242, 1782-1787 (1969)
30. Griffiths, D.E., Wharton, D.C.: Studies of the electron transport system XXXVI. Properties of copper in cytochrome oxidase. J. Biol. Chem. 236, 1857-1862 (1961)
31. Henderson, R., Capaldi, R.A., Leigh, Jr., J.S.: Arrangement of cytochrome oxidase molecules in two dimensional visicle crystals. J. Mol. Biol. 112, 631-648 (1977)
32. Hopfield, J.J.: Electron transfer theory in photosynthesis. In: Tunneling in biological reactions. Chance, B., DeVault, D., Frauenfelder, H., Marcus, R., Schrieffer, R., Sutin, N. (eds.) New York: Academic Press 1978 in press
33. Jacobs, E.E., Sanadi, D.R.: The reversible removal of cytochrome c from mitochondria. J. Biol. Chem. 235, 531-534 (1960)
34. Jortner, J., Discussions. In: Tunneling in biological reactions. Chance, B., DeVault, D., Frauenfelder, H., Marcus, R., Schrieffer, R., Sutin, N. (eds.) New York: Academic Press 1978 in press
35. Junge, W., DeVault, D.: Symmetry, orientation and rotational mobility in the a_3 heme of cytochrome c oxidase in the inner membrane of mitochondria. Biochim. Biophys. Acta 408, 200-214 (1975)
36. Keyhani, E., Chance, B.: Cytochrome biosynthesis under copper-limited conditions in *Candida utilis*. FEBS Lett 17, 127-132 (1971)
37. Kurtin, S.L., McGill, T.C., Mead, C.A.: Direct interelectrode tunneling in GeSe. Phys. Rev. B. 3, 3368-3379 (1971)
38. Oesterhelt, D.: Energy conservation and H^+-pump in the membrane of helobacteria. This symposium

38a. Powers, L.S., Blumberg, W., Chance, B., Barlow, C., Leigh, J.S., Jr., Smith, J., Yonetani, T., Vik, S., Peisack, J.: Biochim. Biophys. Acta. (submitted for publication).

39. Shulman, R.G., Eisenberger, P., Blumberg, W.E., Strombaugh, N.A.: Determination of the iron-sulfur distances in *Ruberdoxin* by X-ray absorption spectroscopy. Proc. Natl. Acad. Sci. U.S.A. 72, 4003-4007 (1975)

40. Vanderkooi, J.M., Landesberg, R., Hayden, G.W., Owen, C.S.: Metal-free and metal-substituted cytochrome *c*. Use in characterization of the cytochrome *c* binding site. Eur. Biochem. J. 81, 339-347 (1977)

41. Vanderkooi, J.M., Chance, B., Waring, A.: Luminescence of Cu-cytochrome *c*. FEBS Lett 88, 273-274 (1978)

42. Wakabayashi, T., Senrio, A.E., Hatase, O., Hayashi, H., Green, D.E.: Conformational changes in membranous preparations of cytochrome oxidase. J. Bioenergetics 3, 339-344 (1972)

43. Wever, R., Van Drooge, J.H., Musjers, A.O., Bakker, E.P., Van Gelder, B.F.: The binding of carbon monoxide to cytochrome *c* oxidase. Eur. J. Biochem. 73, 149-154 (1977)

44. Wikström, M.K.F., Harmon, J., Ingledew, W.J., Chance, B.: A Re-evaluation of the spectral, potentiometric and energy-linked properties of cytochrome *c* oxidase in mitochondria. FEBS Lett 65, 259-277 (1976)

45. Wikström, M.K.F., Krab, K.: Cytochrome *c* oxidase is a proton pump: a rejoiner to recent criticism. FEBS Lett. (1978) in press

Structure and Function of Iron-Sulfur Proteins

L. H. JENSEN

Introduction

The iron-sulfur proteins are widely distributed in both prokaryotic
and eukaryotic cells, occurring both as simple proteins with one or
more active iron-sulfur complexes (such as rubredoxin and ferredoxin)
and complex proteins containing other functional groups in addition
to the iron-sulfur centers (such as oxygenase and nitrogenase). They
have been studied by a number of physical techniques and the three-
dimensional structures of some of the simpler ones have been determined by X-ray analysis. I will describe the known three-dimensional
structures of the iron-sulfur proteins and the analog compounds that
are models for them and then discuss some observations relating the
structure and the function of these intriguing molecules.

Structures of Some Iron-Sulfur Proteins

Rubredoxin: FeS_4^{γ} Complex

The typical rubredoxin has one FeS_4^{γ} complex per molecule of weight
$\sim$ 6000 daltons, four cysteine S atoms being coordinated to the iron
atom. The molecule appears to function in redox reactions with an E_0'
of approximately -60 mV.

The structure of rubredoxin from *Clostridium pasteurianum* has been solved
and the resulting model refined in the crystallographic sense of the
term, first at 1.5 Å resolution and subsequently at 1.2 Å resolution
(Herriott et al., 1970; Watenpaugh et al., 1973; Watenpaugh et al.,
to be published). Figure 1 is a stereo plot of the C^{α} positions showing the course of the polypeptide chain and the position of the FeS_4^{γ}
complex.

The molecule is globular but somewhat flattened laterally as viewed in
Figure 1, approximately 26 Å in its major dimension and 20 Å in its
minor one. Despite its relatively small size, the molecule has a hydrophobic core, composed primarily of Tyr 13, Ile 24, Ile 33, Trp 37,
and Phe 49. The FeS_4^{γ} complex is exposed on the molecular surface,
readily accessible to the surroundings.

Since the bond lenghts and angles are important parameters of the FeS_4^{γ}
complex, it is a matter of considerable importance to determine their
values with high precision. To do so, the model for rubredoxin was
subjected to crystallographic refinement, i.e., the atomic parameters
x,y,z, and B (B is the so-called thermal parameter) are adjusted to
minimize the difference between the calculated diffraction amplitudes,
$|F_c|$, and their observed values, $|F_0|$. Initially, the rubredoxin model
was refined against a 1.5 Å resolution data set consisting of 5033
reflections. The bond lengths and angles from this refinement are
shown in Figure 2 along with their standard deviations (in parentheses

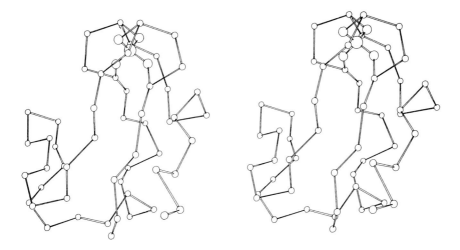

Fig. 1. Stereo view of the C^α atoms and FeS^γ_4 complex of rubredoxin from *C. pasteuri-anum* (from Watenpaugh et al., 1973)

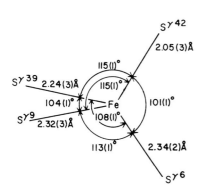

Fig. 2. Bond lenghts and angles in the FeS^γ_4 complex of rubredoxin from 1.5-Å resolution refinement (see text)

following each value). The bond angles range from 101 (1)° to 115 (1)°, deviating significantly from the tetrahedral value, but the deviations are no more than has been observed for tetrahedral iron-sulfur complexes in some small structures.

Three of the four bond lengths do not differ significantly from the average value of 2.28 Å found in many other iron-sulfur structures (Watenpaugh et al., 1972). The Fe-$S^{\gamma 42}$ bond, however, is much shorter, 2.05 (3) Å, or 8 σ less than the average of the other three. This is surprising because the ligands are chemically identical, but the difference is highly significant if the standard deviations are to be taken at face value.

To check the validity of the parameters of the FeS^γ_4 complex, a much more extensive data set extending to 1.2 Å resolution was collected, consisting of 10,964 reflections compared to 5033 in the earlier data set. Refinement of the model with the more extensive data led to bond

lengths and angles close to the values in Figure 2 except for the
Fe-S$^{\gamma 42}$ bond, which is now 2.24(1) Å (Watenpaugh et al., to be published).

The fact that two different data sets lead to significantly different
values for the Fe-S$^{\gamma 42}$ bond lenghts suggests the presence of systematic error in one or both data sets. The most likely sources of error
are in the absorption correction and radiation damage to the crystal.
One should note, however, that both these sources are likely to be more
severe for the 1.2 Å resolution data set because the crystal was considerably larger than that used for the earlier data, and it was exposed to the X-ray beam for a much longer period of time.

The method of X-ray absorption fine structure (EXAFS) has been applied
to rubredoxin from *C. pasteurianum* in order to determine the Fe-S distances in the FeS$_4^{\gamma}$ complex (Schulman et al., 1975; Sayers and Stern,
1976). In one study the average distance found was 2.24 Å with no one
distance differing by more than 0.1 from a fixed value for the other
three. In another study, an average distance of 2.30 Å was found with
an *rs* deviation about this value of 0.06 Å.

Ferredoxins

Chloroplast Type: Fe$_2$S$_2^{*}$ S$_4^{\gamma}$ Complex

The ferredoxins found in the chloroplasts of plant cells typically
have molecular weights of ∿ 12,000 daltons and E$_0'$ values of ∿ -400 mV.
The complex consists of two iron atoms bridged by two inorganic sulfur
atoms, which are labile, with four cysteine sulfur ligands, two coordinated to each iron atom. A model for the complex based on electron
spin resonance in spinach ferredoxin is shown in Figure 3 (Gibson et
al., 1966).

A preliminary account of the structure of the ferredoxin from *Spirulina
platensis* has appeared (Ogawa et al., 1977). The structure was at a
resolution insufficient to reveal the details of the complex, but a
plot of the composite electron density through the complex is consistent with the model shown in Figure 3. The molecular shape in cross
section is approximately 30 Å by 40 Å with the iron complex approximately 8 Å from one surface.

Proteins with the two-iron complex found in plant ferredoxin occur in
other cells, e.g., adrenodoxin found in certain mammalian cells of the
adrenal cortex and putidaredoxin in the bacterium *Pseudomonas putida*.
Both these proteins have E$_0'$ values in the range of -240 to -260 mV,
considerably higher than values for chloroplast ferredoxins.

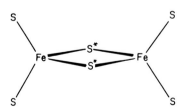

Fig. 3. Model of Fe$_2$S$_2$S$_4^{\gamma}$

"High-Potential Iron Protein" (HiPIP): $Fe_4S_4^* S_4^\gamma$ Complex

An iron-sulfur protein with a high redox potential, +350 mV, was isolated from *Chromatium* by Bartsch (1963). It was reported to contain
four iron atoms, four labile sulfur atoms, and four cysteine residues
per molecule of weight $\sim$ 10,000 daltons.

A 4-$\overset{\circ}{A}$ resolution X-ray analysis led to the conclusion that the four
iron atoms were clustered about a single site, too close to be resolved
and possibly in a tetrahedral arrangement (Kraut et al., 1968). The
authors noted that the clusters were arranged in a way approaching
cubic close packing and that this could be interpreted to mean that
the clusters were near the center of roughly spherical protein molecules
stacked in an almost close packed array. Higher resolution (2.25 $\overset{\circ}{A}$)
revealed an unusual cubic arrangement of four iron and four sulfur
atoms, the iron atoms in the complex being bound to the protein by four
cysteine sulfur atoms (Carter et al., 1972a) as shown in Figure 4.

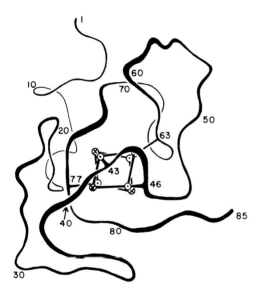

Fig. 4. Schematic diagram of HiPIP main
chain and $Fe_4S_4^*$ core (based on Fig. 2
of Carter et al., 1974b; from Jensen,
1977)

Table 1. Mean distances and angles in iron-
sulfur complex of oxidized and reduced HiPIP
(estimate standard deviations in parentheses)

Distance/angle	HiPIP$_{ox}$	HiPIP$_{red}$
Fe-Fe	2.73(0.04) $\overset{\circ}{A}$	2.81(0.05) $\overset{\circ}{A}$
Fe-S*	2.25(0.06)	2.32(0.08)
Fe-S$^\gamma$	2.21(0.03)	2.22(0.03)
S*-Fe-S*	103(3)$^\circ$	104(3)$^\circ$
S*-Fe-S$^\gamma$	115(5)$^\circ$	114(6)$^\circ$
Fe-S*-Fe	75(2)$^\circ$	75(2)$^\circ$

The model for HiPIP in both the oxidized and reduced states has been extensively refined leading to the dimensions of the iron-sulfur complex appearing in Table 1 (Freer et al., 1975). Distances in the $Fe_4S_4^*$ core suggests that it shrinks slightly on oxidation. In fact, examination of the individual Fe-S[*] bond lengths in the cube-like core indicates a contraction of 0.16 Å in a direction normal to one face, 0.08 Å perpendicular to this, and a slight expansion of 0.03 Å in the third direction (Carter et al., 1974a). The oxidized complex appears to be more nearly of tetrahedral symmetry than is the reduced complex.

The Clostridial Type: Two $Fe_4S_4^*$ S_4^γ Complexes

A low-molecular-weight iron protein was identified as an essential component of the nitrogen-fixing fraction from *C. pasteurianum*, and it was named ferredoxin in view of the iron content and redox function (Mortenson et al., 1962). Tagawa and Arnon (1962) showed that this ferredoxin could be substituted for the iron-sulfur protein that catalyzes photoreactions in chloroplasts. It was clear, therefore, that the iron proteins from anaerobic, nitrogen-fixing bacteria and from chloroplasts were similar, at least at the functional level, and they extended the name to cover both proteins. The clostridial ferredoxins have E_o' values near -400 mV and a much higher iron content than the chloroplast type. In fact, there are eight iron atoms, eight labile sulfur atoms, and at least eight cysteine residues per molecule of weight ∿ 6,000 daltons.

The structure of ferredoxin from *Peptococcus aerogenes* was solved soon after the structure of the iron-sulfur complex in HiPIP was reported (Sieker et al., 1972; Adman et al., 1973). Figure 5 illustrates the course of the polypeptide chain and shows the iron and sulfur atoms to be distributed in two complexes similar to the one in HiPIP. The very different E_o' values for proteins with similar complexes raises some questions of great interest. For example, to what extent does the surrounding protein modify the properties of the iron-sulfur complexes? Are the complexes distorted from some idealized, regular geometry, and if so, in what way are they distorted and to what extent?

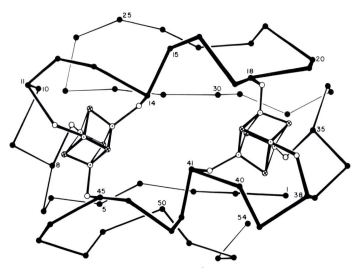

Fig. 5. View of C^α atoms and $Fe_4S_4^*$ S_4^γ complexes of ferredoxin from *P. aerogenes* (based on Fig. 8 of Adman et al., 1973; from Jensen, 1977)

Table 2. Mean distances and angles
in the iron-sulfur complexes of
ferredoxin from *P. aerogenes* (estimated standard deviation in
parenthesis x 2 because of preliminary nature of values listed)

Distance/angle	Value
Fe-Fe	2.74(0.04) Å
Fe-S*	2.26(0.06)
Fe-S$^\gamma$	2.17(0.09)
S*-Fe-S*	103(2)°
S*-Fe-S$^\gamma$	115(3)
Fe-S*-Fe	75(2)

To provide a better basis for answering such questions, the model for
ferredoxin from *P. aerogenes* has been refined at 2-Å resolution, leading to the average bond lenghts and angles listed in Table 2 (Jensen,
1974). In view of the present uncertainty in these values, the complexes in ferredoxin cannot be said to differ significantly in dimensions from the complex in either the oxidized or reduced form of HiPIP.
This is not to say that the complexes do not, in fact, differ, only
that we cannot say in view of the standard deviations in the present
results.

The difference in E_0' values for HiPIP and *P. aerogenes* ferredoxin can
be explained on the basis of a three-state hypothesis (Carter et al.,
1972b). Since magnetic and other properties suggest the close similarity of the complexes in reduced HiPIP and oxidized ferredoxin, they
were assumed to be in the same oxidation state, which was designated
C. Therefore, E_0' for HiPIP applies to the reaction C ⟷ C$^+$ + e, while
that for the clostridial type ferredoxin applies to the reaction
C$^-$ ⟷ C + e, thus accounting in a formal way for the redox potential.

Iron-Sulfur Analog Compounds

Holmes and his colleagues have synthesized a series of compounds that
contain complexes analogous to those found in the proteins described
above. They have crystallized representative compounds with each type
of complex and determined accurate models for each one (Herskovitz et
al., 1972; Mayerle et al., 1975; Lane et al., 1975). Comparison of the
metric parameters in Figure 2 and Tables 1-3 indicates the similarity
of the complexes in the analog compounds with those in the proteins.

Table 3. Mean distances and angles in the iron-sulfur complexes of the
analog compounds

Distance/angle	FeS$_4$ complex	Fe$_2$S$_2^*$S$_4$ complex	Fe$_4$S$_4^*$S complex
Fe-Fe	–	–	2.747 Å
Fe-S*	–	2.202 Å	2.239 Å, 2.310 Å
Fe-S	2.268 Å	2.312 Å	2.251 Å
S*-Fe-S*	–	104.6°	104.1°
S*-Fe-S	–	114.5°	111.7°-117.3°
Fe-S*-Fe	–	75.4°	73.8°
S-Fe-S	105.8°-112.6°	109.5°	–

The accuracy of the parameters for the analog structure is much higher than for the proteins, adding important new information. Thus, the $Fe_4S_4^*$ core has one pair of faces separated by a distance ~ 0.06 Å less than that for the other two sets of faces, resulting in two sets of $Fe-S^*$ bond lengths (Herskovitz et al., 1972).

Structure-Function Relationships

It is a matter of considerable importance to understand how electrons are transferred to and from the iron complexes in redox reactions. In both rubredoxin and *P. aerogenes* ferredoxin, the complexes are near the molecular surface suggesting direct access to them from the out-side. In the case of rubredoxin an indirect route through a remote pathway appears possible in view of the hydrophobic core of the mole-cule. The rings of the aromatic residues Tyr 4, Phe 30, and Phe 49, the latter adjacent internally to the iron-sulfur complex, are stacked together in a staggered array across the inside of the molecule, the side-chain rings being almost parallel. These residues are either part of the hydrophobic core or adjacent to it, and the OH group of Tyr 4 and edge of the Phe 30 ring are exposed at the surface of the molecule. Quantum mechanical tunneling through such a region may provide a path for electron transfer. Indeed, such a mechanism through a hydrophobic region in the structure of plastocyanin has been suggested as an elec-tron pathway in that molecule (Coleman et al., 1978).

Another possible mechanism for electron transfer involves outer sphere interaction in which the metal-ligand bonds of the complexes in the reacting molecules remain unbroken. For the electron to be transferred, this mechanism requires the reacting molecules to be in contact at a point to which an electron in the reductant can be delocalized, and in turn from which it can be localized to the acceptor complex. Such a mechanism may operate in ferredoxin, and the structure suggested electron transfer by way of the two tyrosine residues in the molecule. Each of these residues is located adjacent to one of the complexes and in contact with one of its faces, the OH group and one edge of each tyrosine ring being exposed at the surface of the molecule (Adman et al., 1973). Moreover, spectral evidence suggests that the aromatic residues in *Clostridium acidis-urici* ferredoxin are involved in electron transfer (Packer et al., 1972).

The hypothesis of tyrosine involvement in electron transfer was tested in ferredoxin from *Clostridium M-E* by Rabinowitz and his colleagues, who replaced the only aromatic residue in the molecule, Tyr 2, by leucine (Lode et al., 1974). This substituted ferredoxin, now devoid of any aromatic residue, was still as active in the phosphoclastic enzyme system as the native *C. M-E* ferredoxin or as *C. acidi-urici* fer-redoxin, indicating that one or more aromatic residues are not neces-sary for electron transfer. An outer sphere mechanism may still operate, however, but other pathways for electron transport must be sought.

It is generally assumed that molecular orientation is an important fac-tor in controlling biological electron transport (Moore and Williams, 1976). The specificity of many electron transfer reactions can be ex-plained on the basis of selective molecular orientation between react-ing molecules or between a molecule and a membrane or other ultrastruc-tural components of the cell. Furthermore, specificity may require different electron pathways for the transfer of an electron to the metal complex and from the complex. Indeed, in the case of cytochrome c there is evidence suggesting different pathways (Smith et al., 1973).

As noted above, the three-state hypothesis accounts for the different E_0' values of HiPIP and *P. aerogenes* ferredoxin, but it provides no answer to the question of why the molecules should function between different oxidation states. In seeking an answer, we compare the models for the two proteins for differences which may be suggestive. We note that the complexes in ferredoxin are near the molecular surface, being barely covered by the protein, with a tyrosine side chain adjacent to each. In contrast, the HiPIP complex lies more deeply within the molecule, and it is surrounded on one side by aromatic residues, five of them being within 5 Å of the complex at some point.

Another prominent difference between HiPIP and the *P. aerogenes* structure is in the number and type of N-H...S bond (Adman et al., 1975). Thus HiPIP has five N-H...S bonds distributed among three types (for a description of the bond type see Adman et al., 1975). Ferredoxin, on the other hand, has nine N-H...S bonds per complex, five involving the same three types as HiPIP but with a different distribution and four of a fourth type not present in HiPIP. The marked differences in the N-H...S bonding schemes in the two proteins suggest that such bonds play a key role in determining the oxidation states between which the two proteins function biologically.

A theoretical model for the effects of solvent and protein dielectric on the redox potential of iron-sulfur complexes has been formulated by Kassner and Yang (1977). In their model, one factor contributing to the redox potential that is dependent on the environment of the complex is the difference in charging energies between different redox states. Another factor is the electrostatic free energy within the cluster resulting from the transfer of an electron. The model suggests that the difference in potential between the natural and synthetic systems, when measured in aqueous and nonaqueous solutions, respectively, can be accounted for, at least in part, by electrostatic effects of clusters in environments of different dielectrics. Furthermore, the model may provide an additional basis for understanding the difference in redox potential between HiPIP with its more deeply buried complex and ferredoxin with the complexes near the molecular surface.

Acknowledgments. I wish to acknowledge the contributions of my colleagues, L.C. Sieker, K.D. Watenpaugh, and E.T. Adman and the support of USPHS Grant GM-13366 from the National Institutes of Health.

References

Adman, E.T., Sieker, L.C., Jensen, L.H.: The structure of a bacterial ferredoxin. J. Biol. Chem. <u>248</u>, 3987-3996 (1973)

Adman, E.T., Watenpaugh, K.D., Jensen, L.H.: N-H...S hydrogen bonds in *Peptococcus aerogenes* ferredoxin, *Clostridium pasteurianum* rubredoxin, and *Chromatium* high potential iron protein. Proc. Natl. Acad. Sci. U.S.A. <u>72</u>, 4854-4858 (1975)

Bartsch, R.C.: Nonheme iron proteins and *Chromatium* iron proteins. In: Bacterial Photosynthesis. Gest, H., San Pietro, A., Vernon, L.P. (eds.) Yellow Springs, Ohio: The Antioch Press, 1963, pp. 315-326

Carter, Jr., C.W., Freer, S.T., Xuong, Ng.H., Alden, R.A., Kraut, J.: Structure of the iron-sulfur clusters in the *Chromatium* iron protein at 2.25 Å resolution. Symposia on Quantitative Biology, Vol. XXXVI, Cold Spring Harbor Laboratories (1972a)

Carter, Jr., C.W., Kraut, J., Freer, S.T., Alden, R.A., Sieker, L.C., Adman, E.T., Jensen, L.H.: A comparison of $Fe_4S_4^*$ clusters in high-potential iron protein and in ferredoxin. Proc. Natl. Acad. Sci. U.S.A. <u>69</u>, 3526-3529 (1972b)

Carter, Jr., C.W., Kraut, J., Freer, S.T., Alden, R.A.: Comparison of oxidation-reduction site geometries in oxidized and reduced *Chromatium* high potential iron protein and oxidized *Peptococcus aerogenes* ferredoxin. J. Biol. Chem. <u>249</u>, 6339-6346 (1974a)

Carter, Jr., C.W., Kraut, J., Freer, S.T., Xuong, Ng.H., Alden, R.A., Bartsch, R.G.: Two-Ångstrom crystal structure of oxidized *Chromatium* high potential iron protein. J. Biol. Chem. <u>249</u>, 4212-4225 (1974b)

Coleman, P.M., Freeman, H.C., Guss, J.M., Murata, M., Norris, V.A., Ramshaw, J.A.M., Venkatappa, M.P.: The X-ray crystal structure analysis of plastocyanin, a 'blue' copper-protein at 2.7 Å resolution. In Press, Nature (London) in press (1978)

Freer, S.T., Alden, R.A., Carter, Jr., C.W., Kraut, J.: Crystallographic structure refinement of *Chromatium* high potential iron protein at two Ångstrom resolution. J. Biol. Chem. <u>250</u>, 46-54 (1975)

Gibson, J.F., Hall, D.O., Thornley, J.H.M., Watley, F.R.: The iron complex in spinach ferredoxin. Proc. Natl. Acad. Sci. U.S.A. <u>56</u>, 987-990 (1966)

Herriott, J.R., Sieker, L.C., Jensen, L.H., Lovenberg, W.: Structure of rubredoxin: an X-ray study at 2.5 Å resolution. J. Mol. Biol. <u>50</u>, 391-406 (1970)

Herskovitz, T., Averill, B.A., Holm, R.H., Ibers, J.A., Phillips, W.D., Weiher, J.F.: Structure and properties of a synthetic analogue of bacterial iron-sulfur proteins. Proc. Natl. Acad. Sci. U.S.A. <u>69</u>, 2437-2441 (1972)

Jensen, L.H.: X-ray structural studies of ferredoxin and related electron carriers. Ann. Rev. Biochem. <u>43</u>, 461-474 (1974)

Jensen, L.H.: Contribution of X-ray crystallography to the understanding of the mechanism of redox proteins. In: Proceedings of the Marseille Symposium on Electron Transfer Systems in Microorganisms. Senez, J.C. (ed.) Marseille, 1977

Kassner, R.J., Yang, W.: A theoretical model for the effects of solvent and protein dielectric on the redox potentials of iron-sulfur clusters. J. Am. Chem. Soc. <u>99</u>, 4351-4355 (1977)

Kraut, J., Strahs, G., Freer, S.T.: An X-ray anomalous scattering study of an iron protein from *Chromatium D*. In: Structural Chemistry and Molecular Biology. Rich, A., Davidson, N. (eds.) San Francisco-London: Freeman, 1968, pp. 55-64

Lane, R.W., Ibers, J.A., Frankel, R.B., Holm, R.H.: Synthetic analogs of active sites of iron-sulfur proteins. XII. Proc. Natl. Acad. Sci. U.S.A. <u>72</u>, 2868-2872 (1975)

Lode, E.T., Murray, C.L., Rabinowitz, J.C.: Semisynthetic synthesis and biological activity of a clostridial-type ferredoxin free of aromatic amino acid residues. Biochem. Biophys. Res. Commun. <u>61</u>, 163-169 (1974)

Mayerle, J.J., Denmark, S.E., DePamphilis, B.V., Ibers, J.A., Holm, R.H.: Synthetic analogues of the active site of iron-sulfur proteins. J. Am. Chem. Soc. <u>97</u>, 1032-1045 (1975)

Moore, G.R., Williams, R.J.P.: Electron-transfer proteins. Coord. Chem. Rev. <u>18</u>, 125-197 (1976)

Mortenson, L.E., Valentine, R.C., Cornehan, J.E.: An electron transport factor from *Clostridium pasteurianum*. Biochem. Biophys. Res. Commun. <u>7</u>, 448-452 (1962)

Ogawa, K., Tsukihara, T., Tahara, H., Katsube, Y., Matsu-ura, Y., Tanaka, N., Kakudo, M., Wada, K., Matsubara, H.: Location of the Iron-Sulfur Cluster in *Spirolina platensis* Ferredoxin by X-ray Analysis. J. Biochem. <u>81</u>, 529-531 (1977)

Packer, E.L., Sternlicht, H., Rabinowitz, J.C.: The possible role of aromatic residues of *Clostridium acidi-urici* ferredoxin in electron transport. Proc. Natl. Acad. Sci. U.S.A. <u>69</u>, 3278-3282 (1972)

Sayers, D.E., Stern, E.A.: Measurement of Fe-S bond lengths in rubredoxin using extended X-ray absorption fine structure (EXAFS). J. Chem. Phys. <u>64</u>, 427-428 (1976)

Schulman, R.G., Eisenberg, P., Blumberg, W.E., Stombaugh, N.A.: Determination of iron-sulfur distances in rubredoxin by X-ray absorption spectroscopy. Proc. Acad. Sci. U.S.A. <u>72</u>, 4003-4007 (1975)

Sieker, L.C., Adman, E.T., Jensen, L.H.: Structure of the Fe-S complex in bacterial ferredoxin. Nature (London) <u>235</u>, 40-42 (1972)

Smith, L., Davies, H.C., Reichlin, M., Margoliash, E.: Separate oxidase and reductase reaction sites on cytochrome *c* demonstrated with purified site-specific antibodies. J. Biol. Chem. <u>248</u>, 237-243 (1973)

Tagawa, K., Arnon, D.I.: Ferredoxins as electron carriers in photosynthesis and in biological production and consumption of hydrogen gas. Nature (London) 195, 537-543 (1962)

Watenpaugh, K.D., Sieker, L.C., Herriott, J.R., Jensen, L.H.: The structure of a non-heme iron protein: rubredoxin at 1.5 Å resolution. Symposia on Quantitative Biology, Vol. XXXVI, Cold Spring Harber Laboratory (1972)

Watenpaugh, K.D., Sieker, L.C., Herriott, J.R., Jensen, L.H.: Refinement of the model of a protein: rubredoxin at 1.5 Å resolution. Acta Cryst. B29, 943-956 (1973)

Watenpaugh, K.D., Sieker, L.C., Jensen, L.H.: Refinement of the rubredoxin model at 1.2 Å resolution. To be published, Acta Cryst.

Organization of the Photosynthetic Electron Transport System of Chloroplasts in the Thylakoid Membrane

A. TREBST

The chloroplasts of higher plant leaves contain the enzymic machinery for photosynthesis, defined as light-driven CO_2 assimilation. Surrounded by a double membrane envelope, the matrix (stroma) of a chloroplast contains the water-soluble enzymes of the Calvin cycle for CO_2 fixation and of starch (but not of sucrose) biosynthesis plus other assimilatory activities, such as photosynthetic sulfate and nitrite reduction (see reviews in *The Intact Chloroplast*, (1)). The light-driven primary reactions of photosynthesis take place in the inner, green membrane system of the thylakoid (in grana stacks). The light-dependent reactions of photosynthetic electron flow split water into oxygen and reducing equivalents at a redox potential level, electronegative enough to reduce NADP. This electron flow system against the thermodynamic gradient is furthermore coupled to ATP formation = noncyclic photophosphorylation. In another cyclic photophosphorylation system, only ATP is formed without the net reduction of an electron acceptor. NADPH and ATP, so formed, are consumed in the dark during carbon assimilation by the enzymes in the matrix space of the chloroplast.

Compounds of the Electron Flow System

In photosynthetic NADP reduction, two photosystems, operating in sequence, are needed to absorb and convert light energy in order to bridge the redox potential (and energy) difference between water (+180 mV) and NADP (-320 mV), equivalent to 52 kcal. In cyclic electron flow only one photosystem is required. Each photosystem consists of a light-harvesting pigment (of chlorophyll a and b and carotines, etc.) and a reaction center chlorophyll a. The light energy absorbed in the antenna system is led to the reaction center by excitation energy transfer. In the reaction center the special, excited chlorophyll a (dimer) undergoes a redox reaction to reduce a primary electron acceptor. Several compounds in addition to chlorophyll, undergoing reversible oxidoreduction reactions, connect the two light reactions and participate in electron flow from water to NADP. Most are tightly bound to and in the thylakoid membrane, such as cytochrome f, two cytochromes b_{559} and cytochrome b_{563}, plastoquinone, non-heme iron proteins (FeS), and the manganese protein complex of the oxygen evolution system. Other components may be easily removed from the membrane, e.g., ferredoxin, ferredoxin-NADP reductase (a flavine enzyme), and plastocyanin. Other intermediate electron carriers in the electron transport system - called Q, B or R, X, and Y to indicate that they are not yet identified for sure - have been implicated by fluorescence or other spectroscopic or kinetic evidence (see reviews in Ref. (2)

Hill and Bendall (3) were the first to propose a "zigzag" scheme for photosynthetic electron flow via two photosystems in a redox potential scale (Fig. 1). Their scheme also accounted for coupled ATP formation by postulating a coupling site in the sequence between the two

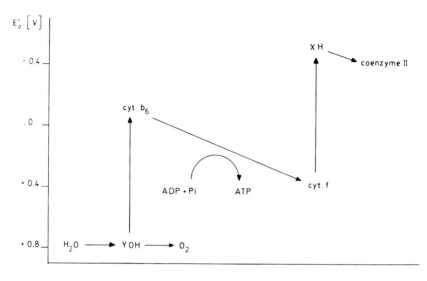

Fig. 1. Concept of Hill and Bendall in 1960 for two light reactions in photosynthetic electron flow and a coupling site located in the thermodynamic gradient (3)

photosystems, in which the electron flow in the thermodynamic gradient
may form ATP analogous to respiratory electron flow in the cytochrome
region. Hill and Bendall did not include plastoquinone, ferredoxin,
or plastocyanin in their scheme in 1960; the later carrier was not
even known then. Instead, they emphasized cytochromes that they themselves had discovered (3). It is interesting that 15 years later the
role of cytochromes in the photosynthetic electron flow scheme of
chloroplasts is still uncertain, and that therefore they are often
left out. Only a small percentage of the electrons pass through cytochrome b_{559} and only 10% through cytochrome f. Thus, other functions
have to be considered. Cytochrome b_{563} participates in cyclic electron
flow. The situation is somewhat different in some algae, in which
plastocyanin may be replaced by a special cytochrome f^4. However, other
compounds have been identified, and at present a scheme of photosynthetic
electron flow might look like the model shown in Figure 2.

The photosynthetic electron flow system may be separated into smaller
sequences either functionally (by inhibitors) or physically (by frag-
mentation). To obtain activity, artificial electron acceptor or donor
systems are added to measure (only) a section of the electron flow
system, the other section being made inoperative (5,6). For example,
ferredoxin and ferredoxin-NADP reductase needed for NADP reduction can
be removed from the thylakoid membrane. However, oxygen evolution and
both photosystems can still be measured by the reduction of an artificial electron acceptor, e.g., K-ferricyanide, or low redox potential
dyes such as methylviologen or anthraquinone. Conversely the oxygen
evolution system may be specifically inactivated by Tris, pH 9.0; the
rest of the chain can be measured by adding an artificial donor system
for photosystem II. By a combination of artificial donor and acceptor
systems, each photosystem and sequences of the electron flow system
may be isolated and measured separately (Fig. 3). The functional site
of an electron carrier can be identified in a specific sequence of the
chain.

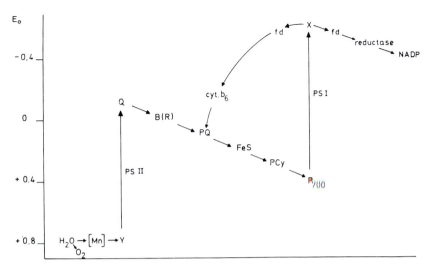

Fig. 2. Likely scheme in 1978 for photosynthetic electron flow from water to NADP and native cyclic electron flow in a redox potential scale. (Mn) = manganese protein complex of the oxygen evolution system; Y = primary donor for PS II; PS II = photosystem II; Q = primary acceptor of PS II, identical with X_{320} and probably a plastosemiquinone; B or R = secondary two-electron acceptor for PS II, possibly plastoquinoldianion; PQ = plastoquinone (main pool); FeS = nonheme iron protein; PC_y = plastocyanin; P_{700} – reaction center of PS I; X = primary acceptor of PS I; fd = ferredoxin; $reductase$ = Fd-NADP-oxidoreductase; $cyt.$ b_6 = cytochrome b_{563}

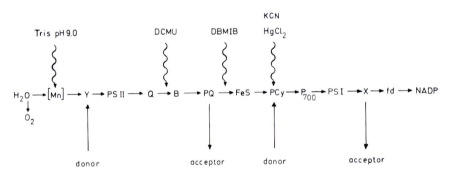

Fig. 3. Artificial donor and acceptor systems permit measurement of sections of the electron flow system when other parts are inhibited. $DCMU$ = dichlorophenyl-dimethyl-urea; $DBMIB$ = dibromothymoquinone

In this way, function and structure of the thylakoid membrane can be correlated. Large particles, seen in the electronmicroscope in freeze-fracture studies on the inner membrane half, have been identified as photosystem II complexes connected to a light-harvesting complex (7, 8,9). Small particles on the inner face of the outer membrane half have been identified as photosystem I complexes and as a cytochrome f + cytochrome b_6 protein complex. Larger protruding knobs on the outer face of the outer half of the membrane (i.e., facing the matrix) are "coupling factors" or the ATP synthetase complex responsible for the

formation of ATP in coupled electron flow. (For more structural data
see reviews by Mühlethaler (7) and Sane (8) and by Staehelin et al.
(9)).

These structural data already indicate a particular organization and
orientation of the components of the electron flow system in the membrane. Some complexes are oriented toward the inside, others are located on the outer side of the membrane. It is important to note that
the thylakoid membrane forms a vesicle with the outer side facing the
matrix of the chloroplast and the inner side facing a distinct additional space, separate from the matrix space. The organization of the
electron flow system in the membrane is of principal importance for
understanding the mechanisms of energy conservation in photophosphorylation, i.e., the mechanisms by which the coupling of electron flow to
ATP formation is accomplished.

Coupling Sites of Electron Flow to ATP Formation

The stoichiometry of the coupling of ATP formation to electron flow
from water to NADP (or in the artificial donor or acceptor systems)
is a matter of debate. Following their discovery in 1957 (10) of noncyclic photophosphorylation, Arnon and his colleagues reported a stoichiometry of one ATP formed per one NADP reduced or half a mole of
oxygen evolved. However, in the twenty years since, the stoichiometry -
the P/2e ratio - slowly crept up above the one mark and even reached
two (see recent reviews (1,12)). Higher P/2e values than one, of
course, indicate more than the one coupling site visualized initially,
and indicated in Figure 1. As will be justified later, two coupling
sites do not necessarily need to yield a P/2e value of two, and the
true stoichiometry in noncyclic photophosphorylation is probably
1.33 ATP per two electrons.

One of the coupling sites is localized between the two photosystems,
as Hill and Bendall proposed, and is now more specifically identified
between plastoquinone and plastocyanin. However, in the last few
years, the second coupling site has also been identified (13,14). By
the technique outlined in Figure 3, i.e., by blocking electron flow
between the two photosystems using the plastoquinone antagonist dibromothymoquinone (DBMIB) or the plastocyanin inhibitors KCN or
$HgCl_2$, Hill reactions driven by photosystem II only, could be measured. A lipophilic, artificial mediator such as phenylenediamine or
benzoquinone, has to be added to connect the lipophilic end of photosystem II (in contrast to the hydrophilic end of photosystem I) onto
a terminal electron acceptor (Fig. 4). In such a photosystem II-de-
pendent reaction, photosystem I is inoperative, but oxygen is nevertheless evolved and an artificial electropositive electron acceptor
reduced. It is important to note that the reaction is still coupled
to ATP formation, but with only half the P/2e value of noncyclic
photophosphorylation (13,14). The stoichiometry obtained is about
0.6 per two electrons moved through photosystem II. In chloroplast
preparations with very high P/2e values of two, the photosystem II
section also yields half this value, i.e., one. This is reviewed in
References (1,12).

It has been known for some time that photosystem I alone is also
coupled to ATP formation, when connected to an artificial donor
system or in a cyclic electron flow system (Fig. 4). The P/2e values
in these systems are complicated because of the overimposed cyclic

88

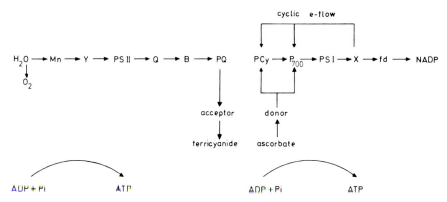

Fig. 4. Two coupling sites in noncyclic electron flow, one connected with photo-system II, the other with photosystem I. An artificial lipophilic acceptor permits measurement of coupled photosystem II reactions; an artificial donor system feeding either into plastocyanin or P_{700} or cyclic electron flow permits measurement of coupled photosystem I reactions, when noncyclic electron flow is interrupted between plastoquinone and plastocyanin

electron flow. The P/2e values tended to creep down from one, and recently corrected values of about 0.6-0.7 for photosystem I coupled electron flow are reported (summarized in (1,12).

In summary, two native coupling sites in photosynthetic electron flow are responsible for ATP formation. Both the electron flow sections with either photosystem II or I are coupled to ATP formation. Either coupling site contributes half the ATP (0.66) of noncyclic electron flow to a total stoichiometry of 1.33.

Artificial ATP Formation

Since the introduction of artificial donors, acceptors, and cofactors of cyclic electron flow, it has been clear that these are artificial electron flow pathways. In addition there are also artificial coupling sites. This concept was introduced (11,15,16) during studies with artificial donor systems in sections of the native electron flow system as discussed above. Such sections were coupled to ATP formation, but the attempted location of the coupling site between the native carriers had become increasingly contradictory and unlikely. The electron pathway included only very few components of the native system and -- most disturbing at first -- did not include the recognized coupling site between plastoquinone and plastocyanin. The contradictions and difficulties dissolved when it became clear that an artificial coupling could be imposed on a section of the electron flow system that does not contain a native coupling site in vivo (11,15,16). An electron flow section can be forced to form ATP. Artificial coupling is possible when the donor or cofactor of cyclic electron flow can translocate hydrogens across the membrane (11,16). The coupling of the donor systems for photosystem I, indicated in Figure 4, is due to such an artificial coupling site, because the native coupling site I between plastoquinone and plastocyanin does not participate in these systems. Artificial coupling sites have now been coupled to either photosystem

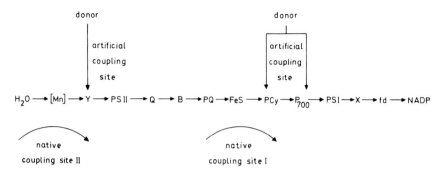

Fig. 5. Two native coupling sites in noncyclic electron flow, one connected to proton liberation in the water-splitting reaction, the other with proton translocation via plastoquinone. An artificial coupling is possible when the donor for either photosystem I or II can carry hydrogens across the membrane

by connecting them with proton translocating electron donor systems (11). In Figure 5, the two native coupling sites are indicated in native electron flow from water to NADP as well as in artificial coupling sites in artificial electron donor systems.

Vectorial Electron Flow

The following problems in the coupling of ATP formation to electron flow had to be resolved: the localization of two native coupling sites and of several artificial coupling sites and the stoichiometry of broken P/2e ratios of 0.6-0.7 for one or of 1.33 for the two sites in noncyclic phosphorylation. The solution to these problems turned out to be possible by, and to provide at the same time strong support for, the chemiosmotic coupling mechanism. In 1961 Mitchell proposed that, according to the chemiosmotic theory, coupling of an electron flow system to ATP formation proceeds via a proton motive force (PMF). A PMF consists of a membrane potential term and a concentration term of a proton gradient across the membrane. The electron flow system was proposed to be vectorial across the membrane via an alternation of an electron and an electron + hydrogen-carrying carrier that would pump protons across. A loop of such an electrogenic + a proton translocating step in the electron flow system would constitute an energy-conserving site in this section. An ATP synthetase, removed spatially in the membrane from the electron flow system, was to drive ATP synthesis at the expense of the PMF. A P/2e ratio is composed of a H^+/e ratio of the electron flow system and a H^+/ATP ratio of the synthetase reaction.

Mitchell had already speculated in 1966 about the orientation of the electron flow components of photosynthesis (17). He had put the watersplitting reaction on the inside of the thylakoid membrane and had both photosystems cross the membrane from the inside to the outside. In this way, he had implicitly proposed two coupling sites at a time, there was as yet no direct indication for them. Jagendorf and his collaborators provided the first direct evidence for a chemiosmotic coupling mechanism when they showed that an artificial proton gradient imposed on chloroplasts would form ATP in the dark in a PH jump experiment (18). The results of Witt and his colleagues later provided the first indication for vectorial electron flow, when a 515-absorption shift was correlated with the build-up of an electric field across the membrane, brought about by the primary charge separation in the two photosystems in the light (19). Because each photosystem contributed equally to the membrane potential, again a second coupling site could have been predicted.

As studies on the topology of the thylakoid membrane progressed, evidence for vectorial electron flow came from investigations on the location of the individual components of the electron flow system in the membrane, and their sidedness became increasingly clear. Studies on the accessibility of electron carriers to membrane-impermeable antibodies - directed against the components - or of impermeable versus permeable chemical probes were particularly indicative. The properties of the lipophilic versus hydrophilic electron donor and acceptor systems, the proof of the second coupling site, the properties of the electric field and of delayed fluorescence contributed to mounting evidence for vectorial electron flow. This evidence (15) was summarized in a review in 1974. Central to the argumentation was the positive,

Table 1. Evidence in 1974 for a sidedness of photosynthetic electron flow in the thylakoid membrane. For references see (15)

Located outside or oriented toward the outside	Located inside or oriented toward the inside
Coupling factor	Low pH in the inner lumen

	Photosystem I Acceptor side	
Hydrophilic acceptors	
Antibody against Fd-NADP-reductase	
Antibody against ferredoxin	
Antibody against S_{L}-eth	
DABS-labeling	
Field formation	

	Donor side	
Polylysine inhibition	Antibody against plastocyanin
	Antibody against cytochrome f
	Lipophilic donor systems
	Precipitation of DAB oxid
	Field formation

	Photosystem II Acceptor side	
Antibody against plastoquinone	Inaccessibility of hydrophilic acceptors
Antibody against chlorophyll	
Antibody against subchloroplasts	
Photoreductions by photosystem II	
Proton uptake by plastoquinone reduction	
Accessibility of cytochrome b_{559}	
Field formation	

	Donor side	
DABS-labeling	Coupling of photosystem II
Antibody against lutein	Proton release in the water- splitting reaction
Antibody against photosystem II particles	Field formation
Trypsin digestion	Delayed fluorescence

though also somewhat inconclusive or even contradictory, evidence for the location or orientation of the donor sites of the two photosystems toward the inside and of the acceptor sites toward the outside of the thylakoid membrane.

Some of the evidence available in 1974 is shown in a condensed form in Table 1. Of course, the data are of different significance. The crucial points are the location of the oxygen evolution system and of plastocyanin inside.

The electron flow scheme of 1974 indicated two loops responsible for two native energy-conserving sites. Site I consisted of the electrogenic photosystem I and proton translocation by plastoquinol, site II of the electrogenic photosystem II and the protons of the water-splitting reaction, liberated inside. Artificial coupling would occur when an artificial proton-translocating donor system would complete a loop with either photosystem, replacing one or both native proton-translocating reactions.

The scheme of Figure 6 would also account for broken P/2e ratios, as observed. Each loop in the electron flow would yield 2 H^+/2e translocated or 4 H^+/2e for the two loops in noncyclic electron flow from water to NADP via both photosystems. If the ATP synthetase reaction in turn would consume 3 H^+/ATP, from the ratio of H^+/2e over 3 H^+/ATP, it follows that a coupling site (energy-conserving site) would have a stoichiometry of 2/3 ATP/2e = 0.66 or for two sites 4/3 ATP/2e = 1.33.

Present evidence on the H^+/e and H^+/ATP ratios provides good support for these stoichiometries in photosynthetic electron flow (see reviews (1,12,20). There is satisfactory agreement for a 4 H^+/2e ratio; a recent report (21) on a higher stoichiometry was refuted (20,22). The H^+/ATP ratio of 3 is not unanimously accepted: values of 2 and 4 have also been reported (see (1,12)). Careful measurements by Pick et al. (23) and by Portis and McCarty (24) did yield a value of 3. This result was reviewed recently by Reeves and Hall (12).

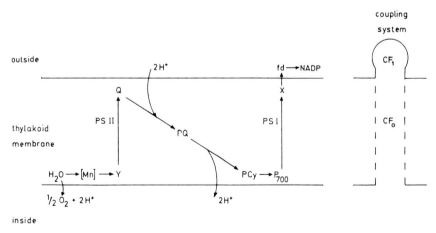

Fig. 6. Vectorial electron flow across the thylakoid membrane with two energy-conserving sites, consisting of an electrogenic photosystem and a proton liberation (water splitting) or translocation (plastoquinol) reaction. A coupling system, driving ATP synthesis at the expense of a pH gradient, is separated in space from the electron flow system

Location of the Donor Sites of the Photosystems in the Membrane

As stated previously, the locations of the oxygen evolution system and of plastocyanin in the membrane are of crucial importance for the concept of vectorial electron flow. Both should be oriented toward the inside and should not be accessible from the outside by impermeable probes. In 1974 the evidence for plastocyanin inside seemed satisfactory, but that for oxygen evolution was not. A reevaluation in 1978 of the available evidence for the siddedness of components of the electron flow system yields further positive support but also some new contradictions. As Table 2 indicates, a comparison with the data available in 1974 (shown in Table 1) shows a much improved situation concerning the location of the water-splitting system on the inside. The effect of internal pH (25-27) or of manganese release into the inside (28), of further studies on proton release inside (29), artificial

Table 2. Location of the donor sides of photosystems I and II in the membrane. For references see text

Evidence	Outside	Inside
Plastocyanin		
1974	Polylysine inhibition	Early antibody studies
		Artificial donor systems
1978	HgCl$_2$ inhibition	
	DABS labeling	
	New antibodies	
Oxygen evolution		
1974	DABS labeling	Proton release inside
	Antibody against lutein	
	Trypsin digestion	
1978	Antibodies against	Control by internal pH
	purified proteins	Release of manganese inside
		Further studies on proton release
		Artificial donor systems
		Lavorel model of moving units

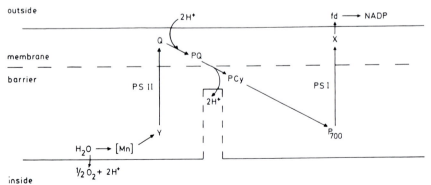

Fig. 7. Hypothetical vectorial electron flow with proton translocation by plastoquinol into a microspace or a channel from the inside lumen to account for plastocyanin accessibility

donor systems for photosystem II (see (6,11)), and the Lavorel model of a moving of the water-splitting system on the inside surface (30) are good evidence for inside location. The results with new antibodies (31-34) (directed against purified proteins from thylakoids) as well as with DABS (35,36), which inhibit oxygen evolution to some extent, seem to contradict, but do not strongly argue against it. Surely antibodies and DABS can react only with accessible components. However, the water-splitting reaction is closely connected to the large photosystem II complex, which in turn is connected with a light-harvesting protein oriented toward the outside (37). This large integral protein complex will relay an attack by the probes or antibodies on the outside through the membrane, which results in (some) inhibition of oxygen evolution on the inside.

The situation concerning the inside location of plastocyanin has, however, not improved. The evidence in 1974 from antibody studies for an inside location seemed satisfactory. In the meantime, however, new antibodies against plastocyanin have been described (38-40), which do agglutinate the membrane and do inhibit plastocyanin function and therefore argue for an orientation or at least for a cleft toward the outside. It has been shown (41) that DABS, KCN, and $HgCl_2$ do react with the active center (copper) of plastocyanin. Therefore the argument for an indirect effect of the inhibitors does not hold. Still, the reported data of plastocyanin antibody effects are not fully convincing, as has been discussed elsewhere (42).

Though, at present , the evidence for the functional site of plastocyanin seems less convincing, the evidence for photosystem I being oriented across the membrane remains strong. P $_{700}$ (43) and more specifically one of the subunits of the P_{700} complex were shown to react from the inside (44), as does an antibody against another purified protein (45). The same is true for the orientation of photosystem II (for a summary see (46)), though there is a recent report that the photosystem itself might not span the whole membrane (47).

Vectorial Electron Flow in Microspace Compartments Within the Membrane

The new results on the accessibility of plastocyanin, as discussed above (as well as for cytochrome f (39) do not require a principal revision of concepts. On the other hand, an adjustment of the certainly oversimplified schemes, such as the one in Figure 6, is possible. The electron carriers between the two light reactions, in particular plastocyanin and cytochrome f, could be arranged so that they are partly accessible from the outside, but so that the principle of vectorial electron flow is preserved: proton translocation across a barrier separating two pH compartments. The barrier could be close to the surface and perhaps in the outer lipid layer only. The inside space might protrude into the membrane in channels toward plastoquinol and plastocyanin oriented toward the outside. Such channels are known to exist in the coupling system: A proton channel CF_O connects the inside space with the ATP synthetase CF_1 on the surface (48). Instead of a channel, one could visualize that the protons from plastoquinol are not translocated across the membrane, but into a microspace within the membrane. The microspace would communicate with the inside lumen of the thylakoid and under steady-state conditions both spaces would be in equilibrium. However, assume that the proton channel of the ATP synthetase (or some of them) connects directly with the microspace. Then ATP formation could proceed before the proton gradient and the membrane

potential are delocalized in the inside space of the thylakoid. Such
a microspace concept would accomodate some recent evidence reported
by Ort et al. (49,50) for localized pH gradients in transient photosynthesis, where ATP formation seems to occur before a massive, delocalized pH gradient is built up. Such a microspace hypothesis would
also help to account for some unexplained data on the acceptor system
of photosystem II, which indicate an outside location, yet governed
by the internal pH (see (42)). Also less strain is put on the mechanism of plastoquinone tumbling across the membrane, as recently summarized (11). The microspace hypothesis might even allow small differences in the coupling mechanisms for coupling sites I and II, as observed, for example, by the effect of $HgCl_2$ as an energy transfer
inhibitor (51-53). Coupling site II and the water-splitting system
might connect always via the inside space, since the evidence for the
oxygen evolution being inside seems reasonable (Table 2), whereas
coupling site I requires a channel or microspace (Fig. 6).

The microspace hypothesis does not necessarily contradict the chemiosmotic mechanisms, but it could do without vectorial electron flow,
as proposed by Williams (54). A proton pump, driven by the redox reactions of the electron transport chain, would replace proton translocation via the hydrogen-carrying redox carrier itself. The stoichiometry
of H^+/e could be greater than in a proton pump mechanism. Though there
is no evidence for this in photosynthesis, results and concepts for
respiratory electron flow favor such proton pumps (55) and should not
simply be ignored in concepts of photosynthesis.

References

1. Barber, J. (ed.): The Intact Chloroplast. Amsterdam: Elsevier 1976
2. Trebst, A., Avron, M. (eds.): Photosynthesis I, Encyclopedia of Plant Physiol.
   New Series. Berlin-Heidelberg-New York: Springer, 1977, Vol. 5
3. Hill, R., Bendall, F.: Nature (London) 186, 136 (1960)
4. Bohnen, H., Böger, P.: FEBS Lett. 85, 337 (1978)
5. Trebst, A.: In: Methods in Enzymology. Pietro, A. (ed.) New York: Academic Press,
   1972, Vol. XXIV, p. 146
6. Hauska, G.: In: Encycl. Plant Physiol. New Series. Trebst, A., Avron, M. (eds.)
   Berlin-Heidelberg-New York: Springer, 1977, Vol. 5, p. 253
7. Mühlethaler, K.: Ibid. p. 503 (1977)
8. Sane, P.V.: ibid. p. 522 (1977)
9. Staehlin, L.A., Armond, P.A., Miller, K.R.: Brookhaven Symp. Biol. 28, 278 (1976)
10. Arnon, D.I., Whatley, F.R., Allen, M.B.: Science 127, 1026 (1958)
11. Hauska, G., Trebst, A.: In: Current Topics in Bioenergetics. Sanadi, D.R. (ed.)
    New York: Academic Press, 1977, Vol. 6, p. 152
12. Reeves, S.G., Hall, D.O.: Biochim. Biophys. Acta 463, 275 (1978)
13. Trebst, A., Reimer, S.: Biochim. Biophys. Acta 305, 129 (1973)
14. Ouitrakul, R., Izawa, S.: Biochim. Biophys. Acta 305, 105 (1973)
15. Trebst, A.: Annu. Rev. Plant Physiol. 25, 423 (1974)
16. Hauska, G., Reimer, S., Trebst, A.: Biochim. Biophys. Acta 357, 1 (1974)
17. Mitchell, P.: Biol. Rev. 41, 445 (1966)
18. Jagendorf, A.T., Uribe, E.: Proc. Natl. Acad. Sci. U.S.A. 55, 170 (1966)
19. Witt, H.T.: Q. Rev. Biophys. 4, 365 (1971)
20. Junge, W.: Annu. Rev. Plant Physiol. 28, 503 (1977)
21. Fowler, C.F., Kok, B.: Biochim. Biophys. Acta 423, 5110 (1876)
22. Saphon, S., Crofts, A.R.: Z. Naturforsch. 32c, 810 (1977)
23. Pick, U., Rottenberg, H., Avron, M.: FEBS Lett. 48, 32 (1974)
24. Portis, A.R., McCarty, R.E.: J. Biol. Chem. 251, 1610 (1976)
25. Harth, E., Reimer, S., Trebst, A.: FEBS Lett. 42, 165 (1974)

26. Briantais, J.M., Vernotte, C., Lavergne, J., Arntzen, C.J.: Biochim. Biophys. Acta 461, 61 (1976)
27. Renger, G., Gläser, M., Buchwald, H.E.: Biochim. Biophys. Acta 461, 392 (1976)
28. Blankenship, R.E., Sauer, K.: Biochim. Biophys. Acta 357, 252 (1974)
29. Junge, W., Renger, G., Ausländer, W.: FEBS Lett. 79, 155 (1977)
30. Lavorel, J.: FEBS Lett. 66, 164 (1976)
31. Radunz, A., Schmid, G.H.: Z. Naturforsch. 30c, 622 (1975)
32. Schmid, G.H., List, H., Radunz, A.: Z. Naturforsch. 32c, 118 (1977)
33. Schmid, G.H., Menke, W., Koenig, F., Radunz, A.: Z. Naturforsch. 31c, 304 (1976)
34. Schmid, G.H., Renger, G., Gläser, M., Koenig, F., Radunz, A., Menke, W.: Z. Naturforsch. 31c, 594 (1976)
35. Dilley, R.A., Giaquinta, R.T.: In: Current Topics in Membranes and Transport. Bronner, F., Kleinzeller, A. (eds.) New York: Academic Press, 1975, Vol. 7, p. 49
36. Dilley, R.A., Giaquinta, R.T., Prochaska, L.J., Ort, D.R.: In: Water Relations in Membrane Transport in Plants and Animals. New York: Academic Press, 1977, p. 55
37. Bose, S., Burke, J.J., Arntzen, C.J.: In: Bioenergetics of Membranes. Packer, L. et al. (eds.) Amsterdam: Elsevier, 1977, p. 245
38. Schmid, G.H., Radunz, A., Menke, W.: Z. Naturforsch. 30c, 201 (1975)
39. Schmid, G.H., Radunz, A., Menke, W.: Z. Naturforsch. 32c, 271 (1977)
40. Böhme, H.: Eur. J. Biochem. 83, 137 (1978)
41. Smith, D.D., Selman, B.R., Voegeli, K.K., Johnson, G., Dilley, R.A.: Biochim. Biophys. Acta 459, 468 (1977)
42. Trebst, A.: In: Methods in Enzymol. San Pietro, A. (ed.) in press
43. Junge, W.: In: Proc. of the 3rd. Intern. Congr. Photosynth. Avron, M. (ed.) Amsterdam: Elsevier, 1975, Vol. 1, p. 273
44. Nelson, N., Notsani, B.E.: In: Bioenergetics of Membranes. Packer, L. et al. (eds.) Amsterdam: Elsevier, 1977, p. 233
45. Menke, W., Koenig, F., Schmid, G.H., Radunz, A.: Z. Naturforsch. 33c, 280 (1978)
46. Crofts, A.R.: In: Photosynthesis 1977. Hall, D.O. et al. (eds.) London: The Biochemical Soc., 1978, p. 223
47. Jursinic, P., Govindjee, Wraight, C.A.: Photochem. Photobiol. 27, 61 (1978)
48. Racker, E.: A New Look at Mechanisms in Bioenergetics. New York: Academic Press, 1976
49. Ort, D.R., Dilley, R.A.:Biochim. Biophys. Acta 449, 95 (1976)
50. Ort, D.R., Dilley, R.A., Good, N.E.: Biochim. Biophys. Acta 449, 108 (1976)
51. Bradeen, D.A., Winget, G.D., Gould, J.M., Ort, D.R.: Plant Physiol. 52, 680 (1973)
52. Bradeen, D.A., Winget, G.D.: Biochim. Biophys. Acta 333, 331 (1974)
53. Gould, J.M.: Biochem. Biophys. Res. Commun. 64, 673 (1975)
54. Williams, R.J.P.: In: Current Topics in Bioenergetics. Sanadi, D.R. (ed.) New York: Academic Press, 1969, Vol. 3, p. 79
55. Papa, S.: Biochim. Biophys. Acta 456, 39 (1976)

Electron Transport Phosphorylation in Anaerobic Bacteria

A. KRÖGER

Ten years ago it was generally believed that anaerobic nonphotosynthetic bacteria could synthesize ATP only by substrate level phosphorylation and that the occurrence of electron transport phosphorylation
was restricted to aerobic bacteria (1). The only exception was the
reduction of nitrate (Table 1), which was known to yield ATP from
electron transport phosphorylation (2). In the meantime it was discovered that certain anaerobic bacteria can derive the ATP required
for growth from the reduction of fumarate, sulfate, and CO_2 (3). The
respective products are succinate, sulfide, and methane. The redox
potentials of these acceptors are considerably lower than that of
nitrate. ATP synthesis coupled to the reduction of these acceptors was
demonstrated by growth of the bacteria with the acceptors in the absence of other energy substrates. Because growth is sustained with molecular hydrogen as the only donor, it can be excluded that substrate
level phosphorylation may be responsible for ATP synthesis. Therefore
the ATP obtained from the reduction of fumarate, sulfate, and CO_2 by
molecular hydrogen must be formed by electron transport phosphorylation.
It is of great interest to know whether or not these systems function
according to the same basic principle as the oxidative phosphorylation
of mitochondria and bacteria, and it can be expected that an investiga-
tion of these systems will contribute to the understanding of the mechanism of electron transport phosphorylation in general.

Table 1. Terminal acceptors of bacterial
electron transport

Acceptor	E_O' (mV) (3)
O_2/H_2O	+815
NO_3^-/NO_2^-	+420
fumarate/succinate	+ 30
$SO_4^=/HS^-$	-220
CO_2/CH_4	-244

Comparison of the Energy-Transducing Systems

The properties of the electron transport phosphorylation systems with
fumarate, sulfate, and CO_2 are compared in Table 2. Bacteria that
reduce fumarate and sulfate as well as methanobacteria contain a membrane-bound magnesium-stimulated ATPase in their membrane (4,5).

Table 2. Properties of anaerobic phosphorylative electron transport (3)

Acceptor	Growth with H_2	Membrane-bound ATPase	Phosphory-lation (in cell-free preparations)	Uncoupling	Electrochem. H^+ potential	Components of electron transport
Fumarate	+	+	+	+	+	MK, b cytochrome
$SO_4^=$	+	+	+	+	?	MK, b c cytochrome
CO_2	+	+	?	?	?	F_{420}, CoM

+ = The respective property has been demonstrated.
? = The respective property has not yet been shown.

Fumarate Reduction

Fumarate reduction is the best investigated system so far. It appears
to be similar to oxidative phosphorylation with respect to nearly all
of the properties (3,6). Phosphorylation that is sensitive to uncouplers
has been demonstrated with a cell-free membrane preparation of a strictly anaerobic bacterium with hydrogen as the donor (7). The yield was
between 0.5 and 0.9 mol ATP per mol of fumarate. The reduction of
fumarate by formate, which is catalyzed by the anaerobic *V. succinogenes*,
is associated with the liberation of protons in the external medium
(8) and with the consumption of protons from the cytoplasm. Measurements indicated that 1-2 $H^+/2e^-$ were produced. The reaction is electric,
since it depends on the presence of potassium ions and is abolished by
uncouplers.

Sulfate Reduction

Less is known about energy transfer in sulfate reduction (3,9,10). The
actual acceptor of electron transport is adenylylsulfate (APS), which
is formed from ATP and sulfate (reaction a):

$$SO_4^2 + ATP \longrightarrow APS + PP_i \qquad\qquad\qquad\qquad (a)$$

$$PP_i + H_2O \longrightarrow 2P_i \qquad\qquad\qquad\qquad (b)$$

$$APS + H_2 \longrightarrow HSO_3^- + AMP + H^+ \quad \Delta G_o' = -16.4 \text{ kcal} \qquad (c)$$

$$HSO_3^- + 3H_2 \longrightarrow HS^- + 3H_2O \quad \Delta G_o' = -41.0 \text{ kcal} \qquad (d)$$

$$SO_4^2 + 4H_2 + H^+ \longrightarrow HS^- + 4H_2O \quad \Delta G_o' = 36.4 \text{ kcal} \qquad (e)$$

Its redox potential of about -60 mV is 160 mV more positive than that
of sulfate. Uncouplers, when applied to intact cells, inhibit not only
phosphorylation of the bacterial adenine nucleotides, but also electron transport. The explanation is that the bacterial concentration of
ATP is decreased by uncoupling to an extent that leads to the inhibition of the activation reaction. Regular uncoupling is found with sulfite as the acceptor, and sulfite is an intermediate of sulfate reduction (reactions c and d). Phosphorylation that is inhibited by uncoupling was shown with a cell-free preparation of *Desulfovibrio* in the

presence of hydrogen as donor and sulfite as the acceptor. The ratio
$P/2e^-$ was maximally 0.2 (11). The activation of sulfate requires two
phosphate bonds, since the pyrophosphate resulting from the synthesis
of adenylphosphosulfate is hydrolyzed subsequently (reaction b). The
activated sulfate is reduced to give sulfite (reaction c), which is a
free intermediate. The reduction of sulfite to sulfide (reaction d),
which requires six reducing equivalents, is either catalyzed by a single
enzyme or by three different reductases. From measurements of the cell
yields of bacteria growing with hydrogen as the donor and sulfate on
the one hand, and sulfite on the other, it was concluded that 1 ATP
per 2 electrons is formed in reaction (d), whereas the reduction of
APS is not coupled to phosphorylation (12). Thus 3 mol of ATP are synthesized per mol of sulfite reduced, that is, per 6 electrons. Since
2 ATP are required for the activation, only 1 mol ATP per mol of sulfate
results from the overall reaction (reaction e).

Desulfovibrio contain menaquinone (MK), *b*, and *c* cytochromes, but it is
not yet known whether these components participate in sulfate reduction (3,9,10).

CO$_2$ Reduction

From the growth of methanobacteria on hydrogen and CO$_2$ it is clear that
CO$_2$ reduction is coupled to some sort of electron transport phosphorylation. However, little is known about the details of the system. Even
the presence of a membrane-bound magnesium-stimulated ATPase (5) does
not prove a relation to oxidative phosphorylation, since this enzyme
was also detected in bacteria, which definitely do not carry out electron transport phosphorylation (13).

Methanobacteria do not contain quinones or cytochromes. Instead F$_{420}$
and coenzyme M (CoM) were detected in all species of methanobacteria
and could not be detected in other bacteria. F$_{420}$ is a redox compound
of low molecular weight. Its structure is not yet known. CoM is thio-
ethane-sulfonate. The reduction of the methylthioether of CoM to methane
and CoM (reaction f)

$$CH_3-CoM + H_2 \rightarrow CH_4 + CoM \qquad\qquad (f)$$

is probably the final step in methane formation. The steps leading to
methyl-CoM are not known. CO$_2$ is activated before reduction, but the
activated form is unknown. The activation seems to require ATP, since
uncouplers inhibit the phosphorylation of the endogenous adenine nucleotides as well as CO$_2$ reduction.

The hydrogenase of methanobacteria catalyzes the reduction of F$_{420}$ by
hydrogen; and it is feasible that F$_{420}$ passes the electrons to the
reductases (see Fig. 6B), which are responsible for the reduction
of the activated CO$_2$ in four different steps (3).

Electron Transport with Fumarate as Acceptor

Whereas the reduction of CO$_2$ and sulfate is regarded as a capability
of highly specialized bacteria, fumarate reduction is a widespread
property of facultative and anaerobic bacteria (6). Most of the entero-
bacteria, like *Escherichia*, *Proteus*, *Salmonella*, and *Klebsiella*, can use
fumarate reduction for synthesizing ATP by electron transport phosphorylation. The same is true for many rumen bacteria like *Propionibacteria*, *Bacteroides*, and *Vibrio succinogenes*. The growth rates of some

Table 3. Redox potentials of unsaturated
carboxylic acids

	E_o' (mV) (3)
Fumarate/succinate	+30
Acrylate/propionate	-30
Crotonate/butyrate	-30

fumarate-reducing bacteria are greater than those of other anaerobic
bacteria, and the specific activities of fumarate reduction are in
many cases as great as those of the respiration of fast-growing aerobic bacteria. Because fumarate can be synthesized from many amino
acids and other metabolites, an environment containing organic matter
appears to be the most favorable biotope for bacteria performing
fumarate reduction.

The redox potential of the fumarate/succinate couple is at least
200 mV more positive than that of most of the other redox couples involved in carbon metabolism. Therefore, fumarate is well suited as an
oxidant in anaerobic metabolism. Unsaturated fatty acids have a redox
potential that is only 60 mV more negative than that of fumarate
(Table 3), and their reduction by NADH would also allow the formation
of 1 ATP/2e⁻. The CoA-esters of acrylate and crotonate, whose redox
potentials are even 15 mV more positive than those of the free acids,
are well-known hydrogen acceptors of anaerobic bacteria that form
fatty acids as end products (3). However, electron transport phosphorylation with unsaturated fatty acids or their CoA-esters has not
yet been demonstrated. The reason for this discrepancy between fumarate
reduction and that of the unsaturated fatty acids is not known.

Propionate formation proceeds via acrylyl-CoA in only a few bacteria
(3,14). In most cases it is formed from succinate via methyl-malonyl-
CoA. A well-known example is the fermentation of lactate by propionibacteria (Fig. 1) (15,16). Here pyruvate is carboxylated to give
oxaloacetate, which is reduced to fumarate. Because fumarate reduction is coupled to phosphorylation, 3 ATP/3 lactate can be gained in
this pathway, whereas only 1 mol ATP per 2 or 3 mol of lactate are
produced from the pathway via acrylyl-CoA.

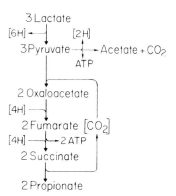

Fig. 1. Metabolic pathway of the fermentation of
lactate by propionibacteria (15,16)

Table 4. Donors of fumarate reduction

Donor	E_O' (mV)	(3)
H_2/H^+	-420	
Formate/HCO_3^-	-416	
NADH/NAD^+	-320	
Lactate/pyruvate	-197	
G-1-P/DAP	-190	
Malate/oxaloacetate	-172	

Fumarate reduction is catalyzed by an electron transport chain that
is localized in the cytoplasmic membrane of the bacteria. Depending on
the organism and on the growth conditions, the substrates listed in
Table 4 can serve as hydrogen donors for fumarate reduction. The chain
consists of at least one membrane-bound dehydrogenase that is specific
for one of the donors, a naphthoquinone, and fumarate reductase. Most
of the systems catalyze fumarate reduction with more than one substrate.
The question as to whether cytochromes are involved is not yet definitely answered (3,6). However, a b cytochrome is present in almost all
of the fumarate-reducing bacteria, even in the strictly anaerobic
species.

The naphthoquinone involved in fumarate reduction is MK in most bacteria, and in some cases desmethylmenaquinone (DMK) (6). They cannot be
replaced by ubiquinone, the redox potential of which is too positive
to allow oxidation by fumarate. By extraction of the quinones and re-
activation of electron transport, it was shown that MK or DMK is a necessary component of the chain. Inspection of the redox reactions of
bacterial MK showed that quinone is converted to hydroquinone upon reduction by the donor substrates and that hydroquinone is reoxidized to
quinone by fumarate (6). The amount of quinone radical measured was
very small (17).

The function of the quinones appears to be similar to that of ubiquinone
in the respiratory chain (18,19). They collect the reducing equivalents
liberated by the various dehydrogenases and transport them to fumarate
reductase or to other acceptor-activating enzymes like nitrate reductase and cytochrome oxidase (Fig. 2). The active sites with respect to
the substrates of both the dehydrogenases and the reductases may face
either the external or the cytoplasmic aspect of the membrane. The electrons are probably transported by diffusion of the quinone in the lipid
phase of the membrane (19). On the basis of this view of the function
of quinone, the electron transport chain may not necessarily be considered a single multienzyme complex. It appears likely that the chain

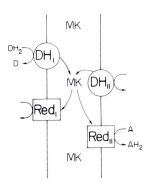

Fig. 2. The function of MK in electron transport. MK
is assumed to diffuse within the membrane and to react
with various donor (DH) and acceptor complexes (Red).
DH_2 = donor substrate; A = acceptor substrate

consists of two structurally independent enzymes: the donor and acceptor complexes. Each of these complexes is comprised of the dehydrogenase or reductase and of other components of electron transport such
as cytochromes and iron-sulfur proteins that catalyze electron transport between the enzymes and the quinone. The donor and acceptor complexes may be randomly distributed on the surface of the membrane,
because the distances could be bridged by the diffusion of the quinone.

The adaptation of bacterial electron transport to varying growth conditions is more easily explained with this concept than with that of
the chain as a structural unit. In the latter case it would be expected that the variation of the species and of the contents of the dehydrogenase and the reductase is limited by structural requirements of
the assumed electron transport complex, and this is not in line with
the actual situation. On the other hand, with a quinone as a diffusing
mediator, both the dehydrogenases and the reductases could be varied
independently and without any restriction. Therefore, we consider
the basic function of the quinones to be that of a diffusing redox
mediator between two structurally independent constituents of the chain.

The validity of this concept with respect to fumarate reduction is confirmed by the results obtained from studying the electron transport
chain of *V. succinogenes* that catalyzes the reduction of fumarate by
formate (Fig. 3). After solubilization of the membrane, the enzyme
catalyzing the dehydrogenation of formate by artificial acceptors was
separated from a fumarate-reducing enzyme with artificial donors (20).

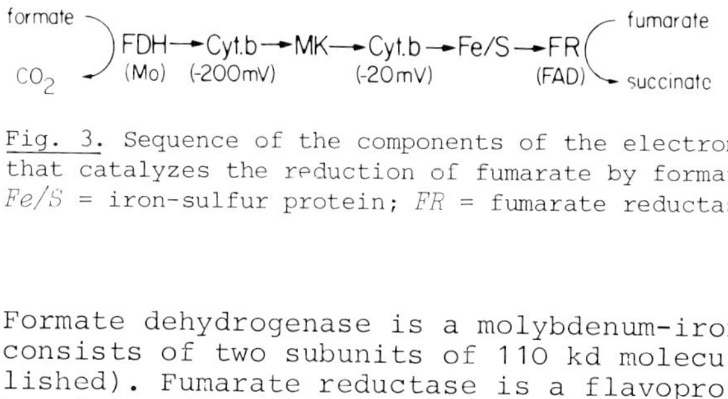

formate $\searrow$ FDH $\rightarrow$ Cyt.b $\rightarrow$ MK $\rightarrow$ Cyt.b $\rightarrow$ Fe/S $\rightarrow$ FR $\nearrow$ fumarate
CO_2 $\nearrow$ (Mo) (-200mV) (-20mV) (FAD) $\searrow$ succinate

Fig. 3. Sequence of the components of the electron transport chain of *V. succinogenes*
that catalyzes the reduction of fumarate by formate (22). *FDH* = formate dehydrogenase;
Fe/S = iron-sulfur protein; *FR* = fumarate reductase

Formate dehydrogenase is a molybdenum-iron-sulfur protein that probably
consists of two subunits of 110 kd molecular weight (Kröger, unpublished). Fumarate reductase is a flavoprotein: It contains covalently
bound FAD (20), acid-labile sulfur and nonheme iron, and consists of
two subunits of 80 and 30 kd molecular weight. Thus it appears to be
structurally similar to the mitochondrial succinate dehydrogenase,
whereas the catalytic properties are those of a fumarate reductase (see
Table 5). The electron transport between formate dehydrogenase and MK
is mediated by a low-potential *b* cytochrome (Fig. 3) and that between
MK and fumarate reductase by a second *b* cytochrome with a midpoint
potential of -20 mV (21, 22). Thus the chain is probably made up of
only five polypeptides, each of which carries a redox-active group.

Generation of the Electrochemical Proton Potential

In view of its relative simplicity, the system of fumarate reduction
may be suited for investigating the mechanism of generation of the
electrochemical proton potential. On the basis of the classical hypothesis that the protons are liberated and taken up by redox reactions
of hydrogen carriers with electron carriers (23, 24), two different
mechanisms are feasible in the formate-fumarate reduction of *V. succinogenes*.

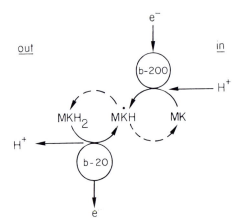

Fig. 4. Possible role of MK and the b cytochromes of the electron transport chain of *V. succinogenes* in electric proton translocation driven by formate-fumarate reduction (6). b-200 = low-potential cytochrome b with E_m = -200 mV; b-20 = high-potential cytochrome b with E_m = -20 mV; $M\dot{K}H$ = neutral radical of MK; MKH_2 = hydroquinone of MK

The first mechanism is based on the assumption that protons are translocated across the membrane by quinone (Fig. 4). The requirements for this mechanism would be met by the chain of *V. succinogenes* in which the donor as well as the acceptor of the reducing equivalents of MK is a cytochrome (Fig. 3). The reduction of MK by the more negative b cytochrome that is associated with the uptake of protons may occur on the inside of the membrane and the quinone radical may be formed. The radical would diffuse within the membrane and dismutate to give quinone and hydroquinone. The oxidation of hydroquinone by the more positive cytochrome could possibly occur on the outside. This would result in a release of protons into the external medium and the radical would be formed again. Thus one proton would be translocated per electron transported, which is in agreement with the experimental result. Although the operation of this mechanism in the formate-fumarate reductase of *V. succinogenes* cannot be excluded, another mechanism is favored by the experimental data.

With the second mechanism, electrons rather than protons are translocated across the membrane (Fig. 5). Protons are liberated on the outside by the reaction of formate dehydrogenase with formate and are taken up on the inside by the reaction of fumarate reductase with fumarate.

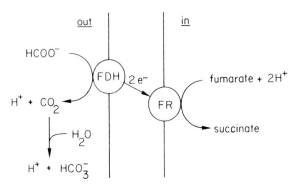

Fig. 5. Possible mechanism of generation of the electrochemical proton potential across the membrane of *V. succinogenes*, which may operate during reduction of fumarate by formate (6). *FDH* = formate dehydrogenase; *FR* = fumarate reductase

The chain serves only in transporting the electrons from formate dehydrogenase to fumarate reductase. Because this mechanism accounts for
the observed H^+/e^- ratio, there would be no need for proton translocation by MK.

The mechanism requires that the active site of formate dehydrogenase
face the outside and that of fumarate reductase the inside of the membrane. This is consistent with the results obtained from accessibility
studies of the two enzymes with intact bacteria and spheroplasts (6).
The enzymes react with a variety of dyes, some of which do not permeate
through the membrane. Formate dehydrogenase is accessible to the impermeable dyes in intact cells or spheroplasts, whereas fumarate reductase reacts only either with permeable dyes or with the impermeable
dyes after disruption of the cytoplasmic membrane.

The accessibilities of the enzymes to their substrates confirm the data
found with the dyes. This was shown by measuring the activities of the
enzymes as functions of the concentrations of their substrates with
intact and broken cells. The relations obtained with both formate
dehydrogenase and fumarate reductase follow the Michaelis equation.
The K_m of formate dehydrogenase for formate is the same when measured
with intact and broken cells (Table 5) and is not affected by the pH
of the external medium (not shown). This indicates that formate does
not pass through the membrane in order to react with formate dehydrogenase. In contrast, the K_m of fumarate reduction for fumarate is
drastically different for intact and broken cells (Table 5). Thus with
the impermeable viologen dye MOV the K_m for fumarate is about twenty
times smaller with lysed than with intact cells. With the permeable
dye methylene blue and succinate as the donor for fumarate reductase
the K_m for succinate is ten times greater with lysed than with intact
cells. This indicates that the K_m of the reductase is measured only
with broken cells, whereas the apparent K_m obtained with intact cells
also implies the transport of fumarate into the cytoplasm. The transport
of fumarate across the membrane, which may operate by exchange against
succinate, was confirmed to occur by measuring the uptake of fumarate
by the cells (Kröger, unpublished). For a final proof of this mechanism,
formate dehydrogenase must be identified as the site of production and
fumarate reductase as the site of comsumption of the protons that are
associated with overall electron transport. However, this has not yet
been achieved.

Table 5. Influence of cell disruption on the K_m of formate dehydrogenase and
fumarate reductase for their substrates. Cells of *V. succinogenes* were lysed
by the addition of Triton X-100. (Kröger, A., unpublished)

	Formate → MOV[a]	
	K_m (mM)	V_{max} (U/mg)
Intact cells	0.55	2.0
Lysed cells	0.55	2.0

	MOV[a] → fumarate		Succinate → methylene blue	
	K_m (mM)	V_{max} (U/mg)	K_m (mM)	V_{max} (U/mg)
Intact cells	1.7	0.6	0.04	0.18
Lysed cells	0.08	8.0	0.40	0.48

[a] 1,1'-Bis(dimethylmorpholinocarbonylmethyl)-4,4'-bipyridylium(dichloride)

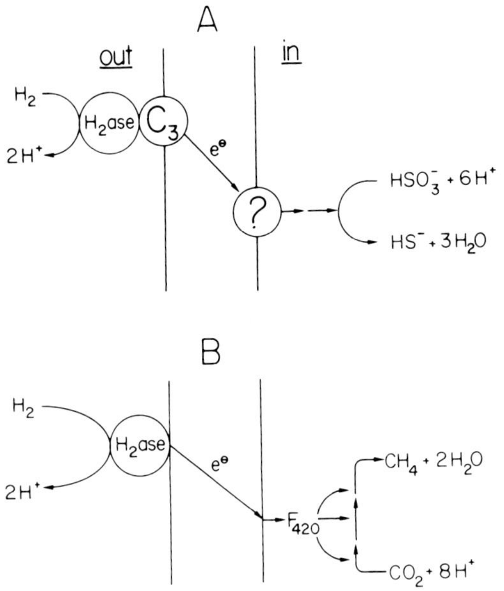

Fig. 6 A and B. Hypothetical mechanisms of generation of electrochemical proton potentials across the membrane of bacteria that reduce sulfate (A) and CO_2 (B) (3)

It may be speculated that a similar mechanism is operative in sulfate and CO_2 reduction by molecular hydrogen (3) (Fig. 6A). Sulfate-reducing bacteria contain hydrogenase, an iron-sulfur protein, together with its primary acceptor cytochrome c_3 in the periplasmic space, whereas the enzyme(s) catalyzing sulfite reduction are present in the cytoplasm (25). Although electron transport from hydrogenase to sulfite reductase is not known, it is feasible that only the electrons of hydrogen are transported across the membrane, whereas the resulting protons are liberated in the external medium. Furthermore the protons required for sulfite reduction are likely to be taken up from the cytoplasm.

The possible generation of a proton potential by methane formation from CO_2 by methanobacteria cannot proceed via the quinone mechanism, since these bacteria do not contain a quinone. However, it is feasible that a mechanism similar to that proposed for fumarate and sulfate reduction may operate in methane formation (3) (Fig. 6B).

From the present state of knowledge it is probable that more than one mechanism of generation of the proton potential operates in electron transport phosphorylation. This seems to be true also for mitochondrial oxidative phosphorylation. Thus proton translocation can probably be coupled directly to the redox reactions of single cytochromes (26 - 30), but neither this nor the mechanisms depicted in Figures 4 and 5 are expected to operate in the mitochondrial reduction of ubiquinone by NADH. Whether the proton potential is a prerequisite of electron transport phosphorylation is discussed by Kawaga (31).

References

1. Decker, K., Jungermann, K., Thauer, R.K.: Angew. Chem. Int. Ed. Engl. 9, 138-158 (1970)
2. Stouthamer, A.H.: Adv. Microb. Physiol. 14, 315-375 (1976)
3. Thauer, R.K., Jungermann, K., Decker, K.: Bacteriol. Rev. 41, 100-180 (1977)

4. Guarraia, L.J., Peck, H.D.: J. Bacteriol. 106, 890-895 (1971)
5. Thauer, R.K.: personal communication
6. Kröger, A.: In: Microbiol. Energetics. Haddock, B.A., Hamilton, W.A. (eds.) London: Cambridge University Press, 1977, pp. 61-93
7. Barton, L.L., Le Gall, J., Peck, H.D.: Biochem. Biophys. Res. Commun. 41, 1036-42 (1970)
8. Kröger, A.: In: Electron Transfer Chains and Oxidative Phosphorylation. Quagliar-iello, E. et al. (eds.) Amsterdam: North-Holland Publ. Co., 1975, pp 265-270
9. Le Gall, J., Postgate, J.R.: Adv. Microb. Physiol. 1o, 81-133 (1973)
10. .Siegel, L.M.: Metab. Pathways 7, 217-286 (1975)
11. Peck, H.D.: Biochem. Biophys. Res. Commun. 22, 112-118 (1966)
12. Badziong, W., Thauer, R.K.: Arch. Microbiol. 117, 209-214 (1978)
13. Riebeling, V., Jungermann, K.: Biochem. Biophys. Acta 430, 434-444 (1976)
14. Anderson, R.L., Wood, W.A.: Annu. Rev. Microbiol. 23, 539-578 (1969)
15. Wood, H.G., Utter, M.F.: Essays Biochem. 1, 1-27 (1965)
16. De Vries, W., van Wyck-Kapteyn, W.M., Stouthamer, A.H.; J. Gen. Microbiol. 76, 31 41 (1973)
17. Weber, M.M., Hollocher, T.C., Rosso, G.: J. Biol. Chem. 240, 1776-87 (1965)
18. Kröger, A., Klingenberg, M.: Eur. J. Biochem. 39, 313-323 (1973)
19. Kröger, A., Klingenberg, M.: Vitam. Horm. 28, 533-574 (197o)
20. Kenney, W.C., Kröger, A.: FEBS Lett. 73, 239-243 (1977)
21. Kröger, A., Innerhofer, A.: Eur. J. Biochem. 69, 487-495 (1976)
22. Kröger, A., Innerhofer, A.: Eur. J. Biochem. 69, 497-506 (1976)
23. Robertson, R.N.: Biol. Rev. 35, 231-264 (1960)
24. Mitchell, P.: Biol. Rev. 41, 445-502 (1966)
25. Bell, G.R., Le Gall, J., Peck, H.D.: J. Bacteriol. 120, 994-997 (1974)
26. Urban, P.F., Klingenberg, M.: Eur. J. Biochem. 9, 519-525 (1969)
27. Papa, S.: Biochim. Biophys. Acta 465, 39-84 (1976)
28. Wikström, M., Krab, K.: this volume, pp. 128-139 (1978)
29. v. Jagow, G., Schägger, H., Engel, W.D., Hackenberg, H., Kolb, H.: this volume pp. 43-52 (1978)
30. v. Jagow, G., Schägger, H., Engel, W.D., Machleidt, W., Machleidt, J., Kolb, H.: FEBS Lett. 91, 121-125 (1978)
31. Kagawa, Y.: this volume, pp. 195-219 (1978)

Primary Mechanisms of Energy Conservation

The Plastoquinone Pool as Possible Hydrogen Pump

H.T. WITT, R.TIEMANN, P. GRÄBER, and G.RENGER

Figure 1 shows the zigzag scheme of the molecular machinery of photosynthesis with the vectorial pathways of electrons, protons and hydrogens derived from pulse spectroscopic studies (1,2). The time sequence
of the reaction patterns has been evaluated as follows. (1) Excitation
of Chl-a$_I$ and Chl-a$_{II}$. (2) Vectorial ejection of electrons from
Chl-a$_I^*$ and Chl-a$_{II}^*$ at the membrane inside to the outside and electric
field generation. (3) Oxidation of H_2O, reduction of $NADP^+$, reduction
of PQ, and reoxidation. (4) Proton translocation into the inner phase
through protolytic reactions with the charges at the outer and inner
surface of the membrane. (5) Discharging of the energized state through
efflux of protons. (6) Formation of free ATP from ADP + P by the energy
released with the efflux of H^+ via the ATP-synthetase. In toto the H^+-
translocation corresponds to a H^+-circulation. This cycle is closed,
however, only if between system I and II a hydrogen (H^+ plus e) is
translocated from the outside to the inside (Fig. 1). Based on the
protolytic properties of redox reactions in quinone systems, the PQ
pool was inferred to be a candidate for a "pump" for hydrogen (H^+ + e)
from the membrane outside to the inside (3). Recently it was discussed
that a pump not coupled to redox reactions may be responsible for this
H-transfer (4). Such a pump may operate as in the manner of the light-
driven H^+ pump of bacteriorhodopsin (see below).

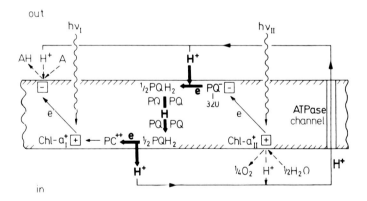

Fig. 1. Scheme of the proton and hydrogen translocations in the thylakoid membrane

---

Abbreviations

Chl-a	Chlorophyll a	ATP	Adenosine triphosphate
PQ	Plastoquinone	ADP	Adenosine diphosphate
PC	Plastocyanine	P	Inorganic phosphate

However, that the PQ pool might operate as a hydrogen pump driven by redox reactions is supported by the following five proofs (Fig.1).

1. The number of active PQ-molecules is about six times larger than the other members of the electron chain. (5,6). This makes the PQ-pool suitable to span the membrane from the outside to the inside. That the pool may be arranged in this way is supported by the following observations:

2. The primary acceptor of light reaction II located at the membrane outside is a special PQ-molecule (X - 320) (6,2). Obviously, also at the inside, at least one PQ-molecule is located, because the time of reoxidation of PQH_2 (20 ms) corresponds to the time of the H^+ release at the inside (7).

3. We have now changed drastically the kinetics of the electron release from the PQH_2 by corresponding change of the turnover of $Chl-a_I$ with appropriate far red light intensities. The kinetics of the simultaneously measured internal H^+ release coincides with the kinetics of the electron release from PQH_2 over a range of 20-fold variation (8).

4. The number, n, of protons, H^+, taken up at system II from the outer phase increases parallel with the number, n, of electrons, e, taken up by the PQ pool (9).

$$n \cdot H^+ \text{ (uptake external)} = n \cdot e \text{ (uptake PQ-pool)} \qquad n = 1 - 7$$

5. Recently, we were able to show that the number, n, of the internal H^+ release (corrected for the H^+ release due to the oxidation of H_2O) increases with the number, n, of electrons, e, released by the PQ pool (8).

$$n \cdot H^+ \text{ (release internal)} = n \cdot e \text{ (release PQ pool)} \qquad n = 1 - 7$$

These results are in consistence with the assumption that the PQ pool might operate as a H - pump.

With respect to the mechanism of the pumping of hydrogens through the PQ pool, two possibilities can be considered: (a) a diffusion-type mechanism and (b) a hopping-type mechanism. In the diffusion-type mechanism each individual molecule (oxidized or reduced) of the PQ pool is assumed to be able to move statistically between the inside and the outside of the membrane. The hopping mechanism envisages the PQ pool as a lattice-like structure. In this case the transport occurs after PQ collision by a statistical hopping of an electron and proton from one PQ to another. If by chance via mechanism (a) or (b) PQH_2 is at the membrane inside next to the oxidized electron acceptor, the electron from PQH_2 is trapped at this place, whereas the remaining proton is released into the inner space due to its much higher affinity to the internal water phase than to the hydrophobic part of the membrane lipids. The internal H^+ release into the inner water phase occurs very likely via protonizable heads of lipids acting as buffer groups at the inner face of the membrane (10).

For both models one has to assume that due to the long hydrophobic side chain of PQ, the PQ molecules are soluble within the lipids of the thylakoid membrane. This is supported by the observation that in artificial bilayers of the lipids plastoquinones are indeed soluble (11).

It is of interest to compare the mechanism of the plastoquinone-pool-hydrogen pump with the light-driven proton pump realized by bacterio-rhodopsin in the membrane of purple bacteria (12). The essential difference between both pumps is as follows: the PQ pump is translocating a neutral atom ($H^+ + e^-$) in the dark, the rhodopsin pump, however, a charged atom (H^+) in the light. The PQ pump works mechanistically relatively simple (see above) because the energy-requiring steps take place by the preceding reduction of PQ at the outside (via electron ejection from $Chl-a_{II}^*$) and by an oxidation via PC at the inside induced by electron ejection from $Chl-a_I^*$ (Fig.1). This means that the overall process of H^+ pumping is subdivided into two sequences, (a) into a vectorial electron transfer at the chlorophyll reaction centers from inside to outside, and (b) into the vectorial H - transfer via the PQ pool from outside to inside.

On the other hand the rhodopsin pump translocates a proton, H^+. Therefore, this translocation must be coupled with the energy-requiring act, i.e., with generation of an electric field and Δ pH formation. Because such a pump has to perform all elementary steps within one highly specialized protein - and, therefore, characterized as a system "beautiful for its simplicity" - the reaction cycle is expected to be rather complex. It is, therefore, not surprising that such a pump includes a photoreaction, conformational change, and a sequence of as yet not well-known different H^+ transfer reaction within the enzyme.

Other models for the H pump have been discussed (4). In contrast to the model outlined above in a so-called Q-cycle model, PQ transfers with the uptake of one electron two H across the membrane instead of one H (13). This is achieved by shuttling one additional electron between PQ and PQH_2 with the help of two cytochromes, Cyt b-6 and Cyt b-559. This model would correspond to a H^+/e-ratio of 3; but the most accepted value is 2. Furthermore, the model includes a relatively slow electron shift from cytochrome b-6 to b-559. This should be detectable as a slow phase in addition to the fast phase (< 20 ns) of the rise of the field-indicating absorption change. However, under our conditions (isolated class-II-chloroplasts excited by flashlight), such a slow phase has not been observed.

References

1. Witt, H.T.: Coupling of Quanta, Electrons, Fields, Ions and Phosphorylation in the Functional Membrane of Photosynthesis. Results by Pulse Spectroscopic Methods. Rev. Biophysics 4, 365-477 (1971)
2. Witt, H.T.: Primary Acts of Energy Conservation in the Functional Membrane of Photosynthesis. In: Bioenergetics of Photosynthesis. Govindjee (ed.) New York-San Francisco-London: Academic Press, 1975, pp. 493-554
3. Mitchell, P.: Chemiosmotic Coupling in Oxidative and Photosynthetic Phosphorylation Biol. Rev. 41, 442-502
4. Hauska, G., Trebst, A.: Proton Translocation in Chloroplasts. In: Current Topics in Bioenergetics. Sanadi, D.R. (ed.) New York-San Francisco-London: Academic Press, 1977, Vol. 6, pp. 151-220
5. Rumberg, B., Schmidt-Mende, P., Weikard, J., Witt, H.T.: Correlation between Absorption Changes and Electron Transport of Photosynthesis. In: Photosynthetic Mechanism of Green Plants. Washington: Acad. Sci. Res. Council, Publ. 1145, 1963, pp. 18-34
6. Stiehl, H.H., Witt, H.T.: Quantitative Treatment of the Function of Plastoquinone in Photosynthesis. Z. Naturforsch. 24b, 1588-1598 (1969)
7. Ausländer, W., Junge, W.: The Electric Generator in the Photosynthesis of Green plants. II. Kinetic Correlation between Protolytic Reactions and Redox Reactions. Biochim. Biophys. Acta 357, 285-298 (1974)

8. Tiemann, R., Renger, G., Gräber, P., Witt, H.T.: The Plastoquinone Pool as Possible Hydrogen Pump in Photosynthesis. Biochim. Biophys. Acta, in press (1978)

9. Reinwald, E., Stiehl. H.H., Rumberg, B.: Correlation between Plastiquinone Reduction, Field Formation and Proton Translocation in Photosynthesis. Z. Naturforsch. $\underline{23b}$, 1616-1617 (1968)

10. Siggel, U.: The Function of Plastoquinone as Electron and Proton Carrier in Photosynthesis. Bioelectrochem. $\underline{3}$, 3o2-318 (1976)

11. Hauska, G.: The Permeability of Quinone through Membranes. In: Bioenergetics of Membranes. Packer, L. et al. (eds.) Amsterdam: Elsevier, North-Holland Biomedical Press, 1977, pp. 177-187

12. Oesterhelt, D.: Bacteriorhodopsin als Beispiel einer lichtgetriebenen Protonen-pumpe. Angew. Chem. $\underline{88}$, 16-24 (1976)

13. Mitchell, P.: Protonmotive Function of Cytochrome Systems. In: Electron Transfer Chains and Oxidative Phosphorylation. Quagliariello et al. (eds.) Amsterdam: North-Holland Publ. Co., 1975, pp. 305-316

Proton Pumping Across the Thylakoid Membrane Resolved in Time and Space

W. JUNGE, A. J. McGEER, W. AUSLÄNDER, and J. KOLLIA

Introduction

For more than a decade the pathway of energy flow in photophosphorylation and in oxidative phosphorylation has been under debate. After the
discussion on whether or not protons are obligatorily involved faded
away [for an "afterglow" see the multiauthored review by Boyer et al.
(1)] in favor of the causality sequence, electrons-protons-ATP, postulated by Mitchell (2,3), another controversy was revived. The question
is whether the protons drive ATP synthesis via the osmolar volumes
[Mitchell (2,3,4,5) in his chemiosmotic hypothesis] or whether they are
confined to special conducting subspaces within the membrane [Williams
(6,7,8,9) in his localized proton hypothesis]. In the public debate
on the matter (4,5,8,9) the weight seems to be on the side of the
chemiosmotic version. Especially in chloroplasts, there is no doubt
that electron transport is coupled to inwardly directed proton pumping
[for reviews, see Refs. [(10,11,12,13)] and that protons placed in the
internal osmolar volume *can* be used for ATP synthesis [for a review,
see Ref. (14)]. However, there are a few observations that seem to shed
doubt on the chemiomotic concept or, at least, make it worthwhile to
take a closer look at localized protons.

1.) Ort et al. (15) studied ATP synthesis under illumination of chloroplasts with short light gates (10 ms - some 100 ms). In the presence of
permeating buffers, which delayed the build-up of a sufficiently high
acidification inside, they found ATP synthesis to be less inhibited
than expected. This led them to suggest "that the protons produced by
electron transport may be used directly for photophosphorylation without ever entering the bulk of the inner aqeous phase of the lamella
system."
2.) Studies on the free energy difference of the proton across the cristae membrane and on the phosphate potential in steadily respiring mitochondria led van Dam et al. (16) to speculate that the proton motive
force seen by the ATP synthase is not the average one existing between
the aqueous bulk phases.
3.) Ausländer and Junge (17) observed the existence of intramembrane
proton sinks at the outer part of the thylakoid membrane at the reducing
sides of both photosystems. The salient point of the controversy on the
pathway of protons (if there is any salient point, because any intermediate version is also conceivable) is whether or not passage of a proton through the internal osmolar volume of thylakoids is a must for
their utilization for ATP synthesis. The problem is a kinetic one. It
may be attacked by posing three different questions:

Abbreviations: DBMIB = dibromothymoquinone; PS II photosystem two.
Buffers: ACES = N-(2-acetamido)-2-aminoethanesulfonic acid;
HEPES = N-2hydroxyethylioerazine-N-2-ethanesulfonic acid;
MES = 2-(N-morpholino)-ethanesulfonic acid; MOPS = morpholinopropane sulfonic acid;
PIPES = piperazine-N,N-Bis(2-ethanesulfonic acid); PP_i = pyrophosphate;
TES = N-Tris(hydroxymethyl)methyl-2-amino-ethanesulfonic acid;
Tricine = N-Tris(hydroxymethyl)methylglycine; Tris = Tris(hydroxymethyl)aminomethane.

1. How fast is the release of protons into the internal osmolar volume by the pumps?

2. How fast is the uptake of protons from this volume by the ATP synthase?

3. How fast is the transverse passage of protons in the internal osmolar volume from the "sources" [at least the pumps associated with photosystem II activity are located in the stacked portions of thylakoids, grossly, the center parts of disks] into the "sinks", the ATP synthases [located in the unstacked portions, according to good evidence (18,19)]?

We investigated questions (1) and (2) by time resolving the internal pH changes with a pH-indicating dye in a rapid kinetic spectrophotometer [for instrumentation, see (20)]. We previously introduced neutral red as an extremely rapid pH indicator for changes occurring inside thylakoids (21-24). Because thylakoids are flat disk-shaped vesicles with a narrow internal phase (thickness ca. 10 nm and diameter ca. 500 nm) there is no way to rapidly record transient pH changes inside other than to use molecular probes. This technique is susceptible to artifacts and therefore perhaps not generally to be trusted. It is sometimes difficult to decide whether absorption changes of a "pH indicating" dye reflect pH changes or some other events (photochemical reactions, redox reactions, shifts of binding equilibrium). In addition, it may be difficult to decide where the supposed pH changes occur, i.e., in an aqueous phase or within some subspace of a membrane. In this communication we review how these difficulties may be overcome. We demonstrate that neutral red is a clean indicator of pH changes inside thylakoids. We further demonstrate that the pH changes indicated by absorption changes of the dye occur in a space that exchanges protons with the internal osmolar volume within less than 100 μs. Both the release of protons into the internal osmolar volume by the pumps and the comsumption of protons from this volume by ATP synthase are time resolved. The results are discussed in light of the above controversy on the mechanism of photophosphorylation.

Experimental details and a more thorough description of the results can be found in two forthcoming papers (25,26).

Dyes as Indicators of pH Changes Inside and Outside the Thylakoids

Figure 1 illustrates how two pH-indicating dyes, a neutral one (neutral red, left) and a charged and less lipid-soluble one (cresol red, right) may be used to measure pH changes in internal and external phases. In the presence of these dyes, flash excitation of chloroplasts produces extra changes of absorption at various wavelengths. These are superimposed on the already complex patterns of intrinsic absorption changes that are mainly attributable to electron transport and to electrochromic affects [for a review, see Witt (27)]. By means of appropriate buffers, e.g., the nonpermeating bovine serum albumin (BSA) or the permeating imidazole and phosphate, it is possible to label the pH_{out} - and the pH_{in} -indicating absorption changes of these dyes [see Ausländer and Junge (21)].

Figure 2 shows transient absorption changes of neutral red (left) and of cresol red (right) under the same layout as in the schematic representation of Figure 1. At t=0, a chloroplast suspension was excited with a short flash of light ($T_{1/2}$ of duration 15 μs), which turned over each photosystem once and induced the uptake of two protons per electron from the outer phase and the release of two protons into the inner one (28,29). The measuring wavelength was tuned to the respective peaks

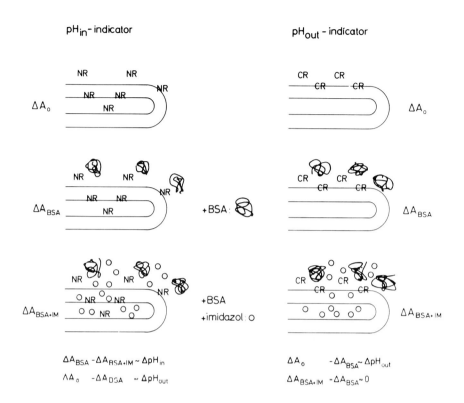

Fig. 1. Schematic illustration of the correlation of absorption changes of two pH-indicating dyes with transient pH-changes in the inner and outer phases of thylakoids, respectively. A section through one thylakoid is simplified by only three phases: the internal phase (ca 10 nm thick), the membrane, and the external phase. Neutral red (*NR*) is a dye that is distributed over all three phases, while cresol red (*CR*) is "confined" to the external phase [why this is so is discussed elsewhere (25)]. Bovine serum albumin (*BSA*) is a buffer that does not enter into the internal phase, while imidazole is distributed over all phases. Activation of the proton pumps in the thylakoid membrane by flashing light produces transient absorption changes of the pH-indicating dyes. Subtraction of transient signals obtained under different buffering conditions yields signals that can be attributed to pH changes in the external and internal phases of thylakoids as given at the bottom of Figure 1

of neutral red (left) and cresol red (right). The traces in the upper row were obtained in the absence of the dyes. The traces below were obtained in the presence of dyes but in the absence of added buffer. The third row shows the situation with dye and with bovine serum albumin, buffering pH changes in the external phase almost completely, and the last row shows the situation in the presence of both the permeating buffer imidazole plus the nonpermeating BSA.

Comparison of the lower two traces on the right side of Figure 2 shows that there are practically no absorption changes of cresol red other than those attributable to pH changes in the outer phase. On the other hand, it is obvious from the traces on the right that both types of buffers act on the absorption changes of neutral red. Under strong

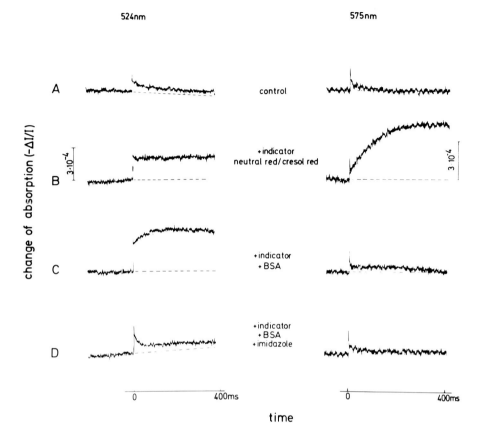

<u>Fig. 2 A-D.</u> Transient absorption changes of chloroplasts in the absence (A) and in the presence of two pH-indicating dyes (neutral red, left; cresol red, right) under different buffering conditions. The situation in the three lower rows (B-D) corresponds to that illustrated schematically in Figure 1. Each trace is the average of 32 signals induced by repetitive flashes (each given at t=0 at a rate of 0.1 Hz). Neutral red (if present), 6.6 μM; cresol red (if present), 30 μM; bovine serum albumin (if present), 1.3 mg/ml; imidazole (if present), 3 mM, pH 7.2. The superimposed electrochromic absorption changes (37,44) were accelerated (and hence virtually eliminated) by addition of nonactin, 1 μM. Both photosystems were active. For further details, see Ausländer (31)

buffering the external-phase, neutral red can be used as a pH indicator for the inside of the thylakoids. However, how clean this indication is, is questionable. Comparison of the traces in the upper and lower left of Figure 2 shows a small difference. This difference was studied in more detail.

How Clean is the pH_{in}-Indication by Neutral Red?

Figure 3 (upper row) shows the flash-induced absorption changes of
chloroplasts in the absence and in the presence of neutral red. The
buffering in the presence of neutral red was such that only negligible
pH changes in the internal phase could be excepted. In the left trace,
no neutral red was present. In the middle only the permeating buffer
imidazole was present. In the right trace, both neutral red and the
buffer were present. All traces were obtained in the presence of BSA.
The difference between the traces in the middle and those on the left
shows the influence of the permeating buffer on the intrinsic absorption changes of chloroplasts (Fig. 2, lower left). The difference between the traces on the right and those in the middle shows all absorption changes that are not sensitive to buffering of the internal phase,
i.e., all but those attributable to pH changes inside (Fig. 3, lower
right). It is obvious that there are no absorption changes that are
not attributable to pH changes inside.

There is a small error if one obtains the pH_{in}-indicating absorption
changes of neutral red by subtracting the transient signal obtained
in the presence of imidazole from the one obtained in its absence [as
we did previously (21)]. The influence of other permeating buffers
(see Fig. 4) on the intrinsic absorption changes is less pronounced
while the pH_{in}-indication is clean if it results from the difference
between transient signals with and without neutral red. The attribution
of the absorption changes of neutral red to transient pH changes inside

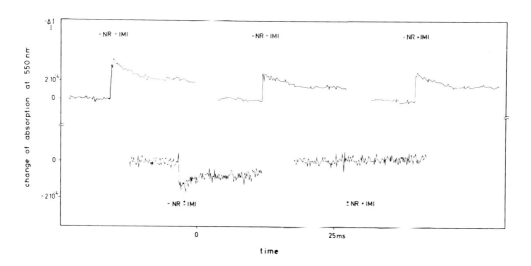

Fig. 3. Flash-induced absorption changes at 550 nm that could serve as controls for
the pH_{in}-indicating absorption changes of neutral red (25) [550 nm is off the peak
of neutral red but avoids a strong contribution of electrochromic underground (37,
44)]. Each trace is the average of 64 signals (flash given at t=0, repetitively at
0.1 Hz), the time resolution was 100 µs per address of the averaging computer. Neutral red (if present), 13 µM; imidazole (if present), 6.8 mM. Only photosystem II
active in proton translocation because of the presence of DBMIB, 3 µM. The presence
and absence of neutral red (NR) and imidazole (IMI) are indicated in the figure. The
lower traces represent the difference between the respective upper ones at twofold
expanded sensivity. [For further details see Ref. (25)]

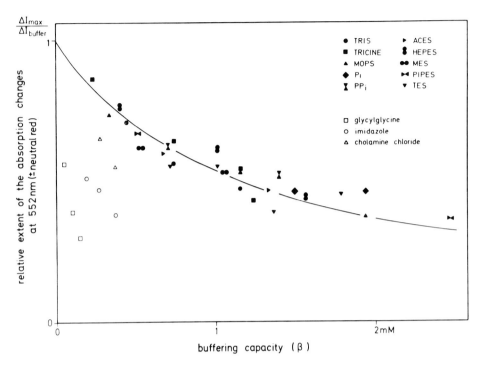

Fig. 4. Extent of the flash-induced pH_{in} -indicating absorption changes of neutral red ($\pm$ neutral red) as a function of the buffering capacity of added buffers (26) The line is calculated according to the equation given in the text (assuming an intrinsic buffering capacity of the thylakoid interior of 1.15 mM). The external phase was strongly buffered by BSA, 1.3 mg/ml. Neutral red at 13 μM. [For details, see Ref. (25)]

is based on the influence of the permeating buffer imidazole. One might argue that this buffer not only acts as such but possibly changes the binding of neutral red to the inner side of the membrane (possibly expels it from its sites). If so, one should not expect chemically different buffers to act in the same way.

Figure 4 shows the extent of the flash-induced absorption changes of neutral red (obtained as the difference ± neutral red in the presence of BSA as a function of the buffering capacity of various buffers. The buffering capacity was calculated according to:

$$\beta = 2.3\ c_b[H^+]K_b / ([H^+] + K_b)^2$$

wherein c_b is the buffer concentration in aqueous solution (the average concentration in the chloroplast suspension), K_b is the dissociation constant of the buffer in aqueous solution as taken from the literature, and $[H^+]$ is the proton concentration in the chloroplast suspension.

The line shows the theoretically expected dependence of the transient pH changes in the internal phase of thylakoids under the assumption of (1) solution of the buffers in the internal aqueous volume, and (2) an intrinsic buffering capacity of the internal space of thylakoids of 1.15 mM. It is obvious that chemically very different buffers (see full

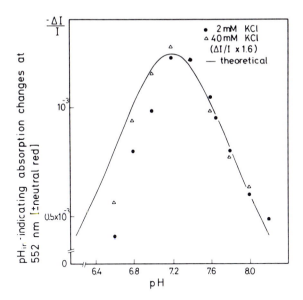

Fig. 5. Extent of the flash-induced pH_{in}-indicating absorption changes of neutral red (± neutral red) as a function of the medium pH at two different ionic strengths (25). Repetitive flash excitation at 0.05 Hz and measuring light intensity (only 3 μW/cm^2) were chosen such that no steady pH difference existed across the thylakoid membrane. Sorbitol, 200 mM reduced the internal osmolar volume to approximately 25 25 l/mol chlorophyll. Neutral red (if present), 13 μM and BSA 1.3 mg/ml. The curve shows the theoretically expected sensitivity of a pH-indicating dye with a pK value of 7.25 to very small pH changes. Note that the signals at higher salt concentrations are smaller by a factor 1.6. [For details, see Ref. (25)]

symbols in Fig. 4), with net charges of the respective base ranging between 0 and 3- act almost identically, while only the more lipophilic buffers imidazole, glycylglycine, and cholamine are more effective. The latter may indicate that these are bound to the internal side of the membrane and hence enriched in the internal phase.

This clearly demonstrates that buffers like ACES, pyrophosphate, or Tris influence the absorption changes of neutral red by buffering and by nothing else. Figure 4 also tells us that these buffers are present inside at the same concentration as outside, which suggests that they are dissolved only in the internal aqueous volume (but see below).

Pick and Avron (30) reported that in dark-adapted chloroplasts neutral red is enriched in the internal volume. This agrees with our own observations (25,31). We asked whether the pK value of the dye is changed when enriched because of binding to the inner side of the thylakoid membrane.

Figure 5 shows the extent of the flash-induced pH_{in}-indicating absorption changes of neutral red as a function of the pH of the suspending medium [for experimental conditions, see Ref. (25)]. (Note that the absolute number of protons released inside per flash is constant in this relatively narrow pH range.) The two sets of points belong to different salt concentrations in the suspending medium. The solid

line shows the theoretically expected indicator sensitivity to small
pH changes: calculated according to [see Ref. (25)]:

$$S(\lambda H^+) = -2.3(\varepsilon^H(\lambda) - \varepsilon(\lambda))[H^+]K_d/([H^+] + K_d)^2$$

wherein $\varepsilon^H(\lambda)$ and $\varepsilon(\lambda)$ are the decadic extinction coefficients of the
protonated and the unprotonated forms of the indicator at a given
wavelength and K_d is the dissociation constant of the indicator dye.

The experimental points agree fairly well with the theoretically
expected sensitivity curve of the dye under the assumption of a
$pK_d = 7.25$. [There is independent evidence that the intrinsic buffering
capacity of the interior of the thylakoids is approximately constant
between pH 6.5 and 8 (25)]. We learn that the apparent pK of the bound
neutral red molecules is not much changed in comparison to its value
in aqueous solution (where it is 6.6 under the composition of our suspension medium, see Ref. (25)). A shift toward alkalinity agrees with what one would
expect for a neutral-base indicator bound to a negatively charged lipid
vesicle [for in vitro studies, see Fernandez and Fromherz (39)].

Figure 5 further suggests that the pK is not sensitive to the salt
concentration in the medium (i.e., the influence of surface potentials
is low). However, it is obvious that there is a strong influence on
the absolute sensitivity. Whether this is via the binding equilibrium
of neutral red at the membrane or via the intrinsic buffering capacity
inside is not known at present.

The flatness of the empirical response of neutral red in the pH range
from 7 to 7.5 tells us that the absorption changes of neutral red are
a linear indicator of pH transients in this domain. We calibrated the
pH transient induced by stimulating both photosystems for a single
turnover to 0.05 pH units (25).

The well-behavedness of neutral red is lost if one tries to resolve
pH changes inside at an external pH below 6.5 or at a considerable
pH difference across the thylakoid membrane [e.g., under continuous
light; see also Ref. (30)]. Then the redistribution of the dye in
response to the pH difference and a marked pH dependence of the intrinsic buffering capacity complicate a quantitative analysis (25).

How Fast Are Proton Release and Proton Consumption Inside Thylakoids?

We previously studied the kinetics of proton release inside under exitation of chloroplasts with single-turnover flashes (21-23). Under
excitation of both photosystems a multiphasic rise of the internal
acidification was observed. This is illustrated in Figure 6 (left). The
slowest phase with a half-rise time of 20 ms matches the expectation
for proton release coupled to the oxidation of plastohydroquinone by
photosystem I under our conditions. This phase accounts for one half
of the total extent. It is eliminated if electron transfer between
the two photosystems is blocked by dibromothymoquinone [for details,
see Ref. (21)]. This is documented in Figure 6 (right). The remaining
proton release, which we attributed to the oxidation of water by photosystem II activity (21), is in itself multiphasic. Exponentials with
half-rise times of less than 100 µs, 300-600 µs, and 1-2 ms could be
discrimimated [(21,23) but see also Fig. 10]. The reason for the
kinetic complecity is not fully understood. We observed no dependence

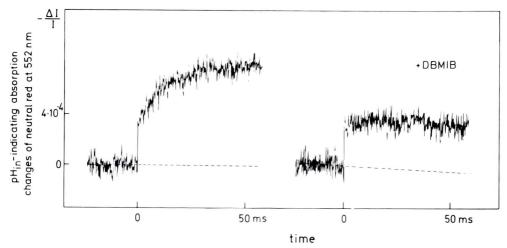

<u>Fig. 6.</u> Time course of the flash-induced pH_{in}-indicating absorption changes of neu-
tral red (± neutral red) with both photosystems active in proton translocation
(*left*) and with only photosystem II active (*right*) due to the presence of DBMIB.
Neutral red (if present), 13 µM; BSA, 1.3 mg/ml; pH = 7.2. [For details, see Ref. (21)]

of the kinetics on the neutral red concentration, which suggests that
the kinetic complexity is not due to different neighborhood relationships between the indicator dye and the water-splitting enzyme systems.
On the other hand, it is well known that the water-oxidizing enzyme
system cycles through 4 oxidation states before liberating one molecule of dioxygen [for a review, see Joliot and Kok, (32)]. Several
authors have observed that protons are liberated during at least three
of the subsequent transitions between these oxifation states (22,23,24,
35). This may indicate that these oxidation states of the manganese-
containing enzyme involve water and its oxidation products (hydroxyl,
peroxide) in a cryptic form as postulated by Renger (33). We observed
that the different kinetic phases dominate different transitions between
the four oxidation states (23). Hence, there is good reason to believe
that the kinetic complexity under repetitive excitation (where all
oxidation states are equally populated) reflects an intrinsic property
of the enzyme system but not an artifact due to the detection method.

In the context of this communication we note that proton release inside
thylakoids by the water-oxidizing enzyme system is rapid with components ranging from 100 µs to some milliseconds. Figure 7 summarizes
the kinetic data on proton release and proton uptake, respectively.
The times given are based on experiments with dyes under excitation
of chloroplasts with single-turnover flashes. That proton uptake from
the outer phase is delayed relative to the reduction of plastoquinone
and of the terminal electron acceptor, respectively, by a removable
diffusion barrier for protons was demonstrated elsewhere (17). The
relaxation time of the flash-induced pH difference across the thylakoid
membrane under an inactive ATP synthase is of the order of 5-10 s
(depending on the chloroplast preparation, at a medium pH of 7-8, in
the absence of a steady pH difference superimposed on the flash-induced
one). Under the given conditions the decay of the pH difference via
the leak permeability for protons is by orders of magnitude slower than
the decay of the electric potential difference [some 100 ms (37)]. This
is changed at increasing acidification of the internal phase (38).

122

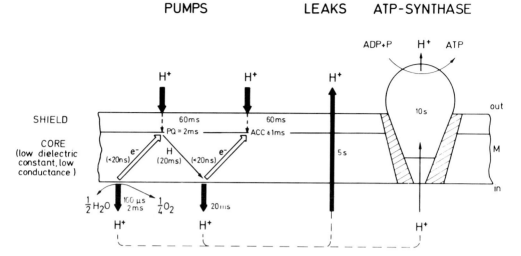

Fig. 7. Schematic representation of proton pumping and of passive proton fluxes across the thylakoid membrane. The *zigzagging arrows* give the path of the electron transport chain with alternating electron transfer *(open arrows)* and hydrogen transfer *(full arrow)*. The electron transport chain is not represented in full detail [for reviews see (27,45)]. Two layers of the thylakoid membrane have to be discriminated, the dielectric core of the membrane (with a very low leak conductivity for protons) and a somewhat lower diffusion barrier for protons, which shield the reducing sites of both photosystems at the outer side of the membrane against the external aqueous phase [for experimental evidence, see Ref. (17)].The half-times for the relaxation of the respective events (under flashing light) are taken from previous work: the electrogenic reactions (36), proton uptake (17,28), proton release (21-23), and proton flux via the ATP synthase (26,31)

We studied the relaxation of the internal acidification under conditions in which the ATP synthase is active. Figure 8 shows the time course of the absorption changes of neutral red in the absence (left) and in the presence (right) of ADP. The stepwise rise of the internal acidification was induced by exciting photosystem II with a series of short flashes. Photosystem I did not contribute to the acidification due to the presence of dibromothymoquinone (DBMIB) [see Ref. (25)]. An accelerated decay is obvious in the presence of ADP. However, the acceleration stops at a certain level. This behavior parallels that which we observed previously for the accelerated decay of the flash-induced electric potential difference across the thylakoid membrane under ATP synthesis (40). There we found the accelerated decay strictly correlated with the synthesis of ATP at a stoichiometric ratio of three elementary charges translocated across the membrane per ATP formed (40,41). We interpreted the stop of the accelerated decay at a given level of the electric potential under the assumption that the enzyme is electrically gated into an active conformation (41,42). Independent evidence for an electric (or electrochemical) gating of the ATP synthase was provided by Gräber et al. (43) in their studies on electrically induced nucleotide release from membrane-bound ATP synthase.

In the context of this paper we note that the flash-induced acidification of the internal phase relaxes through the ATP synthase at a half-relaxation time of about 10 ms (at a membrane energization just above threshold!). This is also illustated in Figure 7.

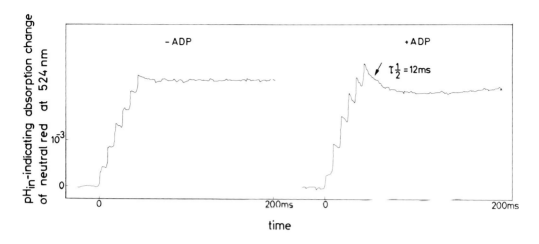

Fig. 8. Time course of the pH$_{in}$-indicating absorption changes of neutral red under exitation with a group of 6 short flashes at 8 ms interval in the absence (*left*) and in the presence of ADP (*right*). Neutral red (if present), 6.6 µM; phosphate, 300 µM; ADP (if present), 600 µM; pH = 7.6. Only photosystem II active in proton trans-location due to the presence of DBMIB, 3 µM. Outer phase strongly buffered by BSA, 1.3 mg/ml. *Left:* average over 20 signals (± neutral red), *right:* average over 24 signals (± neutral red) to correct for different buffering capacity and different internal pH. Repetition rate of flash groups, 0.2 Hz. [For details see Refs. (26,31)]

Do the pH Transients Seen by Neutral Red Occur in the Internal Osmolar Volume?

So far we have used the terms inside and interior to denote the space where neutral red indicates pH changes. We know that neutral red is adsorbed at the inner side of the thylakoid membrane. Therefore the question as to whether it might indicate pH changes in subspaces of the membrane is particularly relevant. Figure 8 shows the extent of the flash-induced absorption changes of neutral red (± neutral red) as a function of the added phosphate buffer for two different osmolar states of thylakoids. The curves are theoretical. They were calculated according to Eq. (A14) in Ref. (26) under the following assumptions: (1) Neutral red is adsorbed at the membrane. Its average concentration in the internal osmolar volume is therefore reciprocal to this volume. (b) The intrinsic buffering capacity of thylakoids is due to a fixed number of titratable groups and therefore is also reciprocal to the osmolar volume. (3) The number of protons released inside the thylakoids per flash is independent of the osmotic state of the thylakoids (this was checked in independent experiments). (4) The added buffer (here the double negatively charged phosphate) does not bind to the membrane. As it is solved in the internal aqueous volume, its concentration equals that in the suspending medium.

A reasonable fit between the experimental points and the theoretical curves was obtained under the assumption of the following internal volumes: 70 l/mol chlorophyll (at 25 mM KCl and 3 mM MgCl$_2$) and 25 l/mol chlorophyll (at the same salts as before but plus 200 mM sorbitol). These volumes are in fair agreement with those determined

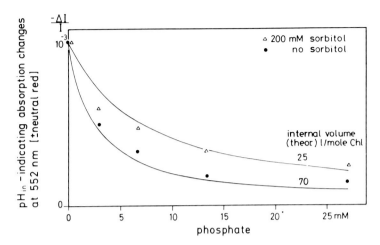

Fig. 9. Extent of the pH_{in}-indicating absorption changes of neutral red as a function of the concentration af added phosphate for two different osmotic states of thylakoids. Neutral red (if present), 13 µM; outer phase buffered by BSA, 1.3 mg/ml; pH = 7.3; sorbitol (if present), 200 mM. The solid curves show the theoretically expected behavior under the assumptions mentioned in the text [see also Refs. (25,26)]. The influence of the magnitude of the internal volume of the buffering efficiency of phosphate demonstrates that neutral red exchanges protons with the internal osmolar volume

by ^3H_2O for the internal osmolar volume at the given osmolarities (15). The fact that the extent of the absorption changes of neutral red changes only slightly as a function of the osmolar state in the absence of added buffer justifies the first two assumptions. More important is the fact that the buffering of the absorption changes of neutral red by phosphate depends on the internal osmolar volume. This shows that neutral red "sees" pH changes in the internal osmolar volume proper — or at least it "sees" pH changes in a subcompartment of the membrane, which rapidly exchanges protons with the internal osmolar volume.

We wished to determine the time constant of this proton exchange. Figure 9 shows the time course of the absorption changes of neutral red under partial buffering with TES (see Fig. 4 for a comparison of the various permeating buffers) for two different osmolar states of thylakoids. The time resolution was 40 µs per address of the averaging computer. Due to the presence of DBMIB, only the water-splitting enzyme system contributed to the internal acidification. With the TES concentration equal for both traces the difference is caused only by the osmolarity of the medium. The greater extent was observed at higher osmolarity when the internal volume was smaller. The reduction of the internal volume partially "squeezes out" the buffer (TES) that is dissolved in this volume. It is important to note that even the most rapid phase of proton release inside is buffered away by TES in a volume-dependent way. This demonstrates that neutral red exchanges protons with the internal osmolar volume in less than 100 µs.

Conclusions

Our data show that neutral red is a clean indicator for pH changes that occur in the internal osmolar volume of thylakoids. The communication time via protons between the neutral red space (at the inner side of the thylakoid membrane) and the internal osmolar volume is less than 100 µs. This is illustrated schematically in Figure 10. Four subspaces are tentatively defined within the thylakoid membrane: one in which netral red is located, another one into which protons are released by the water-oxidizing enzyme system, a third one in which protons are consumed at the reducing side of photosystem II, and a last space from which protons enter into ATP-synthase. The open arrow indicates electron transfer by photosystem II. This reaction step is electrogenic (37) with a half-rise time of less than 20 ns (36). Full arrows indicate proton transfer. The times for proton transfer are minimum relaxation times for a pH difference between the respective subspaces (half-rise). Our data show that proton release into the internal osmolar volume by the water-oxidizing enzyme system occurs within less than 100 µs (fastest component). Even the slowest component of proton release relaxes at 2 ms, which is fast compared to the relaxation of proton uptake by ATP synthase (ca. 10 ms). We cannot discriminate between the following possible reasons for the limitation of the latter by: (1) diffusion of protons from the sources laterally along the thylakoid membrane into the sinks (ATP synthases), (2) the migration of protons from the internal osmolar volume across a barrier into the enzyme, or (3) a limited turnover capacity of the enzyme.

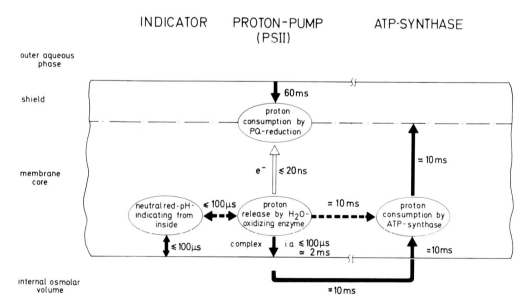

Fig. 10. A closer look at the proton pump of photosystem II and at ATP synthase as already illustrated in Figure 7. The subspaces within the membrane are tentative-ly defined (see text). Electron transport is represented by an *open arrow* and pro-ton transfer by *black arrows*. Although some of the pathways of proton flow are hypothetical, upper limits for the characteristic relaxation times may be attributed, based on our data. The major feature is that there is ample time for protons re-leased by photosystem II to equilibrate with the internal osmolar volume before entering into ATP synthases

With respect to the controversy between Mitchell and Williams our data indicate:

1. The protons liberated by the water-oxidizing enzyme system have ample time to equilibrate with the internal osmolar volume before they enter into ATP synthase. This is in line with Mitchell's chemiosmotic hypothesis.
2. We cannot exclude the possibility that a proton gradient is installed laterally through the internal osmolar volume under continuous illumination (because of a limited proton conductivity of the internal osmolar volume). This possibility, which has been proposed for mitochondria by van Dam et al. (16), is compatible with the chemiosmotic hypothesis (3). If localized protons have any role in photophosphorylation at all, it can only be in the form of localized proton sinks at the outer side of thylakoid membrane, which became evident from our previous studies (17).

Acknowledgments. We wish to thank Ilse Columbus for technical assistance, Angelika Schulze for the graphs, Dr. G. Renger and Dr. P. Fromherz for critical comments. This work was financially supported by the Deutsche Forschungsgemeinschaft and by the European Communities (Solar Energy Program, Project D).

References
========

1. Boyer, P.D., Chance, B., Ernster, L., Mitchell, P., Racker, E., Slater, E.C.: Ann. Rev. Biochem. 46, 955-1026 (1977)
2. Mitchell, P.: Nature (London) 191, 144-148 (1961)
3. Mitchell, P.: Biol. Rev. 41, 445-502 (1966)
4. Mitchell, P.: FEBS Lett. 78, 1-20 (1977)
5. Mitchell, P.: TIBS 3, N58-61 (1978)
6. Williams, R.J.P.: In: The Enzymes. 2nd Ed. Boyer, P.D., Lardy, H., Myrbäck, K. (eds.) New York: Academic Press, 1959, Vol. 1, p. 391
7. Williams, R.J.P.: J. Theor. Biol. 1, 1-13 (1961)
8. Williams, R.J.P.: TIBS 1, 222-224 (1976)
9. Williams, R.J.P.: FEBS Lett. 85, 9-19 (1978)
10. Witt, H.T.: In: Bioenergetics of Photosynthesis. Govindjee (ed.) New York: Academic Press, 1975, pp. 493-554
11. Trebst, A.: Ann. Rev. Plant Physiol. 25, 423-458 (1974)
12. Junge, W.: Ann. Rev. Plant Physiol. 28, 503-536 (1977)
13. Avron, M.: Ann. Rev. Biochem. 46, 143-155 (1977)
14. Jagendorf, A.T.: In: Bioenergetics of Photosynthesis. Govindjee (ed.) New York: Academic Press, 1975, pp. 413-492
15. Ort, D.R., Dilley, D.A., Good, N.E.: Biochem. Biophys. Acta 449, 108-126 (1976)
16. van Dam, K., Wiechmann, A.H.C.A., Hellingwerf, K.J., Arents, J.C., Westerhoff, H.V.: In: Membrane Proteins. Nicholls, P. et al. (eds.) Oxford: Pergamon Press, 1978, pp. 121-132
17. Ausländer, W., Junge, W.: Biochem. Biophys. Acta 357, 285-298 (1974)
18. Berzborn, R.J., Kopp, F., Mühlethaler, K.Z.: Naturforsch. 29C, 694-699 (1974)
19. Miller, K.R., Staehelin, L.A.: J. Cell Biol. 68, 30-47 (1976)
20. Junge, W.: In: Biochemistry of Plant Pigments. Goodwin, T.W. (ed.) New York: Academic Press, 1976, Vol. II, pp. 233-333
21. Ausländer, W., Junge, W.: FEBS Lett. 59, 310-315 (1975)
22. Junge, W., Renger, G., Ausländer, W.: FEBS Lett. 79, 155-159 (1977)
23. Junge, W., Ausländer, W.: In: Photosynthetic Oxygen Evolution. Metzner, H. (ed.) New York: Academic Press, 1978, in press
24. Ausländer, W., Junge, W.: In: The Proton and Calcium Pumps. Azzone, G.F., Avron, M., Metcalfe, J., Quagliarello, E., Siliprandi, N. (eds.) Amsterdam: Elsevier, 1978, pp. 31-44
25. Junge, W., McGeer, A., Ausländer, W., Runge, T.: Biochim. Biophys. Acta, in press (1978)

26. Junge, W., Ausländer, W., McGeer, A. In: Frontiers of Biological Energetics (P.L. Dutton, J.S. Leigh and A. Scarpa, eds.) Academic Press, New York - in press
27. Witt, H.T.: Rev. Biophys. 4, 365-477 (1971)
28. Schliephake, W., Junge, W., Witt, H.T.: Z. Naturforsch. 23B, 1571-1578 (1968)
29. Junge, W., Ausländer, W.: Biochim. Biophys. Acta 333, 59-7o (1974)
30. Pick, U., Avron, M.: FEBS Lett. 65, 48-53 (1976)
31. Ausländer, W.: PhD-Thesis, Technische Universität Berlin, 1977
32. Joliot, P., Kok, B.: In: Bioenergetics of Photosynthesis. Govindjee (ed.) New York: Academic Press, 1975, pp. 387-412
33. Renger, G.: FEBS Lett. 81, 223-228 (1977)
34. Fowler, C.: Biochim, Biophys. Acta 459, 351-363 (1977)
35. Saphon, S., Crofts, A.R.: Z. Naturforsch. 32C, 617-626 (1977)
36. Wolff, C., Buchwald, H.E., Rüppel, H., Witt, K., Witt, H.T.: Z. Naturforsch. 24B, 1038-1041 (1969).
37. Junge, W., Witt, H.T.: Z. Naturforsch. 23B, 244-254 (1968)
38. Gräber, P., Witt, H.T.: Biochem. Biophys. Acta 423, 141-163 (1976)
39. Fernandez, M.S., Fromherz, P.: J. Phys. Chem. 81, 1755-1761 (1977)
40. Junge, W., Rumberg, B., Schröder, H.: Eur. J. Biochem. 14, 575-581 (197o)
41. Junge, W.: Eur. J. Biochem 14, 582-592 (1970)
42. Junge, W.: In: Proc. IInd. Int. Congr. Photosynthesis. Forti, G., Avron, M., Melandri, A. (eds.) Den Haag: Dr W. Junk, N.V., (1972) Vol. II, pp. 1065-1074
43. Gräber, P., Schlodder, E., Witt, H.T.: Biochem. Biophys. Acta 461, 426-440 (1977)
44. Emrich, H.M., Junge, W., Witt, H.T.: Z. Naturforsch. 24B, 1144-1146 (1969)
45. Junge, W.: In: Encyclopedia of Plant Physiol., New Series. vol. 5 Trebst, A., Avron, M. (eds.) Berlin-Heidelberg-New York: Springer, 1977, pp. 59-93

Redox-Linked Proton Pumps in Mitochondria

M. WIKSTRÖM and K. KRAB

Energy Conservation and Transmission in Oxidative Phosphorylation

During the past 15 years it has become evident that the conservation of oxidation-reduction energy, linked to respiratory and photosynthetic processes, is based upon seperation of positive and negative electrical charges across biological membranes. This widely accepted view originated principally with the proposal by Mitchell (1,2) of the chemiosmotic theory of oxidative and photosynthetic phosphorylation.

The general postulate of this theory may be depicted as in Figure 1, showing effective translocation of hydrogen ions across the mitochondrial *cristae* membrane intimately linked to the oxidoreductions in the respiratory chain, and the subsequent influx of protons through the membrane, linked to the function of the proton-translocation ATP synthase. According to this postulate, redox energy is primarily conserved (by the aid of proton translocation) as a proton gradient across the membrane, which is then secondarily utilized for ATP synthesis by proton influx down this gradient.

In this context, the mitochondrial membrane may be thought of as containing two kinds of "proton pumps," using this term in its most general meaning: (1) the exergonic redox complexes of the respiratory chain and (2) the endergonic ATP synthase. In this lecture we will address ourselves entirely to the function of the redox-linked proton

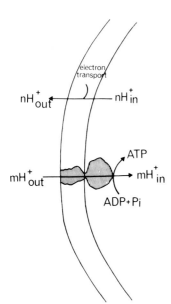

Fig. 1. The main principle of the chemiosmotic theory. n and m refer to the stoichiometric numbers of redox-linked and ATP synthesis-linked proton translocation. According to Mitchell, these numbers are $n=2$ per electron pair transferred (for one redox loop), and $m=2$ per ATP synthesized. For further details, see text

translocating systems in the mitochondrial membrane, while the function of ATP synthase is discussed by other contributors (3,4).

Before leaving these introductory remarks on mitochondrial energy
conservation, it is worth noting that the chemiosmotic theory postulates
that the proton circuitry outlined in Figure 1 involves the aqueous
phases on each side of the membrane. This implies that, strictly speak-
ing, the widely used expression "energization of the mitochondrial membrane" should be replaced by "energization of the aqueous media" [see
also Ref. (5)]. However, as an alternative to this chemiosmotic notion,
one could imagine a more localized proton circuitry, confined to the
membrane itself. Such a concept comes close to the hypothesis proposed
by Williams (6,7) for mitochondrial energy conservation.

The Molecular Principle of Generation of the Electrochemical
Proton Gradient ($\Delta\tilde{\mu}_H+$)

In recent years it has been shown unequivocally that respiratory (and
photosynthetic) redox reactions indeed result in generation of $\Delta\tilde{\mu}_H+$
across the energy-transducing membrane (8-11). Whether or not these
findings are taken as proof for the chemiosmotic theory, they definitely strengthen the more general view that effective proton translocation
(either in a osmotic or a more localized sense) is a fundamental feature of conservation of redox energy. It goes without saying that for
an unravelling of the mechanism of energy conservation, it is therefore of utmost importance to elucidate the mechanism by which H⁺ translocation occurs, and how it is driven by the redox process.

The Redox Loop

In conjunction with formulating the more general concept of the chemiosmotic theory described above, Mitchell also postulated a principle
according to which redox-linked proton translocation would occur. This
postulate, which is today widely known as the "redox-loop," is based
on intermittent vectorial flow of hydrogen atoms (e.g. ubiquinol), and
electrons, in opposite directions across the mitochondrial membrane.
Thus, H atoms are translocated outward and "dissociate" at the external surface into H⁺ ions, which are released into the aqueous phase,
and electrons, which are translocated back across the membrane. On
the inside the electrons react with aqueous H⁺ ions to generate H atoms.
As a sequential process, this will lead to effective translocation of
H⁺ ions across the membrane (and generation of $\Delta\tilde{\mu}H+$). It is important
to note that in this mechanism the coupling between electron flow and
proton translocation is direct, due to the intermittent association of
the two particles (as H atoms) during turnover.

Translocation of H atoms across the membrane may pose mechanistic problems, and it should be realized that the "barrier" across which translocation occurs may not equal the thickness of the entire phospholipid
bilayer. In the regions of the membrane where the redox complexes
reside, there might exist "wells" that make contact with the bulk aqueous phase, so that the actual translocation distance need not be very
large. These more detailed aspects of the redox loop have recently
been discussed in depth (12,13).

Due to the direct character of coupling between electron transfer and
H⁺ translocation, the redox loop mechanism functions with a maximum
stoichiometry of one electrogenically translocated H⁺ per transferred
electron. In some cases (e.g., the Q cycle, see below), more than one

H^+ might be released on the outside per electron transferred, but inspection of the details of the mechanism readily shows that only one charge crosses the membrane per electron transferred. For some time this stoichiometry of 1 was considered as nearly established for the mitochondrial respiratory chain, with translocation of $6H^+$ (3 loops) and $4H^+$ (2 loops) per electron pair in the NADH $-O_2$ and succinate (or ubiquinol)$-O_2$ segments, respectively [(14-17) but see (18,19)]. Recently however, these findings have been seriously challenged [see also (18, 19)] by Brand et al. (20), who reported stoichiometries of 9-12 and 6-8, respectively, for the above-mentioned segments of the respiratory chain. The most obvious consequence of such a finding, if correct, is simply that the redox loop hypothesis would have to be rejected as a model of respiration-linked H^+ translocation, since it cannot accommodate such high stoichiometries.

Proton Pumps
_____

As an alternative to the redox concept, several authors in the past proposed that the redox reactions may be linked to a "true" proton transport process, viz. one that actually translocates H^+ ions as such (and not H or e^-) across the membrane. In this context the term proton pump is used in a more specific sense than merely to express the overall result of the process (cf. introductory paragraph). In this more specific sense, a proton pump functions according to principles fundamentally different from those of the redox loop. Due to the opinions of most proponents of proton pumps [as also emphasized by Mitchell, Ref. (12)], the pumps are described as being coupled to the redox reaction through long-range polypeptide interactions. The model system generally used and often cited in this context is the Bohr effect in hemoglobin (21). Thus, for instance, Papa et al. term their proposed proton pump in the respiratory chain a "vectorial Bohr mechanism" (22). Since the coupling in such models is implied to occur over relatively long distances (cf. Bohr effect), these mechanisms can be termed *indirect* (see above) due to the indirect (*conformational*) linkage between the flows of H^+ and e^-.

One of the often cited differences between "loop" and "pump" mechanisms that can presently be tested experimentally is the H^+/e^- quotient of the H^+-translocating process. While this quotient cannot be higher than 1.0 for a single redox loop (see above), it could, in principle, attain any value in an indirectly coupled (conformational) proton pump, within thermodynamic limits. It is sometimes loosely claimed that proton pumps (as opposed to "loops") may attain *variable* H^+/e^- stoichiometries. However, a simple consideration of this question reveals that any change in H^+/e^- stoichiometry of any proton-translocating device would necessarily imply a change in efficiency of energy transduction. For this reason we find it very difficult to appreciate the proposal of Papa et al. (23) that the "proton pump" in the cytochrome $b-c_1$, segment would show a dramatic *increase* in H^+/e^- quotient upon increasing the pH from 7 to 9. If correct, that would in fact imply that this "pump" worked with very low efficiency at physiologic pH (40%), but would improve energetically at high unphysiologic pH values. It is therefore striking that this finding is the only experimental argument used by these authors in favor of a "pump" mechanism as opposed to a redox loop. The alternative proposal for the cytochrome $b-c$ region is the ubiquinone cycle (Q cycle) model, which works according to the principles of a redox loop (24).

The Proton Pump of Cytochrome c Oxidase

Mitchell's postulate for the function of the terminal segment of the cytochrome chain, the cytochrome c oxidase reaction, is shown in Figure 2. According to this model, the experimental support of which has generally been taken as the best indication in favor of the redox loop type of mechanism [see, e.g., (25-27)], cytochrome c oxidase (EC 1.9.3.1.) merely functions as an electron-translocator with consumption of the H^+ ions required for water formation from the inside (M side) of the mitochondrial inner membrane. Note that no true H^+ transport is associated with the overall reaction, but that the enzyme is furnishing the third redox loop of the chain with its electron-transferring arm. The hydrogen-translocating arm of this third loop is postulated to reside earlier in the chain, in the cytochrome b-c_1/ubiquinone region (24).

We noticed in our earlier work (28) that the heme groups of cytochrome aa_3 appear to "feel" the sum of ΔpH and membrane potential across the mitochondrial membrane, i.e., the entire electrochemical proton gradient ($\Delta\tilde{\mu}H^+$), so that their configuration could be shown to be $\Delta\tilde{\mu}H^+$-dependent (28,29). This suggested that the catalytic redox activity may actually be linked to a true translocation of H^+ ions (proton transport or "pumping") across the membrane, in disagreement with Figure 2. In the following we will briefly review the subsequent tests of this hypothesis [for details, see (13, 29-32)], including some very recent additional findings (31,32).

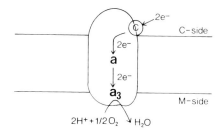

Fig. 2. The arrangement of cytochrome c oxidase as an electron translocator according to Mitchell. For details, see the text

Respiration encompassing only the terminal segment of the respiratory chain may be induced with ferrocyanide, which donates electrons specifically to cytochrome c at the outer surface of the *cristae* membrane (33). As shown in Figure 3D, respiration with ferrocyanide is accompanied by a burst of *acidification of the medium* at an initial rate of 1 H^+/e^- (cf. Figure 3C for respiratory rate). This acidification, which follows an initial alkalinization artifact, is completely abolished if the membrane has been rendered *permeable to H^+* ions by an uncoupling agent (Fig. 3B). In the latter case, oxidation of ferrocyanide is simply accompanied by (after the alkalinization artifact) steady *consumption* of H^+ ions at a rate of 1 H^+/e^- (cf. Fig. 3A). This stoichiometry is exactly that expected from the *overall* reaction, viz.

$$[Fe^{II}(CN)_6]^{4-} + 1/4\ O_2 + H^+ \rightarrow [Fe^{III}(CN)_6]^{3-} + 1/2\ H_2O \tag{1}$$

This simple experiment provides strong evidence for a true H^+-translocating function of cytochrome c oxidase. The burst of acidification is abolished, or nearly so, if no permeant ions are present to balance the net transfer of charge across the membrane. Thus, the acid burst depends on the presence of valinomycin (*plus* K^+), which renders the membrane specifically permeable to potassium ions, or on the presence

132

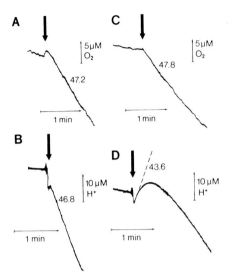

Fig. 3A-D. Proton ejection from rat-liver mitochondria linked to respiration with ferrocyanide. The reaction mixture contained 110 mM HEPES buffer (pH 7.0), 5.5 µM rotenone, 0.118 µg/ml antimycin, 0.055 µg/ml valinomycin, and rat-liver mitochondria (approx. 2.5 mg protein/ml). Temperature 24° C. The *heavy arrows* indicate the points of addition of 0.8 mM potassium ferrocyanide. In (A) and (B) the uncoupling agent FCCP (0.5 µM) was also present. *Numbers* adjected to the traces refer to initial rates in µM electrons or protons per minute. Traces (A) and (C): oxygen consumption. Traces (B) and (D): pH electrode recordings. For further details, see ref. (32)

of Ca^{2+} ions, which penetrate without an exogenously added ionophore. It may therefore be suggested that cytochrome c oxidase is linked to an electrogenic H^+-pumping process (contrast Fig. 2).

The proton ejection from mitochondria may also be demonstrated by several other donor systems of reducing equivalents to cytochrome c, and may even be shown using added ferrocytochrome c, which readily penetrates the outer membrane under the experimental conditions. This is shown in Figure 4, which gives a picture analogous to that using ferrocyanide as the reductant (cf. Fig. 3 B and D). Note especially from Figure 4, that *overall* consumption of H^+ is *independent* of the presence (Fig. 4, right) or absence (left) of uncoupler and agrees closely with the amount of reduced cytochrome c added. Hence any artifactual *net* H^+ production in the system can be excluded as the basis of the H^+ ejection phase, this result being consistent with a true H^+ *transport* process.

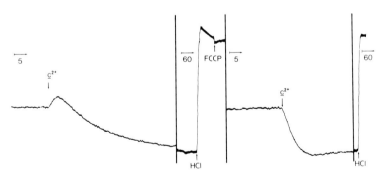

Fig. 4. Proton ejection from rat-liver mitochondria linked to respiration with added ferrocytochrome c. The experimental conditions were similar to those described in the legend to Figure 3. Additions of 4.3 µM ferrocytochrome $c(c^{2+})$, 10 µM HCl, and 1 µM of the uncoupler FCCP were made as indicated. *Horizontal arrows* indicate the time scale in seconds. *Vertical lines* show changes in recorder speed. An *upward deflection* of the pH electrode trace indicates acidification. For further details, see ref.(32)

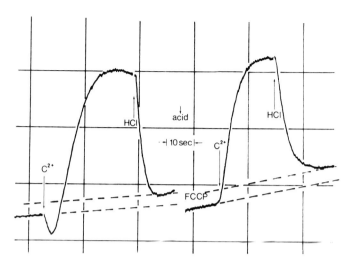

Fig. 5. Proton ejection from cytochrome *c* oxidase vesicles coupled to respiration with added ferrocytochrome *c*. The reaction mixture contained 50 mM K_2SO_4, 1 µg/ml valinomycin, and 40 µl of cytochrome oxidase vesicles in a final volume of 2.0 ml. Temperature 22°. C. pH is recorded electrometrically with respect to time. Acidification is indicated by a *downward deflection* of the trace. The following additions were made as indicated: 8.5 nequiv. ferrocytochrome *c* (c^{2+}) and 7 nequiv. HCl standard solution. In the *right-hand trace*, the experiment is repeated in the presence of 6 µM of FCCP. For futher details, see ref. (31)

Figure 5, shows an entirely analogous experiment, which was performed with artificial phospholipid vesicles inlaid with isolated and purified cytochrome *c* oxidase (30,31). Again H^+ ejection, which is linked to the oxidation of added ferrocytochrome *c*, is abolished by an uncoupler (or by the absence of valinomycin, not shown). We have demonstrated (31) that in this reconstituted system, the stoichiometry of H^+ ejection is also 1 H^+/e^-, as is the case in intact mitochondria. Moreover, the extent of the H^+ ejection can be shown to be dependent upon the intravesicular buffer capacity, proving that it is the consequence of true H^+ transport across the membrane (31).

On the basis of these findings, we conclude that cytochrome *c* oxidase is a proton pump that functions in accordance with the schematic representation shown in Figure 6, (13,34). This scheme predicts that the cytochrome *c* oxidase reaction is associated with translocation of two equivalents of electrical charge per electron (contrast Fig. 2), a prediction that has been experimentally verified (13,36). The H^+-pump has also been demonstrated in sonicated submitochondrial particles (30), in which the polarity of the membrane is inverted as compared to the intact mitochondrion.

It is not surprising, due to the previous widespread agreement on Figure 2, that our concept of the function of cytochrome *c* oxidase will only slowly gain acceptance [cf. Refs. (23,37,38)]. This is so also because it is not consistent with the previously assigned H^+/O quotients of mitochondrial H^+ translocation (see above). In this respect our view is supported by the results of Brand et al. (20) and see below).

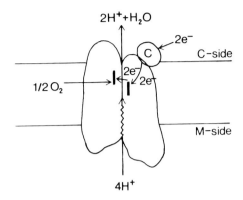

2H⁺+H₂O

2e⁻

C-side

1/2 O₂

M-side

4H⁺

Fig. 6. Schematic representation of proposed proton-pumping function of cytochrome *c* oxidase in the mitochondrial membrane. The two hemes are depicted as *black* rectangulas. The *zigzag line* is intended to represent a proton "channel" connecting the heme region with the M phase of the membrane. The *encircled c* represents cytochrome *c*. See also refs. (13, 34)

Observed Stoichiometry of Respiratory Chain-Linked Proton Translocation

Due to the discrepancy between most of the published H^+/O quotients of mitochondrial proton translocation, and the proton-pumping function of cytochrome *c* oxidase proposed by us, we decided to reinvestigate these quotients using independent techniques.

By measuring *initial rates* of H^+ ejection linked to reduction of added ferricyanide by succinate in rat-liver mitochondria, we found that the $H^+/2e^-$ quotient in this "site 2" segment of the respiratory chain is close to 4.0 (unpublished observations). This is in good agreement with previous measurements by Mitchell and Moyle (39) and others, who used different techniques. However, these authors (see 17 for review) also found the *same* $H^+/2e^-$ (or H^+/O) quotient of 4.0 for the entire segment between succinate (ubiquinol) and O_2. This clearly disagrees with our concept of cytochrome *c* oxidase function, according to which inclusion of the cytochrome c-O_2 segment in the experimental system with succinate (or ubiquinol) should result in an *increase* of the $H^+/2e^-$ quotient by 2.0, from 4.0 to 6.0. This is in agreement with the later findings by Brand et al. (20), although these workers have more recently reproted even higher $H^+/2e^-$ values (up to 8.0), in kinetic measurements of H^+ ejection and O_2 consumption [(40) see also below].

We therefore thought it necessary to reinvestigate the $H^+/2e^-$ quotient in the succinate (ubiquinol) - O_2 span using techniques differing from those previously used.

In Figure 7 the respiration of rat-liver mitochondria with succinate is abruptly blocked by a respiratory chain inhibitor, and the initial rate of H^+ disappearance into the mitochondria is recorded. It may be assumed that during respiration the rate of H^+ efflux equals the rate of H^+ influx (steady state). Then upon respiratory arrest, the initial rate of H^+ influx should be an estimate of the H^+ ejection rate in the aerobic state. Dividing this rate by the rate of consumption of 1/2 O_2 yields the H^+/O quotient. In extensive series of experiments using different respiratory inhibitors (HOQNO, Fig. 7, antimycin and KCN, not shown), Mr. Björn Appelberg in our group has consistently found H^+/O quotients close to 6.0 with this technique.

We further measured *initial rates* of H^+ ejection from rat-liver mitochondria pulsed with duroquinol. The H^+ ejection rates were compared with (1) the initial rates of formation of oxidized quinone, and (2) the rate of O_2 consumption. In typical series of five experiments over a tenfold variation in concentration of duroquinol, we found H^+/O

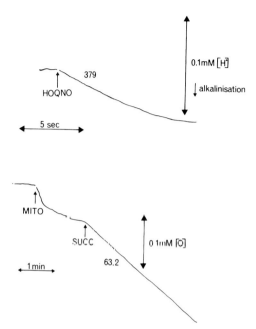

Fig. 7. The kinetics of proton influx into rat-liver mitochondria upon respiratory arrest. The mitochondria (approx. 1.5 mg protein/ml) were suspended in a medium containing 0.2 M sucrose, 20 mM KCl, 5 µM rotenone, 9 µg/ml oligomycin, 0.9 mM EGTA, and 0.25 mM N-ethyl-maleimide. Then 6.7 mM of succinate was added and the respiratory rate recorded polarografically (*lower trace*). In the steady state, the respiratory inhibitor n-heptyl-4-hydroxyquinoline-N-oxide (HOQNO) was added, and the initial rate of H^+ disappearance recorded with a pH meter (*upper trace*). Numbers adjacent to traces indicate the rates of disappearance of H^+ and oxygen consumption, respectively, in µM H^+ or O per min (unpublished experiment by B.C. Appelberg)

quotients of 5.76 ± 0.22 (S.D.) and 6.14 ± 0.71 (S.D.), using methods (1) and (2), respectively, to estimate electron transfer. It is clear that the O_2 electrode does not quite follow the initially very rapid electron transfer. This, we think, is the reason for the H^+/O quotients higher than 3 per "site" reported by Brand et al. using the kinetic approach (40).

Based on these findings, we seriously doubt the validity of the previous findings of a H^+/O quotient of 4.0 for the succinate (ubiquinol) - O_2 segment. We found a quotient close to 6.0, which is in accordance with the values of 2.0 and 4.0 found for the cytochrome c-O_2 and succinate-cytochrome c segments individually. The reason for the lower values has been adequately covered by Brand (41), who suggested that it is due to a fast P_i/H^+ symport leading to a loss of extramitochondrial H^+ ions. Such an effect would be minimized when more turnovers are covered by the experiment, since only a small amount of phosphate (endogenous phosphate that has leaked out of the mitochondria) is present. This is presumably one of the reasons why the kinetic approach gives more reliable stoichiometries than the oxidant or reductant pulse approach. The reason for the agreement between the results of the two methods for the span succinate to cytochrome c might be related to the smaller change in inner pH (the most important contribution to the ΔpH and thus to the motive force on the P_i/H^+ symporter) when electrons are transferred from succinate to cytochrome c compared with that when electrons go all the way to oxygen (-2 $H^+/2e$ compared with -6 $H^+/2e$). Inhibition of the H^+/P_i symporter with N-ethylmaleimide also restores the H^+/O value to 6.0 for the succinate/O_2 span with the O_2-pulse technique (20,41).

Table 1 is a summary of our conclusions on the stoichiometries of H^+ uptake, ejection, and charge transfer in the different segments of the respiratory chain. A comparison is also given with the values predicted by the Q-cycle mechanism for site 2 and the electron-transporting

Table 1. Stoichiometries of proton uptake/release and charge transfer in the mitochondrial cytochrome chain

Segment of respiratory chain	$H^+/2e$ taken up (inside)		$H^+/2e$ released (outside)		Charge/2e transferred ($\oplus$ inside → outside)	
Succinate → (ubiquinol) cytochrome c	2	(2)	4	(4)	2	(2)
cytochrome c → oxygen	4	(2)	2	(0)	4	(2)
succinate → (ubiquinol) oxygen	6	(4)	6	(4)	6	(4)

The values in parentheses are based on the Q-cycle model for site 2 in conjunction with Mitchell's electron-transferring cytochrome c oxidase (Fig. 2)

oxidase model for site 3. Two points are worth noticing. First, the quotients of H^+ ejection and uptake differ for the two sites. This, of course, is the consequence of the oxidation of a hydrogen donor by an electron acceptor in one case, and vice versa in the other. Secondly, the stoichiometry of charge transfer across the membrane is twice as high for "site 3" as for "site 2". This most important conclusion suggests that "site 3" conserves about twice as much energy per electron as does "site 2". This conclusion is in agreement with the fact that the corresponding available redox spans are about 250 mV for "site 2" and nearly 500 mV· for "site 3".

The "site" differences in the quotients of Table 1 suggest that denotations such as "H^+/site" or "$H^+/\sim$" ratio should be avoided. It is, however, equally important to note that H^+/e^- quotients become confusing unless the direction of H^+ shift and the segment of respiratory chain are defined. Note that the H^+/e^- quotients of H^+ *uptake* coincide with those of charge translocation, these quotients being the most relevant ones in an energetic consideration of H^+ transport.

Possible Mechanisms of Linking Proton Transport to Electron Transfer

Our finding of a true proton pumping function of cytochrome c oxidase should not necessarily be taken to imply that the coupling between proton transport and electron transfer is *indirect* in such a way that the energy transmission from the redox center to the proton translocator would have to occur through polypeptide interaction as implied in hypotheses of conformational coupling. It is of great interest in this context to note that oxidoreduction of the heme system in cytochrome c oxidase is linked to H^+ uptake and release as shown by the pH-dependence of measured midpoint redox potentials (42,43). It is unfortunate that this phenomenon has been termed a "Bohr effect," implying proton release and uptake in the apoprotein "conformationally" linked to the redox events by analogy with the Bohr effect in hemoglobin. It should be pointed out that, in contrast to this proposed analogy, the known reasons for the pH-dependent midpoint redox potential in cytochrome c [see ref. (44)] involve protolytic events linked very closely indeed to the heme, by protonation or deprotonation of an axial ligand to heme iron. As long as the protonated group is not known in cytochrome aa_3 (nor in cytochrome b, which interestingly enough behaves similarly), the "cytochrome c model" described above

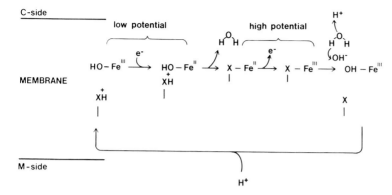

Fig. 8. One possible schematic representation of the mechanism of a directly coupled proton pump in a respiratory complex such as cytochrome c oxidase. In this particular model, H^+ translocation is depicted as being linked directly to one of the two axial ligands of heme iron. X (protonated form, XH^+) is an amino acid side chain (histidine imidazole?), which can attain two positions, either free or liganded axially to the heme. Protonation of X can occur only from the M side of the membrane, the aqueous phase of which is connected to X via a proton-conducting channel (see also Fig. 6)

appears more logical than the hemoglobin model. It is also known from heme model systems that the protonation state of an axial ligand will affect the midpoint redox potential (45,46).

If the protolytic event indeed occurs at the axial (histidine?) ligand of the cytochrome aa_3 hemes, it may be taken as a reasonable model of how proton pumping is achieved by the cytochrome c oxidase complex.

It may be noted that such coupling between electron transfer and proton translocation should be considered to be direct, in contrast to models of "conformational" coupling, but similar to redox loop mechanisms. In considering such possible ways of direct coupling of a respiratory chain-linked proton pump, it may be useful as an initial approach to design a simple model that serves to illustrate what kinds of mechanisms my be proposed and that provides a chemical framework with several experimentally testable details.

The schematic representation in Figure 8, though it is obviously a crude oversimplification, has nevertheless some interesting features. The coupling of H^+ transport with the function of an axial heme ligand (X or OH^- in Fig. 8) is consistent with the orientation of the hemes of cytochrome aa_3 and cytochromes b with their planes perpendicular to the membrane plane (47). The redox system exists in a high- and low-potential form, which is a central requirement of a redox energy transducer, as shown by DeVault (48). Finally, the exchange of a strong (X in Fig. 8) and weak-field axial ligand during catalysis is consistent with our finding that proton translocation may be linked to a spin-state transition in a ferric heme of cytochrome aa_3 (28-30).

Acknowledgments. We are grateful to Ms. Hilkka Vuorenmaa for excellent technical assistance and to Ms. Marja Immonen for help with preparation of the manuscript. One of us (K.K.) acknowledges a long-term postdoctoral fellowship from the European Molecular Biology Organisation.

138

References

1. Mitchell, P.: Nature (London) 191, 144-148 (1961)
2. Mitchell, P.: Chemiosmotic Coupling in Oxidative and Photosynthetic Phosphorylation. Bodmin, U.K.: Glynn Research, 1966
3. Pedersen, P.L.:this volume, pp. 159-194 (1978)
4. Kagawa, Y.: this volume, pp. 195-219 (1978)
5. Mitchell, P.: TIBS 3, N58-61 (1978)
6. Williams, R.J.P.: Curr. Top. Bioenerg. 3, 79-129 (1969)
7. Williams, R.J.P.: Ann. N.Y. Acad. Sci. 227, 98-107 (1974)
8. Mitchell, P., Moyle, J.: Eur. J. Biochem. 7, 471-484 (1969)
9. Rottenberg, H.: Eur. J. Biochem. 15, 22-28 (1970)
10. Nicholls, D.G.: Eur. J. Biochem. 50, 305-315 (1974)
11. Åkerman, K.E.O., Wikström, M.K.F.: FEBS Lett. 68, 191-197 (1976)
12. Mitchell, P.: FEBS Lett. 78, 1-20 (1977)
13. Wikström, M.: In: The Proton and Calcium Pumps. Azzone, G.F. et al. (eds.)
Amsterdam: Elsevier, North-Holland Biomedical Press, 1978, pp. 215-226
14. Mitchell, P., Moyle, J.:Biochem. J. 105, 1147-1162 (1967)
15. Hinkle, P.C., Horstman, L.L.: J. Biol. Chem. 246, 6024-6028 (1971)
16. Lawford, H.G., Garland, P.B.: Biochem. J. 136, 711-720 (1973)
17. Papa, S.: Biochim. Biophys. Acta 456, 39-84 (1976)
18. Cockrell, R.S., Harris, E.J., Pressman, B.C.: Biochemistry 5, 2326-2335 (1966)
19. Rossi, E., Azzone, G.F.: Eur. J. Biochem. 7, 418-426 (1969)
20. Brand, M.D., Reynafarje, B., Lehninger, A.L.: J. Biol. Chem. 251, 5670-5679 (1976)
21. Perutz, M.F.: Nature (London) 228, 726-739 (1970)
22. Papa, S., Lorusso, M., Guerrieri, F., Izzo, G.: In: Electron Transfer Chains and
Oxidative Phosphorylation. Quagliariello, E. et al. (eds.) Amsterdam: North-
Holland, 1975, pp. 317-327
23. Papa, S., Lorusso, M., Guerrieri, F., Izzo, G., Capuano, F.: In:The Proton and
Calcium Pumps. Azzone, G.F. et al. (eds.) Amsterdam: Elsevier, North-Holland Biomedical Press, 1978, pp. 227-237
24. Mitchell, P.: J. Theor. Biol. 62, 327-367 (1976)
25. Mitchell, P., Moyle, J.: In: Electron Transport and Energy Conservation. Tager,
J.M. et al. (eds.) Bari: Adriatica Editrice, 1970, pp. 575-587
26. Skulachev, V.P.: Ann. N.Y. Acad. Sci. 227, 188-202 (1974)
27. Hinkle, P., Mitchell, P.: J. Bioenerg. 1, 45-60 (1970)
28. Wikström, M.K.F.: In: Electron Transfer Chains and Oxidative Phosphorylation.
Quagliariello, E. et al. (eds.) Amsterdam: North-Holland Pupl. Co., 1975, pp.
97-103
29. Wikström, M.K.F.: Nature (London) 266, 271-273 (1977)
30. Wikström, M.K.F., Saari, H.T.: Biochim. Biophys. Acta 462, 347-361 (1977)
31. Krab, K., Wikström, M.: Biochim. Biophys. Acta, in press (1978)
32. Wikström, M., Krab, K.: FEBS Lett. 91, 8-14 (1978)
33. Jacobs, E.E., Sanadi, D.R.: Biochim. Biophys. Acta 38, 12-34 (1960)
34. Wikström, M., Saari, H., Penttilä, T., Saraste, M.: FEBS Symp. Oxford, New York:
Pergamon Press, 1978, Vol. 45, pp. 85-94
35. Hinkle, P.C.: FEBS Symp. Oxford, New York: Pergamon Press, 1978, Vol. 45, pp.
79-83
36. Sigel, E., Carafoli, E.: Hoppe-Seyler's Z. Physiol. Chem. 358, 1284-1285 (1977)
37. Papa, S., Guerrieri, F. Lorusso, M., Izzo, G., Boffoli, D., Stefanelli, R.: FEBS
Symp. Oxford, New York: Pergamon Press, 1978, Vol. 45, pp. 37-48
38. Moyle, J., Mitchell, P.: FEBS Lett. 88, 268-272 (1978)
39. Mitchell, P., Moyle, J.: In: Biochemistry of Mitochondria. Slater, E.C. et al.
(eds.) London, Warsaw: Academic Press and PWN, 1967, pp. 53-74
40. Brand, M.D., Reynafarje, B., Lehninger, A.L.: Proc. Natl. Acad. Sci. U.S.A. 73,
437-441 (1976)
41. Brand, M.D.: Biochem. Soc. Trans. 5, 1615-1620 (1977)
42. Wilson, D.F., Lindsay, J.G., Brocklehurst, E.S.: Biochim. Biophys. Acta 256,
277-286 (1972)
43. Gelder, B.F. van, Rijn, J.L.M.L. van, Schilder, G.J.A., Wilms, J.: In: Structure
and Function of Energy Transducing Membranes. van Dam, K., van Gelder, B.F.,
(eds.) Amsterdam: Elsevier, North-Holland Biomedical Press, 1978, pp. 61-68

44. Dickerson, R.E., Timkovich, R.: In: The Enzymes. Boyer, P.D. (ed.) Part A. New York: Academic Press, 1975, Vol . XI, pp. 397-547

45. Vanderkooi, G., Stotz, E.: J. Biol. Chem. 241, 3316-3323 (1966)

46. Cauquis, G., Marbach, G., Vignais, P.M.: Bioelectrochem. Bioenerg. 1, 23-28 (1974)

47. Erecińska, M., Wilson, D.F., Blasie, J.K.: Biochim. Biophys. Acta 501, 53-62 (1978)

48. DeVault, D.: Biochim. Biophys. Acta 225, 193-199 (1971)

Light-Driven Proton Translocation and Energy Conservation by Halobacteria

D. OESTERHELT, R. HARTMANN, H. MICHEL, and G. WAGNER

Introduction

An important role in bioenergetics is played by enzymatic systems which
are involved in vectorial energy-transducing reactions. Wellknown examples are the phosphotransferase system of sugar transport in bacteria,
the Na^+/K^+ -dependent and the Ca^{2+}-dependent ATPases present in ex-
citable cell membranes and the sarcoplasmic reticulum of muscle cells
respectively. All three enzymes participate in reactions where chemical
energy set free by hydrolysis of ATP is converted into that of a concentration gradient of neutral substances (e.g., a sugar) or charged
substances (e.g., inorganic ions). Energy conservation during respiration and photosynthesis is achieved by a rather complex system of
catalysts. The energy, set free by oxidation of nutrients or by absorption of light, is converted first into an electrochemical proton gradient and in a second separate step into chemical energy by synthesis
of ATP (Fig. 1).

Bacteriorhodopsin in Halobacteria serves a function similar to that
of the electron transport chains of respiration and photosynthesis.
However, no transport of electrons is linked to the action of bacteriorhodopsin. The reaction catalyzed by bacteriorhodopsin can be
written as shown in Figure 2. From this equation an important feature
of the bacteriorhodopsin system becomes evident: its simplicity. A
single polypeptide chain, or possibly several cooperating together,
each containing a retinal molecule, translocates protons from the cytoplasm to the medium upon absorption of light. Not an assembly of
molecules as in the case of the electron transport chains, but one
molecule catalyzes this process and no other ions or substrate molecules participate in this reaction. Therefore the proton translocating
reaction catalyzed by bacteriorhodopsin is a very good system for the
elucidation of a vectorial catalytic process. A further advantage is
that the catalytic cycle of the molecule is attended by a photochemical
cycle, composed of spectrally distinguishable intermediates of the
bacteriorhodopsin chromophore, and which provide useful information
concerning the proton-translocating process.

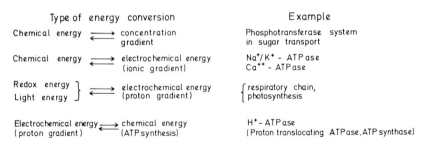

Fig. 1. Enzymes of energy-transducing process

Fig. 2. Reaction catalyzed by bacteriorhodopsin (BR)

Light + H^+_{in} $\xrightarrow{\text{Bacteriorhodopsin}}$ H^+_{out} + heat

$$\Delta \mu_{H^+} = F \Delta \psi + RT \ln \frac{[H^+_{out}]}{[H^+_{in}]}$$

The catalytic cycle of BR comprises absorption changes of the chromophore (purple complex) and proton translocation.

Besides the mechanism of proton translocation, bacteriorhodopsin provides the opportunity to study several other aspects of membrane biology. It occurs in at least two different states of aggregation: crystalline purple membrane and noncrystalline brown membrane. Induction of bacterio-opsin and retinal biosynthesis, insertion and assembly of the molecules into the membrane, modification and crystallization of the chromo-protein into patches and finally patch growth and patch fusion can uniquely be studied in Halobacteria.

The present discussion will be limited strictly to the phenomenon of light energy conversion by bacteriorhodopsin in the intact cell and special emphasis will be put on the relationship between proton translocation, ATP synthesis and cation exchange processes.

General Features of Bacteriorhodopsin and the Purple Membrane

General reviews on the purple membrane system have been published recently (1,2,3). Bacteriorhodopsin in the cell membrane consists of a single polypeptide chain with a molecular weight of 26.ooo. Most of its primary structure has been elucidated (4). From the combination of electron microscopic and X-ray experimental results (5) we know that the polypeptide chain traverses the membrane seven times, mostly as an α helix and that the carboxy terminus projects into the cytoplasm (6). Trimers of bacteriorhodopsin form a hexagonal crystalline array called the purple membrane which can be isolated from the mixture of cell membrane fragments obtained after lysis of the cells by suspension in water. Only 25% of the membrane's dry weight is lipid, corresponding to about 1o lipid molecules per bacteriorhodopsin molecule. Retinal is specifically bound to the protein with a stoichiometry of 1:1, producing the chromophore called purple complex that has a dramatically red shifted absorption maximum (λmax 560 nm) and a high extinction coefficient (ε_M 63,0OO) compared to retinal. The long axis of the retinal molecule is tilted away from the plane of the membrane by 23° (see 1,2,3).

Upon light absorption, the chromophore of the molecule undergoes a photochemical cycle composed of spectral changes which occur in the time range from ps to ms. The photoproduct which was the first to be observed in bacteriorhodopsin is the 412 nm chromophore. It is the longest-lived intermediate of the photochemical cycle (7). Upon formation of the 412 nm chromophore a proton is released into the medium and upon regeneration of the purple complex a proton is taken up from the cytoplasm. Thus the catalytic cycle involves a coupling of the photochemical cycle and vectorial proton release and uptake.

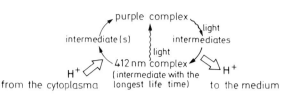

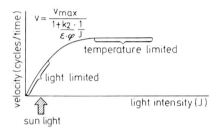

Fig. 3. Overall kinetics of
the photochemical cycle in
bacteriorhodopsin (BR)

The cycles per time (v) are proportional to the steady state concentration of "412"
1 Formation of "412" : $\varepsilon \cdot \varphi \cdot J$
2 Decay of "412" : k_2 (slowest dark reaction)

For the description of the action of bacteriorhodopsin in the intact
cell it is sufficient to understand the overall kinetics of the catalytic cycle and its dependence on light intensity. This relationship
is shown in Figure 3. The rate of formation of the 412 nm chromophore
is equal to the product of light intensity (irradiance) (J), extinction coefficient of the purple complex (ε) and quantum yield of the
primary photochemical reaction (φ). The regeneration of the purple
complex occurs at a constant rate (k_2) of about 200 s^{-1} at room
temperature (turnover number). The frequency of the photochemical cycle
is proportional to the steady state concentration of the 412 nm
chromophore and therefore is given as a function of light intensity as

$$\nu = \nu_{max} \cdot \frac{1}{1 + \dfrac{k_2}{\varepsilon \cdot \varphi} \dfrac{1}{J}}$$

Under conditions of light saturation the maximal frequency ($\nu_{max} = k_2$ BR)
of the cycle is temperature limited whereas at low and medium light
intensities, e.g., solar radiation of the natural habitat of the bacteria, the frequency depends on light intensity. An estimation of the
efficiency of bacteriorhodopsin in the intact cell is not solely dependent on light intensity but also on various parameters such as the
electric capacity of the membrane, the volume of the halobacterial
cell, and the number of bacteriorhodopsin molecules per cell. Since
all these parameters have been determined we know that under light-
saturating conditions a cell could maximally eject 7 × 10^7 protons/s
(8).

If a cell membrane were totally impermeable to ions, thus acting as
a perfect insulator, this rate of proton translocation would produce
an electric field (membrane potential $\Delta\psi$) of more than 300 mV within
the time of one photochemical cycle of bacteriorhodopsin. This rate of
proton translocation exceeds the rates of all other ion movements summarized in the scheme of Figure 4.

Since Halobacteria are aerobic organisms, they also translocate protons to the medium via their respiratory chain. In the presence of

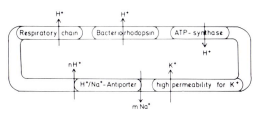

Fig. 4. Main ion transport process in Halobacteria

both oxygen and light, the respiratory chain is in competition with bacteriorhodopsin, thus light can lead to the inhibition of respiration. There exist at least two systems in the halobacterial cell which translocate protons back into the cell. Firstly the ATP-synthase system which is very well characterized in other bacteria and is described later in this book, secondly a proton/sodium antiporter molecule proposed by Lanyi to be present in the cell membrane, which exchanges protons for sodium ions with an unequal stoichiometry (9). A special feature of the cell membrane is the high permeability to potassium ions. K^+ uptake by the cells described in more detail below is not directly coupled to proton outflow or ATP hydrolysis. The high permeability is either an intrinsic property of the membrane or due to an endogenous ionophore yet unidentified. The driving force of potassium uptake is the membrane potential created either by bacteriorhodopsin or respiration. Beside these main ion fluxes, many more transport processes take place in the cell, however, under suitable conditions of incubation they do not interfere with measurements of proton motive force (pmf), photophosphorylation, or potassium transport.

Light Induced Enhancement of the Proton Motive Force (pmf)

It is important to note that under illumination, when bacteriorhodopsin pumps protons, and an increase in membrane potential ($\Delta\psi$) is expected only small amounts of protons can be translocated, and therefore the increase in ΔpH is small unless other charge compensating ions (Na^+, K^+) flow through the membrane. The experimental method for determination of changes in $\Delta\psi$ and ΔpH has to be an indirect one because the bacterial cell is too small for insertion of an electrode. The lipophilic ion triphenyl methyl phosphonium ($TPMP^+$) and the lipophilic weak acid dimethyl oxazolidine-dione (DMO) distribute between cytoplasm and the medium according to $\Delta\psi$ and ΔpH respectively when added to a cell suspension. Cells can be separated from the medium by centrifugation after equilibration of radioactively labeled $TPMP^+$ or DMO and from the distribution $\Delta\psi$ and ΔpH can be calculated (10).

The concentration of the indicator ion used is critical, because at higher concentrations $TPMP^+$ acts as a membrane potential buffer instead of as an indicator. This is illustrated in Figure 5. No significant accumulation of $TPMP^+$ is found when applied in mM concentrations. This buffering effect is similar to that experienced with pH indicator substances such as DMO when used at high concentrations.

As mentioned above, the increase of the membrane potential upon illumination is too fast for a time resolved measurement by the ion distribution method. However, it can be shown that in dicyclohexylcarbodiimide (DCCD) treated cells the initially high $\Delta\psi$ under continous illumination decreases concomitantly with an increase in ΔpH due to charge

144

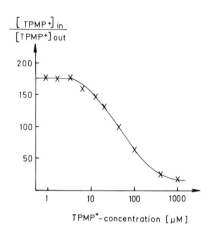

Fig. 5. Dependence of accumulation of TPMP⁺
on its concentration in the medium (pH 8.0).
The experiment was carried out at pH 8 as
described by Michel and Oesterheld (10)

compensating slower processes (12). The total change in pmf ($\Delta\psi$ + $Z\Delta$pH)
remains constant, in accordance with the concept that the proton pump
provides its full thermodynamic potential in the form of an electric
field with low capacity very quickly, and then gradually enhances the
capacity of the system via secondary ion fluxes leading to a large Δ
pH. The size of the change in pmf depends on the level of light intensity within that range where the frequency of the photochemical cycle
is also dependent (12). An interesting fact, discussed in more detail
below, is that maximal pmf in the cell is reached at a level of light
intensity which saturates the photochemical cycle of the isolated PM
to only 1% - 2% (8,12). The maximal value of pmf is found in cells where
ATP synthesis and possibly other "pmf-consuming" processes are blocked
by reaction of the respective catalysts with dicyclohexylcarbodiimide
(DCCD). The maximal pmf value can be used for an estimate of the light
energy converted by bacteriorhodopsin into electrochemical energy. From
the 284 mV found a free energy of 6.5 kcal is calculated which corresponds to about 15 % of the absorbed light energy. The difference between light and dark is 169 mV under these conditions. Thus it seems
that the halobacterial photoconversion system is less efficient than
classical photosynthesis. In cells which are capable of photophosphorylation the total change of pmf induced by light is only 35 mV. The proton
motive force is not abolished completely but values around 60 to 100 mV
remain depending on the state of the cells. This is presumably due
mainly to the residual potassium ion gradient which contributes to the
membrane potential via its diffusion potential.

Photophosphorylation and Oxidative Phosphorylation

The most prominent effect of bacteriorhodopsin is the change of nucleotide levels upon illumination of cell suspensions. This type of photophosphorylation, together with oxidative phosphorylation, is documented
in Figure 6. In both cases, either under nitrogen in light, or with
oxygen in darkness, all adenine nucleotides are converted into ATP and
as anticipated, their sum remains constant. The characteristic features
of halobacterial photophosphorylation have been investigated in detail
(13, 14,15,16) and can be summarized by the following statements:

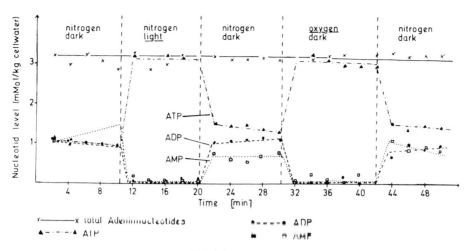

Fig. 6. Nucleotide levels in *H. halobium* in oxidative and photo-phosphorylation

Phosphorylation in Halobacteria is:

1. Not affected by electron transport inhibitors
2. Inhibited by uncoupling reagents
3. Inhibited after blockade of the reversible proton translocating ATPase (ATPsynthase) with DCCD or phloretin
4. Artificially induced by sudden changes of external pH
5. Inhibited by lipophilic cations affecting the membrane potential
6. Induced by addition of monactin (change in potassium diffusion potential)

The inhibition of photophosphorylation by high concentrations of TPMP$^+$ raises the question of whether $\Delta\psi$ or ΔpH or both are the driving force for phosphorylation.

The membrane potential as a driving force for phosphorylation is illustrated in Figure 7. Whether in the presence or absence of DMO the rate and extent of ATP formation is equal. At the concentrations of DMO used no pH gradient could exist. Furthermore, TPMP$^+$, which greatly diminishes the membrane potential at concentration around 1 mM (see Fig. 5), causes a dramatic decrease in the rate of phosphorylation.

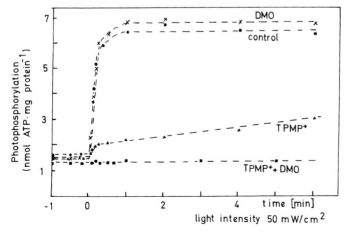

Fig. 7. Influence of DMO and TPMP$^+$ on photophosphorylation. DMO and TPMP$^+$ concentrations were 100 mM and 2 mM respectively. The experiment was carried out at pH 6.5. Data taken from Hartmann and Oesterhelt (16)

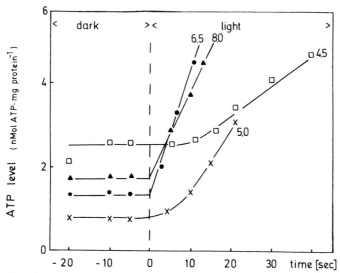

Fig. 8. pH - dependence of the onset of photophosphorylation. Data from Hartmann and Oesterhelt (16)

Since the influx of TPMP$^+$ buffers $\Delta\psi$ an increase in ΔpH will occur which in turn could slowly enhance the ATP level. Thus it becomes clear that a combination of both inhibitors will completely abolish photo-phosphorylation.

The inhibition of photophosphorylation by TPMP$^+$ is strongly pH-depen-dent. The inhibition is 100% at pH 8.5 and only 10%-15% at pH 5 (16). These results strongly suggest a changing contribution of $\Delta\psi$ and ΔpH to the pmf at different external pH values. As a consequence of the decreasing contribution of the membrane potential with decreasing pH one might further expect pH-dependent differences in the onset of ATP synthesis upon illumination for the following reason: as an electrical potential difference across a membrane develops much faster than a pH difference, a lag phase in ATP synthesis should be observable if ΔpH is the main driving force. This is shown for *H. halobium* cell suspensions in Figure 8 where a considerable lag phase in ATP synthesis occurs upon lowering the external pH to pH 5.

Quantitation Between Light Absorption, Respiration, Proton Extrusion and ATP Synthesis

Quantitation of light energy conversion by Halobacteria is possible by application of standard methods of cell physiology (17,8). According to the scheme of Figure 9, which simplifies the actual bioenergetic processes in Halobacteria, one can predict the relationships between quantal absorption, rate of respiration, proton extrusion, and ATP synthesis, and compare the result with experimentally obtained data. The quantum requirement of proton translocation can be determined from experiments in which the rate of proton extrusion is measured as a func-tion of irradiance. A value of 2 is found for cells blocked in ATP-synthesis by phloretin, which compares well with 1.7 found by Bogomolni (18) in the absence of this drug. Inhibition of respiration by light was used as the first experimental approach to prove light energy con-version by Halobacteria (17). A minimum of 24 quanta need to be absorbed by the aqueous phase surrounding the membrane sheets: the intact cell

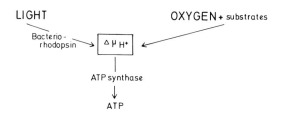

Relationship	Method	Experimentally found value
1. $\frac{Photons}{O_2}$	light inhibition of respiration	24
2. $\frac{Photons}{ATP}$	quantum requirement of photophosphorylation	22
3. $\frac{ATP}{O_2}$	O_2 pulse for oxidative phosphorylation	1.0, predicted from 1 and 2: 1.1
4. $\frac{Photons}{H^+}$	proton extrusion with blocked ATP synthesis	2
5. $\frac{H^+}{ATP}$	initial rates of proton extrusion and photo-phosphorylation	1.0, predicted from 2 and 4: 1.1

Fig. 9. Quantitative relationships between light absorption (H^+), respiration (O_2), phosphorylation (ATP) and $\Delta\mu_{H^+}$. Data from Wagner et al. (19)

establishes an electrochemical gradient which counterbalances the proton motive force of the pump.

Potassium Ion Accumulation as an Energy-Storing Process

Illuminated cells of *H. halobium* accumulate K^+ to a considerable extent. The action spectrum of this uptake process coincides with the absorption spectrum of bacteriorhodopsin showing that potassium ion transport is coupled to the proton pump. Starvation of the cells in the dark leads to a slow leakage of K^+ which takes more than 24 h for total depletion. Therefore the net uptake of K^+ in light following a dark period depends on the extent of starvation and ranges between 600 mmol/kg cell water and 2000 mmol/kg cell water (19). The net uptake of K^+ is accompanied by a release of Na^+ which chargebalances mostly the K^+ uptake (Fig. 10). As can be seen from Figure 11, concentration changes of H^+ and divalent cations like Mg^{2+} and Ca^{2+} are small.

Two alternative mechanisms for a large-scale Na^+/K^+ exchange process will now be considered. In the first case a Na^+/K^+-ATPase, typically found in eukaryotic cells, creates a membrane potential due to its uneven Na^+/K^+ exchange stoichiometry (3:2). The membrane potential then leads to an uptake of K^+ by an electrophoretic process.

$$ATP \xrightarrow{\frac{3Na^+_{in}/2K^+_{out} \text{ antiport}}{\text{catylyzed by } Na^+/K^+\text{-ATPase}}} \Delta\psi \tag{1}$$

$$\Delta\psi \xrightarrow{K^+_{out}, \text{ electrophoresis}} (-)\Delta pNa \ , \quad (+)\Delta pk \tag{2}$$

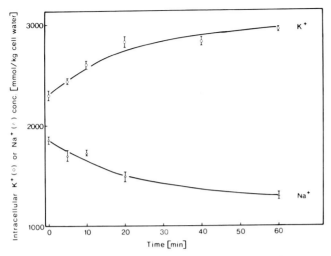

Fig. 10. Changes in the intracellular K⁺ and Na⁺ concentrations induced by illumination. Data from Wagner et al. (19)

K⁺	Na⁺	H⁺	Ca⁺⁺	Mg⁺⁺
+ 637 ± 7	-534 ± 5	- 29 ± 4	+ 24 ± 7	+ 18 ± 6

Changes are given in mMol/kg cell water, + and - indicate
uptake and release

Fig. 11. Changes in the intracellular cationic concentrations during illumination for 60 min (25m W/cm²). Data from Wagner et al. (19)

In and out designate the cytoplasmic and extracellular ion pools respectively (+) and (-) indicate higher and lower concentrations of the respective ions in the cytoplasm as compared with the medium.

As in the case of many other ion pumps the Na⁺/K⁺ ATPase could not catalyze the development of an appreciable ΔpNa and ΔpK of opposite sign unless the K⁺ inflow of reaction (2) charge-balances reaction (1).

In the alternative mechanism the membrane potential created by the proton pump bacteriorhodopsin drives an electrogenic H⁺/Na⁺ exchange process (H⁺ > Na⁺) catalyzed by an antiporter. This partial reaction converts the electrochemical proton gradient ($\Delta\mu H^+$) into that of the sodium ion gradient ($\Delta\mu Na^+$).

$$\text{light energy} \xrightarrow{\dfrac{H^+_{in} \text{ translocation}}{\text{catalyzed by bacteriorhodopsin}}} \Delta\psi\,(\Delta\mu H^+) \qquad (3)$$

$$\Delta\psi \xrightarrow{H^+/Na^+(H^+_>Na^+)\text{ antiport}} (+)\Delta pH, \;(-)\Delta pNa \qquad (4)$$

$$\Delta\psi \xrightarrow{K^+_{out}\text{ electrophoresis}} (-)\Delta pH, \;(+)\Delta pK \qquad (5)$$

The total reactions above lead to the development of a large Na^+/K^+ gradient. However, the pH-gradient created by reactions (4) and (5) are of opposite sign, thereby cancelling large net pH changes (Fig.11) which could otherwise damage the cell. Since Halobacteria are able to synthesize ATP at the expense of $\Delta\psi$ created by the action of bacteriorhodopsin, it is not excluded a priori that K^+ uptake is driven by an ATP-dependent process. Experiments which compared the cellular ATP level and the rate of potassium uptake have yielded evidence against an ATP-driven K^+ uptake and pointed towards a $\Delta\psi$-induced K^+ inflow (K^+_{out} electrophoresis) rather than the existence of a Na^+/K^+-stimulated ATPase in Halobacteria(19). If the selective K^+ uptake is an electrophoretic process, the membrane permeability to K^+ should be high compared to other ions present in the medium. This can be experimentally tested by measuring the K^+ diffusion potential of whole cells under conditions where other charge-transport processes are not possible i.e., under nitrogen in darkness. Beside lipophilic ions such as $TPMP^+$, the membrane potential can be measured by using membrane-permeable dyes which change their fluorescence intensity upon change of $\Delta\psi$. Figure 11 shows that the fluorescence change of such a dye depended linearly on the change in potassium ion gradient. For calibration the same type of experiment was repeated in the presence of valinomycin which leads to membrane which is fully permeable toward K^+, yielding about 60 mV per decade of K^+ gradients according to the Nernst equation. By comparison of the slopes in Figure 11 it follows that in the absense of valinomycin the cell membrane has a value of about 30 mV, which indicates a high selectivity for K^+ by the membrane. We conclude that the K^+ uptake process in the presence of light is a uniport driven by the membrane potential.

This result is also relevant to the bioenergetics of Halobacteria. According to the chemiosmotic hypothesis (20), the free energy change in ATP hydrolysis (ΔG_{ATP}) is in equilibrium with Δpmf ($\Delta\psi + 59 \Delta pH$) of the cell. When it is in starving conditions and in darkness, $\Delta\psi$ is mainly determined by the K^+ gradient. Therefore under these conditions the ATP level should be a function of the K^+ gradient, thus allowing

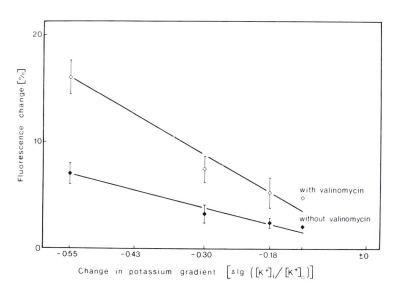

Fig. 12. Selective permeability of the membrane to K^+. Data from Wagner et al. (19)

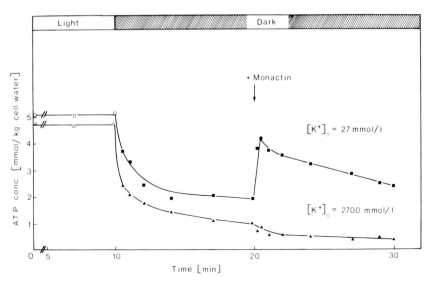

Fig. 13. K - gradient - driven phosphorylation. Monactin was added anaeriobically (final concentration mmol/1). Data from Wagner et al. (19)

the light-induced accumulation of molar concentrations of potassium ions to serve as chemical energy pool for the cell.

The cellular level of ATP is at its maximum in light and decreases in the dark. This decrease is highly dependent on the external potassium concentration, as shown in Figure 13. The K^+_{out} concentration of 2700 mmol/1 which corresponds to a K^+ gradient of 1 causes a more rapid and more extensive decay of the ATP level compared with a K^+_{out} concentration of 27 mmol/1, which is the standard concentration of the growth medium. Direct evidence for the potassium ion diffusion potential as a driving force for ATP synthesis is obtained by the experiment shown in Figure 13. A sudden and considerable increase in cellular concentration of ATP is seen upon the addition of monactin at low (27 mmol/1) but not at high (2700 mmol/1) external K^+ concentration. The correlation of the phosporylation potential and the potassium ion gradient before and after addition of monactin yields straight lines which have slopes differing by a factor of two. This finding supports the result of Figure 11 which revealed a two fold increase in K^+ diffusion potential upon addition of an K^+-specific carrier substance.

Conclusion

The light energy converted by bacteriorhodopsin into electrochemical energy of a proton gradient can be used for various bioenergetic processes such as amino acid uptake, ATP synthesis, ion exchange, and flagella movement. Energy storage, however, is a problem because the ATP pool is small and no metabolic reactions have been found so far in Halobacteria which could serve for chemical energy storage on a large scale. Therefore it is most relevant to the bioenergetics of *H. halobium* that under nonrespiratory conditions in darkness the cell uses the potassium gradient established in light to maintain a certain cellular concentration of ATP. Hence, the potassium gradient serves as the ultimate source of energy when oxidative and photophosphorylation fail. This energy pool decays with a half-life in the order of hours which is long enough to overcome dark periods e.g., the night.

In terms of the chemiosmotic theory, the halobacterial cell transforms light energy into a chemical gradient, rather than into an electrical gradient. This way, the cell increases its effective electrical capacity up to 1 kF/m^2 as compared to the electrical membrane capacity of about 10 mF/m^2 that is usually found in biological membranes (20).

Acknowledgments. We are grateful to Dr. P. Towner for help on improving the style of the English. Work reported from this laboratory was supported by the Deutsche Forschungsgemeinschaft.

References

1. Henderson, R.: Ann. Rev. Biophys. Bioengineer. 6, 87-109 (1977)
2. Lanyi, J.K.: In: Membrane Proteins in energy transduction Capaldi, R.A. (ed) New York: Marcel Dekker, in press, 1978
3. Stoeckenius, W., Lozier, R.H., Bogomolni, R.A.: Biochim, Biophys. Acta, in press (1978)
4. Ovchinnikov, Y.A., Abdulaev, N.G., Feigina, M.U., Kischev, A.V., Lobanov, N.A.: FEBS Lett. 84, 1-4 (1977)
5. Henderson, R., Unwin, P.N.T.: Nature (London) 257, 28-32 (1975)
6. Gerber, G.E., Gray, C.P., Wildenauer, D., Khorana, H.G.: Proc. Nat. Acad. Sci. USA 74, 5426-5430 (1977)
7. Oesterhelt, D., Hess, B.: Eur. J. Biochem. 37, 316-326 (1973)
8. Hartmann, R., Sickinger, H.-D., Oesterhelt,D.: FEBS Lett. 82, 1-6 (1977)
9. Lanyi, J.K., Mac Donald, R.E.: Biochemistry 16, 4608-4613 (1976)
10. Michel, H., Oesterhelt, D.: FEBS Lett. 65, 175-178 (1976)
11. Bakker, E.P., Rottenberg, H., Caplan, S.R.: Biochim. Biophys. Acta 440, 557-572 (1976)
12. Michel, H.: Ph. D. thesis, University of Würzburg, 1977
13. Danon, A., Stoeckenius, W.: Proc. Nat. Acad. Sci. USA 71, 1234-1238 (1974)
14. Oesterhelt, D.: In: Energy Transformations in Biological Systems. Ciba Foundation, Symposium 31, Amsterdam: Elsevier, 1975, pp 147-167
15. Danon, A., Caplan, S.R.: Biochim. Biophys. Acta 423, 133-140 (1976)
16. Hartmann, R., Oesterhelt, D.: Eur. J. Biochem. 77, 325-335 (1977)
17. Oesterhelt, D., Krippahl, G.: FEBS Lett. 36, 72-76 (1973)
18. Bogomolni, R.A.: Fed. Proc. Fed. Am. Soc. Exp. Biol. 1833-1839 (1977)
19. Wagner, G., Hartmann, R., Oesterhelt, D.: Eur. J. Biochem. in press (1978)
20. Mitchell, P.: Chemiosmotic Coupling and Energy Transduction. Cornwall, England Glynn. Research, Bod (1968)

Conformational Changes and Cooperation of Bacteriorhodopsin

B. HESS, R. KORENSTEIN, and D. KUSCHMITZ

The mechanism of coupling between the photocycle and the vectorial proton transfer process of bacteriorhodopsin (BR) of the purple membrane of *Halobacterium halobium* implies an oriented and tight interaction between the chromophore and the protein conformation of BR within the structure of the purple membrane. Recently, the orientation of the transition dipole moment of the retinal chromophore in its dark state was analyzed by linear dichroism at 568 nm in thin layers of purple membrane of less than 1 μm to 4 μm thickness. We found an angle between the transition moment and the plane of the membrane of $\leqslant 27^\circ$ (1), which is schematically drawn in the computed sequence model recently published by Ovchinnikov et al. (2) (see Fig. 1). It should be mentioned here that the direction of the chromophore (to the left or to the right) is not yet clarified. However, the picture illustrates the orientation of the chromophore at its approximate angle within a distribution of polar and nonpolar groups along the folds of the amino acid sequence thus far resolved.

The degree of mobility of bacteriorhodopsin was analyzed in the thin layers of the purple membrane following the transition dichroism of the slow $BR^{LA} \to BR^{DA}$ thermal transition from its light- (LA) to its dark- (DA) adapted state with a relaxation time constant of 14 min at $40^\circ C$ corresponding to an all-trans-13-cis isomerization of the retinal chromophore. Our analysis shows that the anisotropy factor remains constant over a time range of 25 min at $40^\circ C$, indicating that no rotational freedom is observable. It can be concluded that bacteriorhodopsin is completely immobilized within the purple membrane matrix. This result eliminates the possibility for rotational freedom of the bacteriorhodopsin monomer, trimer, or even a big cluster of trimers. The results also exclude the possibility that the chromophore can rotate

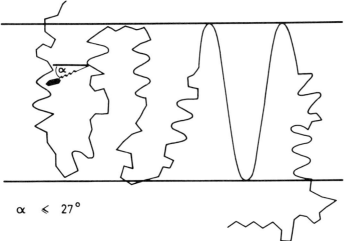

$\alpha \leqslant 27^\circ$

Fig. 1. Backbone-Model of the bacteriorhodopsin polypeptide chain in the purle membrane [according to (2)]

about those axes, which would lead to a dichroic decay, suggesting a strong retinal-opsin interaction. Our results as well as independent studies of the microviscosity of lipid domains in the purple membrane (3) show that the microviscosity that bacteriorhodopsin experiences is due mainly to protein-protein interactions and not to lipid-protein interactions.

The light-triggered series of interactions between the chromophore and the protein are reflected in the spectral changes during the photocycle (4) as well as in the accompanying protein fluorescene changes, indicating the conformational state of the opsin moiety (5). In order to define conformational states of the opsin moiety more directly, we titrated the acid and basic groups of purple membrane in the dark and found a heterogeneous population of acidic groups, the pK values of which are a function of ionic strength in agreement with the Gouy-Chapman theory (6). Upon illumination a drastic change in the pH titration curve is observed: At high ionic strength, an increase in the buffer capacity at pH > 6.8 and a decreased change at pH < 6.8 coupled to the appearance of an apparent pK around 7.8 as a function of light, are detected. This light-induced change is shown in the difference-pH titration curve in Figure 2. The quantitative evaluation of the change in buffer capacity (in percent) at approximately 25% light saturation yields the disappearance of approximately two equivalents and the appearence of 2.3 equivalents per bacteriorhodopsin with an excess of approximately 0.3 equivalents. It is interesting to note that the pK of 7.8, not detectable in dark bacteriorhodopsin, can also be detected in an apomembrane preparation. We are currently investigating the nature of this pH-titratable conformation change, mainly with respect to the vectorial transfer pathway and its sidedness.

The experiments clearly show that the purple membrane responds to the illumination not only by a photocycle and the interaction of its chromophore with neighboring trytophane, tyrosine, and/or phenyl alanine residues, but also by a rearrangement of its pH-titratable groups. The results suggest a pK cascade through the purple membrane molecule as part of the overall mechanism of proton transport relevant to the structural model reproduced in Figure 1. A possible sequence of events in the light-triggered onset of the reactions of the system is illustrated in Figure 3. For each of the signals summarized on the right side of the figure, evidence in model experiments and biochemical as well as biophysical studies of the purple membrane is available. We especially like to point to the time domain in which a conformational change in the bacteriorhodopsin molecule could occur. Recently, we found that

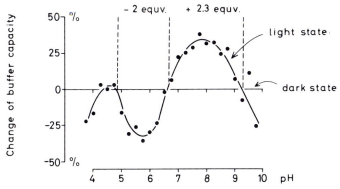

Fig. 2. Light-induced changes in buffer capacity

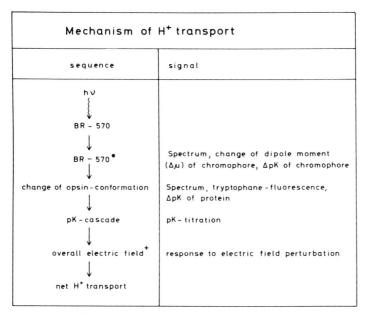

Fig. 3. Sequence of events leading to a stationary light-dependent proton transport

the protein conformation of bacteriorhodopsin, also in the absence of of the chromophore in the form of its apomembrane, detected by its absorbance at 300 nm using a polarization angle of 54°, reacts to an electric field pulse perturbation within a few microseconds with a lag time in the same order of magnitude (7). This responsiveness is well within the time range of the photocycle events leading to deprotonation. We conclude that the conformational changes observed by pH titration might well be an essential part of the pumping mechanism of the purple membrane, as shown in Figure 3.

The complete immobilization of bacteriorhodopsin, even under condition of reduced microviscosity, reflects the exceedingly strong protein-protein interactions within the purple membrane matrix. These strong interprotein forces should play a key role in any cooperative transition that the bacteriorhodopsin undergoes. Thus, such forces can introduce cooperativity into the kinetics of the photocycle of bacteriorhodopsin, i.e., even into the proton transfer processes. Structurally this cooperativity is reflected in the lattice organization of the purple membrane where bacteriorhodopsin molecules are arranged in a two-dimensional array forming an almost perfect crystal lattice of a space group P3 (8). In addition, an interaction between the retinal chromophores was found from the dichroic (CD) spectrum of the purple membrane. The visible CD spectrum of the purple membrane consists of intense positive and negative bands with a crossover at the wavelength of the absorption maximum. Two factors contribute to the overall CD spectrum: a positive band due to a bond between the retinal chromophores and the protein and a positive and negative band due to the exciton interaction between chromophores of neighboring protein (9, 10). Additional evidence for cooperativity results from the observation that the stoichiometry between the protons liberated per occupancy of the 412-nm state is a function of that very state

Table 1. Evidence for cooperative BR unit

Method	Result	Reference
X-ray analysis	Trimetric unit	4,8
CD-spectrometry	Exciton-interaction within trimer	9,10
Rate constants of photocycle	$k_2 = f\,[412]$	6
Proton liberation at pH 7.5	$\dfrac{[H^+]}{[412]} = f\,[412]$	11
Cemical reactivity	Retinal-oxim formation within trimers	12

Table 2. Dependence of '412' decay from the photosteady state on the relative occupancy of '412'

[BR],μM	[412],μM	relative occupancy	τ_1^{-1}, sec^{-1}	τ_2^{-1}, sec^{-1}
22.2	12.7	57%	2.88 ± 0.28 (0.15 ± 0.02)	0.41 ± 0.01 (0.86 ± 0.01)
5.5	3.1	57%	2.96 ± 0.54 (0.20 ± 0.04)	0.40 ± 0.03 (0.79 ± 0.04)
32.4	3.1	10%	0.89 ± 0.16 (0.34 ± 0.09)	0.32 ± 0.02 (0.65 ± 0.10)

as observed in open membrane sheets (11) and furthermore from the
nonlinear chemical reactivity of the system with respect to the retinal-
oxim formation within the trimers (12). In a more detailed analysis
of photocycle kinetics, we observed recently a kinetic coupling among
the three bacteriorhodopsin molecules that constitute a trimeric
cluster. It was found that the photocycle kinetics of a bacteriorhodopsin molecule is dependent on the conformational state of its
nearest neighbor (6) (for a summary of experimental evidence of cooperativity see Table 1). The kinetic coupling among bacteriorhodopsin
molecules is illustrated in the experiments reported in Table 2, in
which the two reciprocal relaxation times of the decay of the cycle,
as analyzed in terms of a sum of exponentials, are given for three
different states of the purple membrane. The table demonstrates that
it is the relative occupancy of the 412-nm state (3rd column) that
determines the decay rates and not the absolute concentration of
bacteriorhodopsin (1st column) or of the 412-nm state itself (2nd
column). These results fit a cooperative model based on subunit-
subunit interactions within the trimer, which is crudely illustrated
in Figure 4. A detailed analysis of the model, which cannot be discussed here, fits our experimental results as shown in Figure 5 (in collaboration with M. Markus).

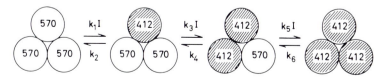

Fig. 4. Schematic model for the cooperation of bacteriorhodopsin molecules in the
photocycle of the purple membrane. 570 = all-trans state of bacteriorhodopsin in the
dark; 412 = photostationary state of bacteriorhodopsin the presence of light
(I = intensity)

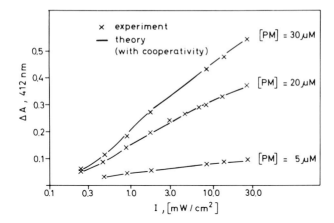

Fig. 5. Dependence of the photostationary concentration of the 412-nm state on light intensity. *PM* = purple membrane. (In collaboration with M. Markus)

Indeed, structural and kinetic evidence of a cooperative function of bacteriorhodopsin, as shown here, add a new property to membrane-bound systems in general. It is obvious that energy conservation mechanisms do not rely only on a maximum energy-trapping device (1) but also on a most efficient molecular energy conversion mechanism as reflected in a cooperative function of bacteriorhodopsin and further-more on a highly efficient cellular energy-storing mechanism in the form of ionic gradients, as reported in this symposium (13).

References

1. Korenstein, R., Hess, B.: FEBS Lett. 89, 15-20 (1978)
2. Ovchinnikov, Y.A., Abdulaev, N., Feigina, M., Kislelev, A., Lobanov, N.: FEBS Lett. 84, 1-4 (1977)
3. Korenstein, R., Chermann, W.V., Caplan, S.: Biophys. Struct. Mech. 2, 267-276 (1976)
4. Henderson, R.: Ann. Rev. Biophys. Bioeng. 6, 87-109 (1977)
5. Oesterhelt, D., Hess, B.: Eur. J. Biochem. 37, 316-326 (1973)
6. Hess, B., Korenstein, R., Kuschmitz, D.: Hoppe-Seyler's Z. Physiol. Chem. 359, 275 (1978)
7. Hess, B., Korenstein, R.: Proceedings of the Solvay Conference 1975, Adv. Chem. Phys., p. 224-227 (1975)
8. Henderson, R.: J. Mol. Biol. 93, 123-138 (1975)
9. Heyn, M., Bauer, P., Dencher, N.: Biochem. Biophys. Res. Commun. 67, 897-963 (1975)
10. Reed, T., Hess, B.: Fed. Proc. 35, Abstract 1227, 1599 (1976)
11. Kuschmitz, D., Hess, B.: Abstract of the 11th FEBS Meeting Copenhagen A 4/13/708, 1977
12. Becher, B., Cassim, J.: Biophys. J. 19, 285-297 (1977)
13. Oesterhelt, D., Hartmann, R., Michel, H., Wagner, G.: Light-Driven Proton Translocation and Energy Conservation by Halobacteria. This volume, pp. 140-151 (1978)

Structure and Function of ATP-Synthesizing Systems

Structure, Function, and Regulation of the Mitochondrial Adenosine Triphosphatase Complex of Rat Liver – A Progress Report

P. L. PEDERSEN, L. M. AMZEL, J. W. SOPER, N. CINTRÓN, and J. HULLIHEN

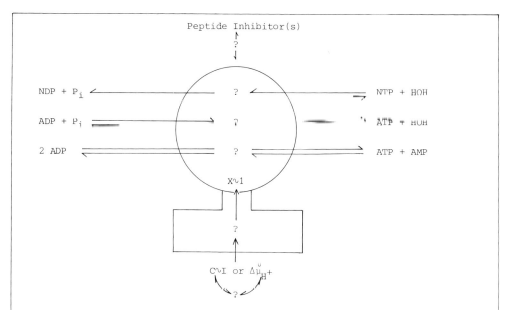

As of this date very little is known about the molecular mechanism(s) of oxidative phosphorylation and ATP-dependent functions of mitochondria, bacteria, and chloroplasts. Also, very little is known about the mechanism(s) or regulation of these processes. However, the enzyme complex responsible for catalyzing ATP synthesis and ATP-dependent functions has been purified in a number of laboratories, and the F_1-ATPase component has been crystallized. Moreover, ATPase inhibitor peptides thought to be involved in the regulation of these enzyme complexes have been isolated in a number of laboratories. Therefore, we can finally state that a concerted and vigorous attack on the problem is under way at the molecular level.

Prolog

There are those who have walked down the path of oxidative phosphorylation and become totally lost; there are those who followed the same path, became lost but did not know it; and finally, there are those who thought they were near the end of the path but retreated in the face of new experimental facts.

Only in recent years have fairly complete ATP synthetase complexes been purified and studied with respect to their nucleotide and P_i binding sites. Only in the past year have X-ray crystallographic studies been initiated on such complexes. *Therefore, the walk down the path leading to the eventual resolution of the problem of oxidative phosphorylation at a molecular level has just begun.* It would be scientifically naive and misleading to state otherwise.

The path leading to the resolution of this problem does not lead necessarily to the resolution of the mechanism of action of ATP synthetases. Such complexes are at the minimum bifunctional enzymes, catalyzing both ATP synthesis and ATP-dependent functions. The final answer to the mechanism of action of this class of enzymes must consider their bifunctional (or multifunctional) nature and those physiological effectors (be they chemical, electrochemical, or both) which direct their active center(s) toward either ATP synthesis or ATP-dependent functions.

Introduction and Background

The mitochondrial ATPase complex and similar ATPase complexes from chloroplasts and bacteria appear to be the most complex enzyme systems constructed on this planet. The two important functions of this enzyme, i.e., its involvement in ATP synthesis and in ATP-dependent processes, have been preserved throughout evolution and are essential for most life processes. For these reasons, the elucidation of the mechanism(s) of action of mitochondrial and similar ATPase complexes represents one of the most challenging and important problems in molecular biology today.

Resolution of this problem is of great interest to the professional bioenergetist who would like to understand how mitochondria, bacteria, and chloroplasts use bifunctional enzyme systems to make ATP under some conditions and utilize ATP under other conditions to do useful work. It is likewise of great interest to the membrane enzymologist who would like to understand in some detail how vectorial enzymes work; to the pathologist who would like to know whether critical alterations occur in energy metabolism in some diseased states, in particular cancer, ischemia, and muscle diseases; and finally, to the energy conservationist who would like to know how biological systems have constructed what appears to be more efficient systems than man for conserving and transducing energy.

As illustrated in Figure 1, mitochondrial ATPase is located exclusively in the inner membrane where it faces the matrix space. In rat liver the inner membrane comprises about 20% of the mitochondrial mass and the ATPase complex about 10% of the mass of the inner membrane. The "complete" mitochondrial ATPase complex is depicted in many review articles as consisting of as many as four major components: A *headpiece* called F_1 (factor 1) which is water soluble, catalyzes the hydrolysis of nucleoside triphosphates, and binds the inhibitor aurovertin; a *basepiece or membrane sector*, which is detergent soluble and binds the inhibitors oligomycin and dicyclohexylcarbodiimide (DCCD); a *stalk* connecting the headpiece and basepiece; and one or more small *peptide inhibitors* of ATP hydrolytic activity. The stalk and basepiece are referred to collectively in animal systems as F_o.

As indicated in Table 1, mitochondrial ATPase complexes are referred to by at least seven different names and to date a name acceptable to all or even most workers in the field has not been agreed upon. In this lecture the mitochondiral ATPase complex will be referred to as OS-ATPase (oligomycin-sensitive ATPase) because this name has been used more than any other to describe the animal enzyme.

To date complete OS-ATPase complexes have not been purified to homogeneity from animal systems and therefore their molecular weight is not known. The F_1 component of the rat liver enzyme purified in this laboratory

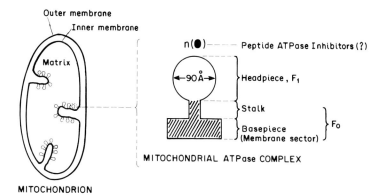

Fig. 1. Inner membrane location and structure usually depcited for the "complete"
mitochondrial ATPase complex. The F_1 component of the rat liver complex isolated
in this laboratory has a molecular weight of 384,000 ± 31,000. The molecular weight
of the complete mitochondrial ATPase complex from animal cells is not known because
what constitutes a homogeneous preparation has not yet been established. The rat
liver enzyme has been purified to a form essentially free of cytochromes (see Tables
2 and 3), but still contains some flavoprotein

Table 1. Current nomenclature for mitochondrial ATPase

Mitochondrial ATPase Complex

Oligomycin-sensitive ATPase

H^+-translocating ATPase

Energy-Transducing ATPase

Complex V (ATP-P_i Exchange Complex)

ATP Synthetase

F_OF_1-ATPase

has a molecular weight of 384,000 ± 31,000, and is comprised of five
or six different polypeptide chains (see below).

In this lecture recent work that we have done at Johns Hopkins School
of Medicine on the "complete" OS-ATPase complex of rat liver, its F_1
component, and its peptide inhibitor will be briefly reviewed. Earlier
work from these laboratories on the rat liver enzyme has been reviewed
elsewhere (Catterall and Pedersen, 1974).

Gross Structure and Inner Membrane Arrangement of OS-ATPase

Gross Structure of OS-ATPase — An Outline of Possibilities

Figure 2 depicts three possible structures of the complete OS-ATPase
complex that have appeared in the literature. Although the "headpiece-
stalk-basepiece" structure shown in Figure 2A is the most common re-

A) Headpiece, stalk, basepiece arrangement.

B) Headpiece with stalk continuing through membrane; no basepiece.

C) Headpiece with basepiece; no stalk.

Fig. 2A-C. Possible gross structures for OS-ATPase. Although the structure of the "complete" mitochondrial ATPase complex (OS-ATPase) is usually depicted as shown in (A), structures of the types shown in (B) and (C) have also been suggested for the enzyme. As shown in Figure 4, the cytochrome-deficient, rat liver OS-ATPase purified in this laboratory does shown in some electron-microscopic fields a structure of the type presented in (A)

presentation, to our knowledge OS-ATPase preparations isolated to date have not clearly revealed this structural arrangement. In fact, electron micrographs of most OS-ATPase preparations are rather difficult to interpret. Some preparations when negatively stained show vesicular-type structures that are not remarkably different in appearance from very small negatively stained, inner mitochondrial membrane vesicles (Kagawa et al., 1976; Sadler et al., 1974; Ryrie, 1975); in other cases negative staining reveals dispersed material with definite suggestions of molecular forms (Tzagoloff and Meager, 1971; Stiggall et al., 1978); and in still other cases electron micrographs have not been published (Serrano et al., 1976). Because there has been little published agreement to date regarding what OS-ATPase preparations should look like, structural arrangements of the type shown in Figure 2A ("headpiece-stalk" arrangement) and in Figure 2B ("headpiece-basepiece" arrangement) have not been ruled out.

Gross Structure of OS-ATPase Extracted with Deoxycholate

We reported earlier (Soper and Pedersen, 1976) that Triton X-100 or deoxycholate (DOC) can be used to solubilize OS-ATPase activity from a P_i-washed inner membrane fraction of rat liver mitochondria. Of these two detergents, DOC proved superior provided that a low concentration (0.1 mg DOC/mg protein) was employed. A higher concentration of DOC solubilizes more ATPase activity but results in a reduced sensitivity of this activity to inhibition by oligomycin (high detergent concentrations may tend to sever the noncovalent linkage between F_0 and F_1).

Electron micrographs of inner membrane vesicles and of OS-ATPase preparations solubilized with DOC are presented in Figures 3 and 4. Figure 3A shows the inner membrane vesicle fraction, which serves as

Fig. 3A and B. (A) Inner membrane vesicle preparation of rat liver mitochondria negatively stained with 1% ammonium molybdate. The inner membrane vesicle preparation shown was prepared by treating mitochondria sequentially with digitonin and Lubrol WX, and then washing 5 x with KP$_i$ as described by Soper and Pedersen (1976). Electron microscopy by Mr. Glenn Decker. x 106,000. (B) Dispersed OS-ATPase Complex (37 µg/ml) prepared as described in Table 2 and negatively stained with 1% phosphotunstate. Electron microscopy by Mr. Glenn Decker. Note that the preparation is not completely dispersed. Some aggregate forms of OS-ATPase can be seen. x 70,000

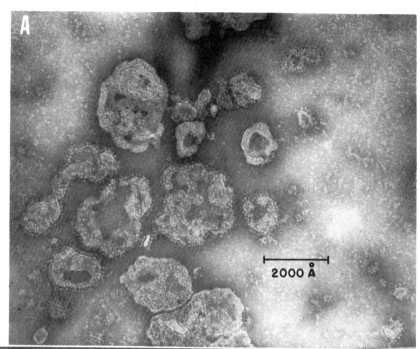

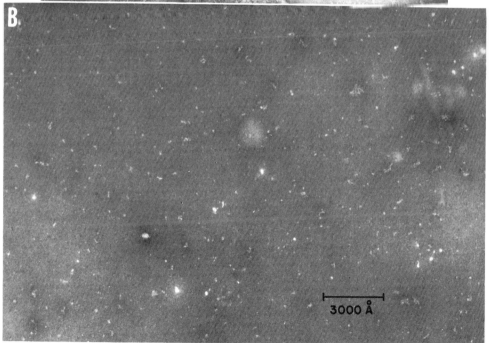

Table 2. <u>Solubilization of OS-ATPase from P_i-washed inner membrane vesicles</u>[a]

Fraction	Protein (mg)	ATPase activity (μmoles min^{-1} mg^{-1})	Inhibition by oligomycin[d] (%)
Inner membrane vesicles (5x P_i washed)	100	21-33	83
Vesicles treated with deoxycholate (DOC)[b]	100	16-28	77
DOC extract (centrifuged and filtered)[c,e]	4.3	9-15	71

[a]This is a slightly modified version of a previously described preparation (Soper and Pedersen, 1976).

[b]DOC concentration = 0.1 mg/mg protein.

[c]Centrifugation was carried out at 48,000 rpm in the Spinco Model L ultracentrifuge; the supernatant was removed and centrifuged again for 4 h at the same speed. The final supernatant was filtered through a 1000-Å Millipore filter pretreated with 10 ml, 1 mg/ml bovine serum albumin.

[d]Oligomycin concentration = 1 μM.

[e]The reader may be interested in the SDS gel pattern of the OS-ATPase preparation of rat liver. The following major bands are observed: The 6 bands characteristic of rat liver F_1 (Fig. 6); two predominant bands in the range 18,000-25,000 molecular weight (one of which is most likely OSCP); and two bands larger than A and B, which may represent aggregates of denatured F_1 subunits contaminating the preparation. [Bands larger than A and B have been reported for the bovine heart preparation of Serrano et al. (1976) and "Complex V" of bovine heart (Stiggall et al., 1978).] Attempts are currently underway in this laboratory to ascertain whether these larger subunits can be removed, and whether a very small nonstainable, hydrophobic subunit (corresponding to what is referred to as "subunit" 9 by investigators working on the yeast OS-ATPase) can be detected in the rat liver OS-ATPase. We believe it important to answer these questions before publishing the SDS gel pattern of the OS-ATPase of rat liver. Otherwise, there might be a tendency for the reader to assume that what is published is completely purified.

starting material for OS-ATPase preparations. The "headpiece-stalk" arrangement of OS-ATPase molecules on the periphery of these inner membrane vesicles is quite prominent. (Presumably a basepiece is integrated in the membrane.) Figure 3B shows a dilute, negatively stained preparation of OS-ATPase (37 μg protein/ml); (OS-ATPase was extracted from P_i-washed inner membranes with DOC and further purified as indicated in Table 2). The preparation appears highly dispersed, although at the magnification shown small molecular aggregates of OS-ATPase dimers and trimers cannot be ruled out. It is important to emphasize that (in our hands) to observe highly dispersed preparations of OS-ATPase molecules, the protein concentration must be low (< 100 μg/ml). Dilute OS-ATPase preparations concentrated to give a protein concen-

Fig. 4A and B. Selected field of OS-ATPase complex (58 μg/ml) prepared as described in Table 2 and subjected to high magnification. (A) x 380,000. (B) Note the "headpiece-stalk-basepiece" arrangement of isolated OS-ATPase molecules. The molecule on the *far left* may represent a side view whereas the two molecules on the *right* may represent tilted or end views. Electron microscopy by Mr. Glenn Decker. x 470,000

165

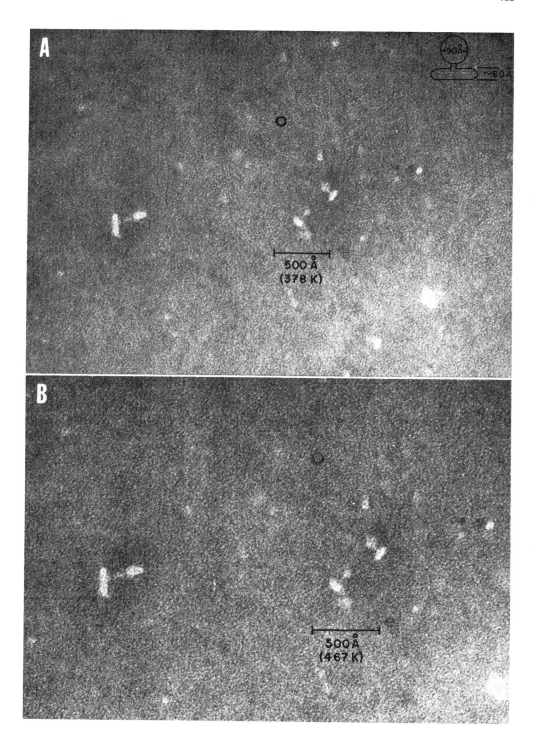

tration of > 100 µg/ml and then negatively stained frequently show
large, rather amorphous aggregates.

Of interest are the electron micrographs presented in Figure 4A and B,
which show dilute preparations of OS-ATPase negatively stained and
examined at high magnification. Molecular details of some individual
OS-ATPase molecules are brought out rather clearly with the definite
suggestion of a "headpiece-stalk-basepiece" arrangement (Fig. 2A)[1].
At this stage the purity of the OS-ATPase preparation containing these
structures became very relevant. Thus, it became important to establish
whether the basepiece (membrane sector) shown in these electron micrographs consisted of "bits and pieces" of the inner membrane containing
both F_O and electron transport chain components, or whether it was an
electron transport-free entity.

Some Properties of OS-ATPase Complex Extracted with DOC[2]

Table 2 shows that an OS-ATPase complex can be extracted from P_i-
washed membrane vesicles of rat liver with 0.1 mg DOC/mg protein and
further centrifuged and filtered to remove membrane fragments. The
overall yield is nearly 4%. P_i-washed membranes reported previously
(Soper and Pedersen, 1976) have a high specific ATPase activity (21-
33 µmol min^{-1} mg^{-1}), whereas the OS-ATPase complex removed from these
membranes with DOC is much lower in activity (10-15 µmol min^{-1} mg^{-1}).
To date we have not found conditions whereby the ATPase activity of
the isolated OS-ATPase complex can be activated to the level characteristic of the starting membranes. Nevertheless, the isolated enzyme
has a specific ATPase activity comparable to that reported for a heart
preparation solubilized with cholate (Serrano et al., 1976), and it
is markedly inhibited by low concentrations of oligomycin.

Table 3 is a summary of the cytochrome and flavoprotein content of the
rat liver OS-ATPase complex purified as described in Table 2. Less
than 1% of the total cytochrome content of the starting P_i-washed inner membrane vesicle fraction remains in the OS-ATPase preparation.
Moreover, neither cytochrome oxidase nor succinic dehydrogenase activity can be detected. There is some flavin present (perhaps flavoprotein), indicating that the preparation is not entirely free of oxidation-reduction components. As yet, we have not analyzed the rat liver
OS-ATPase preparation for iron-sulfur-protein content.

The data presented in Tables 2 and 3, taken together with information
obtained from micrographs (Figs. 3 and 4), support the view that a
"headpiece-stalk-basepiece" arrangement of the OS-ATPase complex does
exist in rat-liver mitochondria. However, structures of the type shown
in Figure 2A ("headpiece-stalk") and Figure 2B ("headpiece-basepiece")
cannot be ruled out. This is because we remain ignorant of the types
of structural rearrangements that the complete OS-ATPase complex might
assume in different energy states of the inner membrane.

---

[1]In depicting models for ATP synthesis and ATP-dependent functions it is important
to know whether or not the F_O unit of the OS-ATPase spans the inner mitochondrial
membrane. The thickness of the basepiece of OS-ATPase (Fig. 2A) is ∿60 Å, indicating
that the total thickness of F_O (basepiece + stalk) is more than sufficient to span
the inner membrane.

[2]The rat liver OS-ATPase described here can be incorporated into azolect in phospholipid vesicles. The reconstituted system catalyzes an ATP-Pi exchange reaction.

Table 3. <u>Amount of electron transport chain components recovered in DOC-solubilized OS-ATPase preparation</u>

Electron transport chain component	Amount recovered in OS-ATPase preparation (%)[a]
Cytochrome a	<1
Cytochrome a_3	<1
Cytochrome b	<1
Cytochrome c	<1
Flavoprotein	<7
Cytochrome oxidase acitivity	Not detectable
Succinic dehydrogenase activity	Not detectable

[a] Recovery is based on the total amount of these components found in 4.3 mg OS-ATPase relative to the amount found in 100 mg of the P_i-washed, starting inner membrane fraction (see Table 2). Cytochrome content was calculated from absorption measurements as described by Estabrook and Holowinsky (1961). Cytochrome oxidase activity and succinic dehydrogenase activity were assayed essentially as described by Schnaitman and Greenawalt (1968).

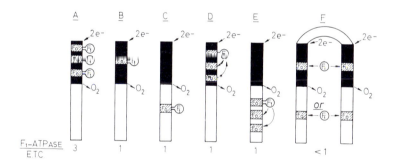

Fig. 5A–F. <u>Possible arrangements of OS-ATPase in the inner mitochondrial membrane.</u> A stoichiometry of 3 mol F_1-ATPase/mol electron transport chain (A) is not consistent with recent experimental data, whereas stoichiometries of <1 (or no greater than 1) seem most likely. (B–F) Whether or not the OS-ATPase is associated directly with the electron transport chain (B, D) or separated from it (C, E) has not been established. The stoichiometry of F_0/F_1 in the inner membrane is not known either. Should it prove to be less than 1 or should the stoichiometry of F_1-ATPase/electron transport chain prove to be less than 1, models of the type shown in (E) and (F) will have to be considered

Possible Arrangements of the OS-ATPase Complex in the Inner Membrane

Figure 5 is an attempt to summarize our current state of ignorance about the relationship between the OS-ATPase complex and the electron transport chain. It would seem that there are several questions that need to be answered correctly before investigators can begin to start making reasonably intelligent guesses about the interaction of the electron transport chain of mitochondria with the ATP-synthesizing center(s). First, there is the question of the stoichiometric relationship of the OS-ATPase complex to the electron transport chain. Secondly, there is the question of whether or not the OS-ATPase complex is physically separated from the electron transport chain or tightly associated with it.

The "stoichiometry" question is relevant to early chemical views of oxidative phosphorylation that depict an ATP-synthesizing center at each of the three energy conserving sites, suggesting an OS-ATPase/ electron transport chain stoichiometry of 3 (Fig. 5A). Both the stoichiometry" question and the "physical separation or association" question are germane to the chemiosmotic view of oxidative phosphorylation. This view does not require a stoichiometry of OS-ATPase/electron transport chain of greater than 1, nor does it require that the electron transport chain be closely associated with the OS-ATPase complex. In fact, most representations of the chemiosmotic view of oxidative phosphorylation depict the OS-ATPase as being physically separated from the electron transport chain (Mitchell, 1966; Hinkle and McCarty, 1978). It should be mentioned, however, that the chemiosmotic view of oxidative phosphorylation could still survive with 1 (or more) OS-ATPase molecule directly associated with each electron transport chain as indicated in Figure 5B.

To date we have very little information bearing on the question of interactions (or lack of interactions) between the OS-ATPase complex and the electron transport chain. We have been able to prepare an OS-ATPase complex from rat liver in a form free of most cytochromes (Table 3), but some oxidation-reduction components (predominantly flavin) still remain. Whether this residual redox material represents a "real" contaminant or an expression of a previous interaction of the OS-ATPase complex with electron transport chain components remains to be established.

The question concerning the OS-ATPase/electron transport chain stoichiometry has been addressed in this laboratory (see below).

Stoichiometric Relationship of ATPase to the Electron Transport Chain

A number of investigators have indicated that the F_1 component of mitochondrial ATPase comprises about 10% of the mass of the inner membrane (for reference see Ferguson et al., 1976). In agreement with these investigators we calculate from either specific activity data, or from gel electrophoresis data that F_1-ATPase comprises about 10% of purified inner membrane vesicles of rat liver (Table 4). Calculations based on reconstitution of purified F_1-ATPase with F_1-depleted urea particles result in a somewhat higher value of 16%, but are more subject to error.

It is known that the inner mitochondrial membrane comprises about 20% of the total protein content of rat liver mitochondria (Chan et al., 1970). Moreover, the cytochrome contents of this organelle are documented (Estabrook and Holowinsky, 1961). With this information at hand, and the additional information that the rat liver F_1-ATPase has a molecular weight of about 384,000 (Catterall and Pedersen, 1971), the F_1-ATPase/cytochrome stoichiometry can be calculated. As shown in Table 4 this stoichiometry ranges from 0.24 - 0.43 when based on cytochrome c and from 0.48 - 0.86 when based on cytochrome b. It would appear, therefore, that in rat-liver mitochondria the stoichiometric relationship of F_1-ATPase to individual cytochromes may be no greater than one and that the ratio may vary from one cytochrome to the other. (Although we have not calculated the stoichiometric ratio of ATPase to complex 1 or coenzyme Q, it can be predicted that these values would differ from values for the cytochromes.)

Table 4. Stoichiometric ratio of F_1-ATPase to cytochromes in rat liver mitochondria

Method of estimation	Amount of F_1 comprising the inner membrane (%)	F_1/ Cytochrome c[d]	F_1/ Cytochrome b[d]
From specific activity data[a]	9	0.24	0.48
From SDS gel data[b]	10	0.26	0.52
From reconstitution data[c]	16	0.43	0.86

[a]Catterall and Pedersen (1971).

[b]Watt and Pedersen, unpublished data. Based on amount of α and β bands comprising the inner membrane.

[c]Pedersen and Hullihen (1978a and b).

[d]Values for cytochrome content were taken from Estabrook and Holowinsky (1961). The molecular weight of the rat liver F_1 was taken as 384,000 (Catterall and Pedersen, 1971), and the amount of inner membrane comprising mitochondria of rat liver was taken as 20% (Chan et al., 1970).

Although earlier work from Slater's lab suggested an ATPase/electron transport chain stoichiometry of 1 on the basis of aurovertin and an- timycin A binding data (Bertina et al., 1973), this same lab has re- cently indicated that the aurovertin data may be difficult to inter- pret (Muller et al., 1977). This leaves the data of Ferguson et al. (1976), which are based on antimycin A and 4-chloro-7-nitrobenzofurazan titrations, as the only remaining evidence to our knowledge for a stoi- chiometry of ATPase/electron transport chain (or more specifically ATPase/cytochrome bc_1 complex) of 1.

Conclusions and Unanswered Questions

In summary, we have purified an OS-ATPase complex from rat liver. The enzyme complex has a very low (<1%) cytochrome content but still con- tains a significant amount of flavin (<7%). At low protein concentra- tion (<100 µg/ml) negatively stained OS-ATPase preparations appear highly dispersed whereas more concentrated OS-ATPase preparations (>100 µg/ml) appear aggregated. The rat liver OS-ATPase preparation when viewed at high magnification clearly shows some molecules that have a headpiece-stalk-basepiece arrangement. The basepiece-stalk or F_0 unit (assuming near homogeneity) is of sufficient thickness to span the inner mitochondrial membrane. The stoichiometry of the F_1-compon- ent of OS-ATPase per cytochrome c or per cytochrome b is less than 1 in rat liver mitochondria. It is suggested that the stoichiometric ra- tio of F_1-ATPase/e^- transport chain component may vary from one compo- nent to the other.

Important questions remaining unanswered at this time concern: (1) A better knowledge of what constitutes a structurally "complete" OS- ATPase complex in animal cells; (2) information about the physical in- teraction (or lack of such an interaction) of OS-ATPase complexes with the electron transport chain; and (3) the stoichiometric relationship of F_1 to F_0 in the membrane.

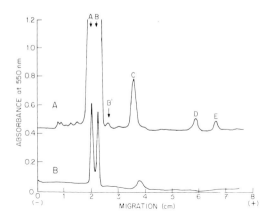

Fig. 6. Electrophoretic analysis of F₁-ATPase of rat liver in SDS-mercaptoethanol.
(From Catterall et al., 1973). F₁-ATPase (*trace A*, 80 μg; *trace B*, 4 μg) was treated
with 8 M urea, 1% SDS, 1% mercaptoethanol and subjected to electrophoresis at 8 mamp
per tube in the direction of the anode. At high protein concentration 6 bands (A,
B, B', C, D, and E) are detected. The B' band has been largely ignored in the past
but its consistent appearance from preparation to preparation merits consideration.
The B' band (or a band of mobility similar to it) has been noted also in heart F₁
(Sadler et al., 1974; Serrano et al., 1976; Gomez-Puyou and Gomez-Puyou, 1977)

Structure of the F₁-ATPase Component of OS-ATPase

Number of Polypeptide Chains

Figure 6 shows that the F₁-ATPase preparation of rat liver is composed
of six different polypeptide chains as revealed by SDS gel electrophoresis. These chains are designated A, B, B', C, D, and E and have molecular weights of about 62,000, 57,000, 53,000, 36,000, 12,500, and
7,500, respectively. In the past the B' band, which comprises only a
small amount of the total staining intensity of liver F₁, has been
largely ignored by us and by other investigators working on animal enzymes. However, we now believe that its consistent appearance in almost one hundred F₁-ATPase preparations (when high concentrations of
protein are placed on electrophoretic gel columns) is reason for concern. We are currently investigating the B' band to establish whether
it is a contaminant, a breakdown product of subunits A or B, or a poly-
meric form of the three small subunits. Significantly, it has been reported recently that the B' band is present also in bovine heart F₁-
ATPase prepared by affinity chromatography (Gomez-Puyou and Gomez-Puyou,
1977). Moreover, a band migrating with a mobility similar to that of B'
has been observed in two different OS-ATPase preparations of heart
(Sadler et al., 1974; Serrano et al., 1976).

Some bovine heart ATPase preparations are characterized by an additional polypeptide chain which migrates between the D and E bands in SDS
gels (Brooks and Senior, 1971; Moudrianakis and Barnes, 1978). This
polypeptide chain has been shown to be a potent inhibitor of ATPase
activity (Brooks and Senior, 1971; Moudrianakis and Barnes, 1978). As
yet we have not isolated a form of the liver F₁-ATPase containing a
polypeptide chain with a mobility equal to that of the ATPase inhibitor
of heart F₁.[3]

---

[3]The ATPase inhibitor peptide prepared from this laboratory from rat liver mitochondria
(Fig. 15) has an amino acid composition very similar to the Pullman and Monroy inhibitor. However, it is slightly larger than the latter inhibitor and comigrates with
the D subunit (12,500 dalton) of rat liver F₁.

Table 5. Stoichiometries suggested for F_1 or F_1-like ATPase preparations[a]

A$_3$B$_3$CDE

A$_2$B$_2$C$_2$DE

A$_2$B$_2$C$_2$E$_2$D$_{1-2}$

A$_2$B$_2$C$_2$D$_x$E$_2$

A$_2$B$_2$CDE$_2$

A$_2$B$_x$C$_2$D$_x$E$_2$

[a]For references refer to Amzel and Pedersen (1978).

Stoichiometries of These Chains Suggested to Data

Table 5 is a summary of a number of stoichiometries suggested for F_1 or F_1-like ATPases to date. All investigators are in agreement that there are 2 or 3 of the larger chains and 1 or 2 of the smaller chains. We proposed a stoichiometry of A$_3$B$_3$CDE for the rat liver enzyme prepared in this laboratory (Catterall and Pedersen, 1971; Catterall et al., 1973). This stoichiometry had been proposed about the same time by Senior and Brooks (1971). The A$_3$B$_3$CDE stoichiometry has been found

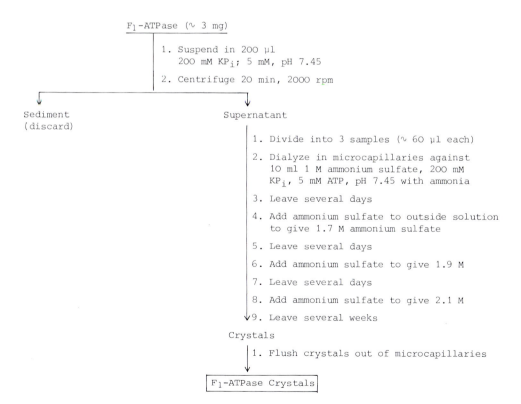

Fig. 7. Method used by Dr. Mario Amzel to crystallize the F_1-ATPase preparation of rat liver (Catterall and Pedersen, 1971) in a form suitable for X-ray diffraction studies. (See also Amzel and Pedersen, 1978)

172

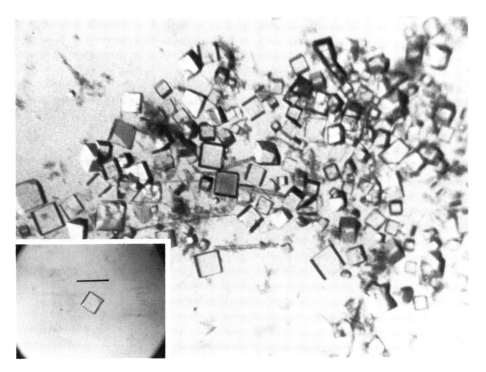

Fig. 8. Photomicrographs of crystals of F₁-ATPase from rat-liver mitochondria. The *inset* shows an enlargement of a large crystal. The *bar* shown represents 1 mm in the scale of the *inset*. (From Amzel and Pedersen, 1978)

acceptable by some workers, whereas other workers including Senior (1975) have proposed modifications on the basis of more recent experimental evidence.

Because most techniques to establish the stoichiometric relationship among subunits of proteins have a number of shortcomings, we decided to resort to the most acceptable and reliable method for obtaining structural information about proteins, namely, single crystal X-ray diffraction studies.

Crystallization of F₁-ATPase of Rat Liver

F₁-ATPase of rat-liver mitochondria (preparation of Catterall and Pedersen, 1971) has been crystallized by an ammonium sulfate precipitation technique exactly as described in Figure 7 (see Amzel and Pedersen, 1978). It is perhaps important to note that during the initial phases of the crystallization process the enzyme was maintained in an environment containing KP_i and ATP. These two compounds are also components of reconstitution assays that result in maximal coupling of ATP synthesis to electron transport (Pedersen and Hullihen, 1978a,b). Thus, the enzyme has been crystallized under conditions where it is known to be maximally active as a coupling factor.

Properties of F₁-ATPase Crystals

Figure 8 shows that the crystals of rat-liver F₁-ATPase are small cubes of approximately 0.3 - 0.6 mm per side. However, some of the crystals

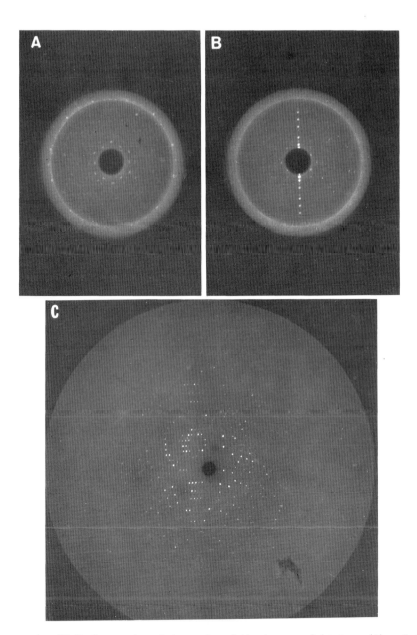

Fig. 9A-C. Precession photographs of the hexagonal hα zone (A) and the hhl zone (B). Photographs were recorded using nickel-filtered CuKa radiation obtained from a sealed X-ray tube. The precession angles were 7°. The hh0 reflections lie in the equatorial line. Observed reflections in both photographs satisfy the condition -h + k + l = 3n. This condition and the presence of the 2-fold axes define the space groups as R32. The space group was confirmed by recording several rhombohedral zones. (C) Oscillation diffraction diagram of F_1-ATPase. The photograph was recorded using an Enraf-Nonius oscillation camera. The crystal-to-film distance was 8.8 cm and the oscillation angle was 0.6°. The X-ray source was an Elliott GX6 rotating anode generator operated at 40 kV and 67mA with a 200-μm focal spot. The beam was focused using two bent mirrors. The total time of exposure was 6 h. The diffraction pattern has measurable data recorded up to a resolution of 3.5 Å. (From Amzel and Pedersen, 1978)

174

Table 6. Some properties of crystals of F$_1$-ATPase of Rat Liver[a]

───

1. Formed in a solution containing ATP and P$_i$ by precipitating with ammonium
    sulfate

2. Crystal size: 0.5 - 1 mm

3. Space group (R32) = rhombohedral

4. Hexagonal "unit" cell dimensions: a = 147.6 Å, C = 365.9 Å

5. Diffract to at least 3.5 Å resolution

6. Molecular weight of the asymmetric unit is 190,000 or about half the molecular
    weigth (384,000) of rat liver F$_1$.

───

[a]From Amzel and Pedersen (1978).

have one dimension smaller than the other two and the corresponding
face is not perpendicular to the other faces. Unlike microcrystalline
preparations reported for F$_1$-ATPase of bovine heart mitochondria
(Spitsberg and Haworth, 1977) and the TF$_1$ portion of the equivalent
enzyme from thermophilic bacterium (Kagawa et al., 1976), the crystals
of rat-liver F$_1$ are large and therefore suitable for X-ray diffraction
studies.

X-ray precession photographs (Fig. 9A and 9B) show that the crystals
are rhombohedral, space group R32 (D$_3$^{7} N° 155) with hexagonal cell
dimensions a$_{hex}$ = 148 Å, c$_{hex}$ = 368 Å. The ratio (c/a)$_{hex}$ in the ATP-
ase crystals is close to $\sqrt{6}$ (2.482). A hexagonal lattice with (c/a) =
$\sqrt{6}$ corresponds to a rhombohedral lattice with α = 60° and a$_{rhom}$ = a$_{hex}$.
The lattice points of such a rhombohedral crystal coincide with those
of a face-centered cubic (fcc) crystal with a$_{fcc}$ = $\sqrt{2}$ a$_{rhom}$. For F$_1$-
ATPase the predominant faces of the crystal corresponds to the "pseudo"
(100)$_{fcc}$ planes. Precession photographs with the X-ray beam perpendicular to these faces show the expected distribution of lattice points
with an interaxial angle of 89° 20'.

To determine the content of the asymmetric unit, the density of the
crystals was determined using several techniques (Amzel and Pedersen,
1978). The density values obtained from the different determinations
for crystals equilibrated with water (D$_o$) range from 1.210 g/cm^3 to
1.219 g/cm^3 with an average value of 1.216 g/cm^3. Using a value of
0.74 cm^3 for the partial molar volume of F$_1$ (Catterall and Pedersen,
1971), a value of 190,000 is obtained for the molecular weight of the
asymmetric unit. The molecular weight of F$_1$-ATPase from rat liver mitochondria was estimated to be 384,000 (Catterall and Pedersen, 1971).
These values seem to indicate that the asymmetric unit of the F$_1$-ATPase
crystals contains one-half of a complete F$_1$ complex, indicating that
a molecular twofold symmetry axis coincides with the crystallographic
twofold axis of space group R32.

A summary of the properties of rat liver F$_1$-ATPase crystals is presented in Table 6.

Conclusions and Unanswered Questions

We do not know the exact number of polypeptide chains which characterize animal F$_1$-ATPases, nor do we know with certainty the stoichiometric
relationships among subunits. Most workers agree that subunits A, B, C,
D, and E exist. However, what about B'? Is it a real subunit intrinsic

Table 7. <u>Functions ascribed to mitochondrial ATPase (Mitochondrial ATPase is at minimum a bifunctional enzyme)</u>

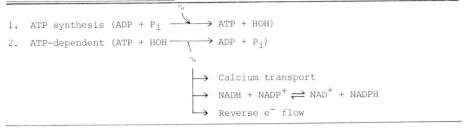

1. ATP synthesis (ADP + P_i ⟶ ATP + HOH)

2. ATP-dependent (ATP + HOH ⟶ ADP + P_i)

                ⟶ Calcium transport

                ⟶ NADH + $NADP^+$ ⇌ NAD^+ + NADPH

                ⟶ Reverse e^- flow

to F_1, or is it a contaminant or breakdown product? Why is B' observed in a number of animal ATPase preparations? Also, are some of the smaller subunits derived from the larger subunits, perhaps by specific prote-ases? Finally, is the ATPase inhibitor peptide a component of F_1 under physiological conditions? These questions must be answered before we can establish with accuracy the structure of F_1-ATPase.

It is of interest that the molecular weight of the asymmetric unit of F_1-ATPase of rat liver is 190,000, indicating that the asymmetric unit of the enzyme contains one-half of a complete F_1 complex. Although the implications of this symmetry for the subunit stoichiometry of F_1-ATPase may appear obvious, we believe it may be premature to speculate about possible subunit stoichiometries at this time. This is both be-cause of the questions raised above and because of the seemingly con-tradictory experimental evidence in the literature for the stoichio-metry of F_1 or F_1-like ATPases (Table 5).

Functions of the F_1-ATPase Component of OS-ATPase

OS-ATPase Is at the Minimum a Bifunctional Enzyme

In an earlier review article (Pedersen, 1975a) emphasis was placed on the fact that OS-ATPase complexes of animal cells are at the minimum bifunctional enzymes; i.e., they can participate not only in ATP syn-thesis but in ATP-dependent functions as well. In mitochondria, for example, *the OS-ATPase complex can participate in at least four known functions: ATP synthesis; ATP-dependent reduction of $NADP^+$ by NADH (transhydrogenase); ATP-dependent Ca^{2+} uptake; and ATP-dependent reverse e^- flow* (Table 7). There may be other hitherto unknown functions of OS-ATPases that have not been identified to date. One case in point may relate to *the ATP + AMP ⇌ 2ADP transphosphorylation reaction* catalyzed by this class of enzymes (Roy and Moudrianakis, 1971; Moudrianakis and Tiefert, 1976; Tiefert et al., 1977). [It should be noted that the word "bifunctional" is used here in a general sense to indicate that OS-ATPases catalyze both ATP syn-thesis (one function) and ATP-dependent activities (taken collectively as a second type of function). Perhaps even a more appropriate descrip-tion of OS-ATPases than "bifunctional" would be "bidirectional, multi-functional complexes".]

Convincing evidence that the F_1 component of rat-liver OS-ATPase par-ticipates in both ATP synthesis and in ATP-dependent functions has been published only recently (Pedersen and Hullihen, 1978a,b). F_1 can be removed from purified inner membrane vesicles by treatment with

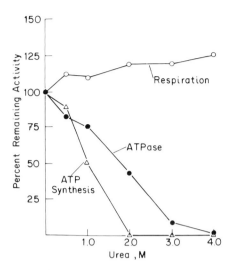

Fig. 10. Loss of ATPase and ATP synthetic activities in inner membrane vesicles after urea treatment. A concentration of urea in 1 ml of water was added to 60 mg of inner membrane vesicles in 1 ml of isolation medium to give the final urea concentrations indicated in the figure. The mixture, after sitting on ice for 5 min, was diluted to 10 ml with isolation medium and centrifuged at 48,000 rpm for 30 min in the Spinco 65 rotor. The sediment was suspended in 8.0 ml of isolation medium and centrifuged again at 48,000 rpm. The final sediment was then suspended in 1 ml of isolation medium and assayed for ATPase activity and ATP synthetic activity by the spectrophotometric procedures. Respiration supported by succinate was assayed in the presence of 33 mM DNP in the medium. The specific ATPase activity of the starting inner membrane vesicle preparation was 2.4 μmol of ATP hydrolyzed min^{-1} mg^{-1}; the specific ATP synthetic activity was 120 nmol of ATP synthesized min^{-1} mg^{-1}; and the specific rate of respiration in the presence of DNP was 125 natoms of oxygen consumed min^{-1} mg^{-1}. Experimental data have been expressed in the figure as a per cent of these original values. All data points represent averages of duplicate experiments. At 4.0 M urea 15% of the total starting inner membrane protein is solubilized and 85% remains sedimentable at 48,000 rpm. (From Pedersen and Hullihen, 1978a,b)

urea without altering the electron transport capacity of the vesicles. Loss of F_1 is accompanied by loss of ATPase activity, ATP synthetic activity, and ATP-dependent activities (Figs. 10 and 11). The purified rat-liver F_1-ATPase preparation of Catterall and Pedersen (1971) readily binds to the F_1-depleted membrane vesicles and restores both ATP synthetic capacity and the capacity of the vesicles to participate in ATP-dependent functions (Table 8). These activities are restored optimally when reconstitutions of F_1 with urea particles is carried out in the presence of ATP or AMP-PNP.

Because AMP-PNP is a potent inhibitor of ATPase activity and because ATP can serve as a negative effector of this activity (see Table 11 and below), it is possible that the form of F_1 most active in ATP synthesis is a "low ATPase activity" form of the enzyme. This point was emphasized earlier by Sanadi and co-workers (Andreoli et al., 1965) in their work on F_A (factor A), a "low ATPase activity" form of F_1. Additional support for this view in more recent years has come from the isolation of other low ATPase activity forms of F_1, which contain an ATPase inhibitor peptide in addition to the five different polypeptides usually characteristic of F_1 preparations (Brooks and Senior, 1971; Moudrianakis and Barnes, 1978). As will be noted below, the action of AMP-PNP tends to mimic the action of ATPase peptide inhibitors.

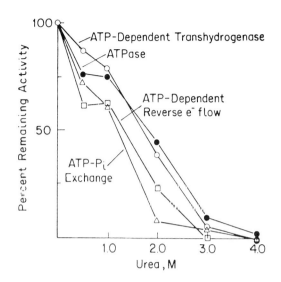

Fig. 11. Loss of ATP-dependent activities in inner membrane vesicles after urea treatment. Inner membrane vesicles were treated with urea at the concentrations indicated, exactly as described in the legend to Figure 10. After sedimentating the membranes at 48,000 rpm for 30 min and washing once with 8.0 ml of isolation medium, the sediment was suspended in 1.0 ml of isolation medium and assayed for ATP dependent activities. Specific activities of starting inner membrane vesicles were as follows. ATPase, 2.5 μmol of ATP hydrolyzed $min^{-1} mg^{-1}$; ATP-dependent trans-hydrogenase, 66 nmol of NADPH formed $min^{-1} mg^{-1}$; ATP-dependent reverse electron flow, 59 nmol of NADH formed $min^{-1} mg^{-1}$; ATP-P_i exchange, 120 nmol $min^{-1} mg^{-1}$. Experimental data have been expressed in the figure as a per cent of these original values. All data points represent averages of duplicate experiments. (From Pedersen and Hullihen, 1978)

Table 8. Evidence that F_1-ATPase of rat liver (Catterall and Pedersen, 1971, preparation) is competent in recoupling ATP synthesis and ATP-dependent functions.

Urea particles containing only 8% of the ATPase activity of the starting inner membrane vesicles were prepared exactly as described by Pedersen and Hullihen (1978a). F_1-restored urea particles were prepared exactly as described in the same reference using 0.5 mg urea particles and 130 μg purified F_1-ATPase. ATP (or AMP-PNP), 4 mM, was included in the reconstitution system. For assay details see reference noted above.

Function	Specific activity (nmol $min^{-1} mg^{-1}$)[a]		
	Starting inner membrane vesicles	Urea particles	F_1-restored urea particles
ATP synthesis	221–291	1	183–220
ATP-dependent trans-hydrogenase	62	0	44
ATP-dependent reverse e⁻ flow	80	1	52

[a] From Pedersen and Hullihen (1978a).

Table 9. <u>Nucleotide triphosphate specificity of some ATP-dependent reactions</u> of rat liver inner membrane vesicles[a]

Reaction	NTP specificity
ATP-P_i exchange	ATP[b]
ATP hydrolysis	ATP > ITP $\cong$ GTP
ATP-dependent transhydrogenase	ATP $\cong$ ITP > GTP

[a]From Pedersen (1976); All assays were conducted on inverted "Lubrol WX" inner membrane vesicles.

[b]ITP and GTP were less than 10% as effective as ATP in the assay employed.

Table 10. K_m(ATP) values for some ATP-dependent reactions in rat liver inner membrane vesicles[a]

Reaction	K_m(ATP)	
	TrisCl buffer	TrisHCO buffer
ATP-P_i exchange	1.0 mM	1.4 mM
ATP hydrolysis	0.068 mM, 0.21 mM	0.076 mM
ATP-dependent transhydrogenase	0.034 mM	0.054 mM

[a]From Pedersen (1976); All assays were conducted on "Lubrol WX" inner membrane vesicles.

AMP-PNP (or ATP), by binding at a regulatory site, may maintain the F_1-like molecule in a form maximally active in ATP synthesis and minimally active in ATP hydrolysis.

Some Experimental Data That Must Be Explained Within the Context of Any Functionally Complete Model for the Mechanism of Action of OS-ATPase

The first steps in the resolution of the molecular mechanism of action of OS-ATPase is to *recognize* that such enzymes are at the minimum *bifunctional* in nature. They participate both in ATP synthesis and in ATP-dependent functions. This point is reemphasized because most workers in the field who depict models for the mechanism of action of OS-ATPases focus exclusively on ATP synthesis with the underlying implication that the site(s) involved in ATP-dependent functions (or ATP hydrolysis) *are identical in every respect* to the sites involved in ATP synthesis. Another assumption frequently made by the same set of investigators is that the site catalyzing ATP hydrolysis on isolated F_1 or F_1-like ATPases is identical to the site involved in ATP synthesis. These implications or assumptions may be correct *but in no way have they been proven*. In fact, as will be emphasized below, there is much data accumulating in the literature which makes it exceptionally difficult to assume that a single, reversible site on OS-ATPases can account for both ATP synthesis and ATP-dependent functions without invoking some special type of regulatory mechanism.

Table 9 shows that the ATP-P_i exchange reaction (a partial reaction of oxidative phosphorylation) catalyzed by inner membrane vesicles of rat liver is highly specific for ATP whereas both CTP and ITP will substi-

Table 11. <u>Additional experimental data that must be explained within the context</u>
<u>of any functional model for mitochondrial ATPase</u>

1. A phosphorylated form of mitochondrial ATPase has never been isolated.

2. Added Mg^{2+} is not essential for intact liver mitochondria (or mitoplasts)
    to catalyze significant rates of ATP synthesis, ATP-P_i exchange, and ATP-ADP
    exchange (suggesting a role for tightly bound Mg^{2+}).[a]

3. ADP is required for maximal ATP-P_i[b] and P_i-H_2O exchange rates[c,d] in submito-
    chondrial particles. IDP does not substitute for ADP.[c]

4.[*] ATP hydrolysis by F_1 is inhibited by ADP but not by other nucleoside diphos-
    phates.[e] ADP inhibits ITP hydrolysis, but IDP does not.[e]

5. Some ATPase inhibitor peptides inhibit ATPase activity, ATP-dependent functions,
    and ATP-P_i exchange activity, but do not inhibit ATP synthesis.[f,g,h]

6. AMP-PNP is analogous in its actions to some ATPase inhibitor preparations (see
    above)[i,j]

7. IMP-PNP inhibits ATPase activity but does not inhibit ATP-P_i exchange or ATP
    synthesis.[k]

8. GMP-PNP inhibits ATP, ITP, and GTP hydrolysis as well as ATP synthesis.[l]

9. Aurovertin (at low concentrations) inhibits ATP synthesis while the ATPase re-
    action and ATP-dependent reverse electron transport reaction remain unchanged.
    High concentrations of aurovertin inhibit the ATPase reaction.[m]

10. HCO_3^- markedly activates ATP hydrolysis but has little effect on the hydrolysis
    of ITP or GTP;[n,o] or on ATP-dependent transhydrogenase and ATP-P_i exchange.[o]

---

[a]Pedersen and Schnaitman (1969); [b] Cooper and Lehninger (1957); [c]Cooper (1965);
[d]Jones and Boyer (1969); [e]Pullman et al. (1960); [f]Pullman and Monroy (1963), [g]Asami
et al. (1970); [h]Cintrón and Pedersen (1978); [i]Penefsky (1974b); [j]Pedersen et al.
(1974); [k]Shuster et al. (1976); [l]Lardy et al. (1975); [m]See Koslov and Skulachev
(1977) for references; [n]Ebel and Lardy (1975); [o]Pedersen (1976).

[*]Under conditions where high concentrations of P_i are present GDP inhibits ATP
hydrolysis (Mitchell and Moyle, 1971).

tute for ATP in assays for ATP-dependent transhydrogenase and ATPase
activity (Pedersen, 1976). Table 10 shows that in bicarbonate buffer
(where all of the above activities and/or functions are characterized
by typical Michaelis-Menten kinetics) the K_m (ATP) for ATP-P_i exchange
is 1.4 mM whereas ATP-dependent transhydrogenase activity and ATPase
activity are characterized by much lower K_m (ATP) values, i.e., 0.076
and 0.054 mM, respectively (Pedersen, 1976).

Finally, in Table 11 are listed a number of additional pieces of data
(obtained in this laboratory, Lardy's laboratory, and other laboratories) which must be explained within the context of any functionally
"complete" model for the mechanism of action of OS-ATPases. Of interest are those experiments which show that some agents are capable of
either activating or inhibiting ATP hydrolysis wihout altering ATP
synthesis (or vice versa).

Possible Models to Explain Bifunctionally [Single Site (with regulator), Separate Site, or Site Sharing]

With the above experimental data in mind and the acceptance of the fact
that OS-ATPases are at the minimum bifunctional enzyme systems, models

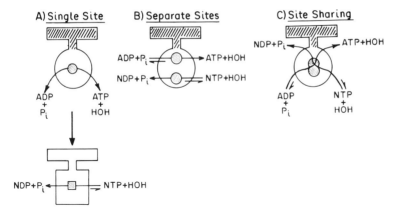

Fig. 12A-C. Models depicting how the F_1-component of OS-ATPase might participate in both ATP synthesis and ATP-dependent functions. In (A) a *single site* is seen to catalyze both activities, but a reversible change in conformational state is suggested to alter the specificity of the site involved. In (B) *two separate sites* are specialized respectively for ATP synthesis and ATP-dependent functions. Such sites could be independent but an interacting site model would be more realistic. Finally, in (C) *two overlapping nucleotide binding sites (site sharing)* are seen to be responsible for ATP synthesis and ATP-dependent activities. At this time sufficient data are not available to decide which of the three models (or some other model) is most likely. It should be noted that the models depicted here are the simplest possible. They assume for example that the sites involved in the three ATP-dependent functions (Ca^{2+} uptake, transhydrogenase, and reverse e^- flow) are the same. If this is not the case, the models will necessarily become more complex. Finally, it should be appreciated that to fully understand the mechanism of action of OS-ATPases, we must account not only for their capacity to participate in ATP synthesis but for their capacity to participate in ATP-dependent functions as well. We must also fully understand the significance of the ATP + AMP $\rightleftharpoons$ 2ADP transphosphorylation reaction characteristic of certain bacterial, chloroplast, and animal preparations (Roy and Moudrianakis, 1971; Moudrianakis and Tiefert, 1976; Tiefert et al., 1977).

It is perhaps important to note that, in addition to the references noted in the text, Leimgruber and Senior (1976) interpret recent data on nucleotide binding to bovine heart F_1 as follows: "Our results are relevant to the question of whether the ATP synthesis site(s) and the ATPase site on F_1 are structurally different. Our results indicate that they are different. At the very least, the ATP synthesis "site" or the site involved in ATP-driven coupled reactions seems to involve two regions on the F_1 which bind ADP moieties (tightly bound ADP) and one region where ATP is hydrolyzed. In this regard the term "site" is probably too restrictive, rather a "cluster" or "series" of sites seems to be involved in energy-coupled ATP synthesis or hydrolysis, whereas a single site may be all that is required for non-energy-coupled ATP hydrolysis." (See also Penefsky, 1974; Pedersen, 1975a)

can be depicted to account for their bifunctionality. Three such models are illustrated in Figure 12 and do not differ markedly from similar models emphasized by one of us (PLP) in 1975a. One model illustrates how OS-ATPases may have a single site specialized for catalyzing both ATP synthesis and NTP-dependent functions. It is important to note, however, that this model necessitates a special regulatory feature (perhaps an allosteric effector-induced conformational change to account for differences in specificity, K_m (ATP), and effect of inhibitors and activators on ATP synthesis and NTP-dependent functions (or NTP hydrolysis). A second model illustrates how separate sites might be involved, respectively, in ATP synthesis and NTP-dependent functions

(a possibility also suggested earlier by Penefsky (1974b)); and finally, a third model illustrates "site sharing" whereby the two sites
are neither completely separate nor are they completely identical. The
site-shearing model for OS-ATPase is similar (if not identical) to the
model proposed recently by Koslov and Skulachev (1977).

At this point in time, it is much too early to state which of the above
models is most likely, but we can state that such models seem to provide a rational framework on which to base future experimentation.

Conclusions and Unanswered Questions

Resolution and reconstitution experiments carried out in this laboratory have provided the first direct evidence that the F_1 component of
OS-ATPase of rat liver is at the minimum a bifunctional enzyme capable
of participating in both ATP synthesis and in ATP-dependent functions
(Pedersen and Hullihen, 1978a,b; Figs. 10 and 11, Table 3). The form
of the enzyme maximally active in ATP synthesis may be a "low ATPase"
form of F_1. This suggestion is derived from the fact that reconstitution of a maximally active ATP synthetic system is formed when the ATP-
ase inhibitor, AMP-PNP, is present (Pedersen and Hullihen, 1978a,
Table 8). Finally, the kinetic, specificity, activator, and inhibitor
data obtained in this laboratory (Pedersen, 1976; Tables 9-11) and in
other laboratories (Table 11) do not readily support the view that a
single, reversible, unregulated site on OS-ATPase of rat liver is responsible for its bifunctional nature. Rather, three models, a single-
site model subject to some special type of regulation; a separate-
site model, or a site-sharing (or overlapping site) model are suggested
as possible alternatives more consistent with available data. A chal
lenging and important aspect of future work on OS-ATPases will entail
experiments to distinguish among these models (or eliminate all of
them) in order establish a complete functional view of the mechanism
of action of this class of enzyme.

[In a recent review article Koslov and Skulachev (1977) *incorrectly state* that a
separate site model specialized respectively for ATP synthesis and ATP-dependent
functions (or ATP hydrolysis) is ruled out because the law of microreversibility
is violated. However, as emphasized in Figure 12, two separate sites can be specialized for different functions on an enzyme with each site being reversible. Also,
in a recent review article, Boyer (1977) has suggested that ATP synthesis may involve two sites, alternating in time sequence in the synthesis of ATP. See also an
earlier report by Adolfsen and Moudrianakis (1976).) The three models noted above
and depicted in Figure 12 are not at all inconsistent with an alternating catalytic
site view and could be expanded to include this view. Our purpose in depicting the
models summarized in Figure 12 is simply to emphasize that to date the vast number
of models presented to explain the capacity of OS-ATPases to catalyze ATP synthesis
have failed to account for the additional function of these enzymes in catalyzing
NTP-dependent activities (and NTP hydrolysis). A possible exception is the recent
model of Koslov and Skulachev (1977)].

Current Thoughts About the Active Site(s) of OS-ATPase of Rat Liver

Nucleotide Binding Sites — Number, Type, and Affinity

Table 12 and Figure 13 summarizes our current views about nucleotide
binding sites on the F_1 component of OS-ATPase of rat liver. Suffice
it to say here that we believe there are at least four-five distinct

Table 12. Summary of nucleotide binding properties of F_1-ATPase of rat liver

Nucleotide	Comment
ADP and ATP bound to F_1 as isolated.[a,b]	The sum (ADP + ATP) usually does not exceed 2 mol/mol F_1. Most of the bound ATP can be removed by precipitating F_1 2-3 times with ammonium sulfate.
ADP that binds *reversibly* to the isolated F_1 (1 or 2x ammonium sulfate-precipitated enzyme)[c,d]	1 mol/mol F_1. Added Mg^{2+} is not required. Assays can be conducted in the presence of EDTA. $K_d = 0.9-2.1$ μM. Other NDPs do not readily substitute for ADP. The same results are obtained with an ammonium sulfate precipitation assay or an equilibrium dialysis assay. Inhibition of ATPase activity does not prevent ADP binding. AMP-PNP does not remove all bound ADP, but almost completely inhibits ATPase activity. Bound ADP is not converted to AMP or ATP.
ATP that binds to the isolated F_1 (1 or 2x ammonium sulfate-precipitated enzyme)[c]	Exact measurements are difficult to make because, even in the absence of added Mg^{2+}, ATP is hydrolyzed to some extent. Nevertheless, in the range of 0.2 - 15 μM ATP (Mg^{2+} absent, EDTA present), some ATP is bound and about half is hydrolyzed to ADP.
AMP-PNP that binds *reversibly* to the isolated enzyme (1 or 2x ammonium sulfate-precipitated enzyme)[b]	1 mol/mol F_1 when Mg^{2+} is present and an ammonium sulfate precipitation assay is used. $K_d = 1 - 2$ μM. This value is in the same range as the K_i(AMP-PNP) for inhibition of ATPase activity. Some AMP-PNP binds in the absence of Mg^{2+}, but Mg^{2+} is required for maximal binding. Equilibrium dialysis experiments detect ∿2 more binding sites with higher K_d values. $K_d < 50$ μM.
Other ADP and ATP sites on F_1 ATPase[e,f]	ADP inhibits F_1-ATPase activity competitively with a K_i = 240 - 310 μM. ATP hydrolysis proceeds with a K_m(ATP) of ∿1 mM in TrisCl buffer and 0.12 mM in TrisHCO₃ buffer. The sites involved in the two processes appear to be looser than the sites described above. It should be noted also that in reconstitution assays (see Table 8) ATP or AMP-PNP is required to obtain maximal rates of ATP synthesis. The K_m for this effect is ∿1 mM ATP or AMP-PNP.
Conservative estimate of the total number of nucleotide binding sites on rat liver F_1	At least 4 - 5.

[a]Pedersen (1975b); [b]Pedersen (Unpublished experiments); [c]Catterall and Pedersen (1972); [d]Pedersen (1975c); [e]Catterall and Pedersen (1974); [f]Pedersen (1976).

nucleotide binding sites on rat liver F_1 which we refer to as *Site 1*, NTP hydrolytic site ; *Site 2*, ATP Mg site ; *Site 3*, Free ADP Site (ADP at this site may interact with tightly bound Mg^{2+} on F_1); *Site 4*, Very tightly bound ADP site ; and *Site 5*, ADP inhibitor site . We have not ruled out the possibility that sites 3 and 4 may be identical. Also, the "free ADP site" may bind free ATP and result in its partial hydrolysis.

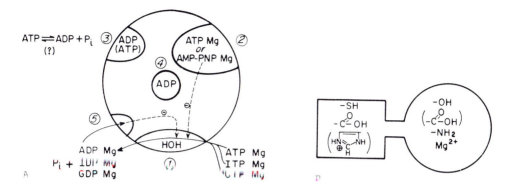

Fig. 13A and B. (A) Minimal view of nucleotide binding sites on rat liver F_1. Rat
liver F_1 hydrolyzes ATP, ITP, and GTP. In bicarbonate buffer, kinetics of hydrolysis
obey typical Michaelis-Menten kinetics for all three nucleoside triphosphates. Thus,
we suggest there is *one kinetic type* of NTP hydrolytic site designated Site 1 .
(We do not rule out the possibility that more than one equivalent type of hydrolytic
site can exist on F_1.) As isolated, the enzyme has associated with it tightly bound
ADP and ATP. The sum of ATP + ADP usually does not exceed 2 mol/mol enzyme. The ATP
can be removed by ammonium sulfate precipitation; the ADP cannot. The ATP can be replaced reversibly as AMP-PNPMg (1 mol/mol enzyme; ammonium sulfate precipitation
assay). We suggest therefore a reversible ATP Mg site designated Site 2 . Because
AMP-PNP is a potent inhibitor of ATP hydrolytic activity at concentration levels
which bind to site 2, and because velocity vs. ATP hydrolysis curves conducted in
TrisCl buffer are biphasic (suggestive of negative cooperativity), we suggest that
ATPMg or AMP-PNPMg binding at site 2 negatively effects the hydrolytic site. In
the absence of added Mg^{2+}, free ADP binds reversibly to rat liver F_1 (1 mol/mol enzyme). Addition of AMP-PNP does not readily remove bound ADP. Moreover, inhibition
of the hydrolytic site by several procedures does not alter binding of ADP. Thus,
we suggest a reversible free ADP site designated Site 3 . This site is highly
specific for ADP (or dADP). ADP may interact through tightly bound Mg^{2+} on F_1. Free
ATP also binds to F_1 in the absence of added Mg^{2+} in the same concentration range
as ADP. About half the ATP is hydrolyzed, as detected by very sensitive labeling
techniques. Site 1 therefore may not be the only site on F_1 capable of hydrolyzing
ATP. Because tightly bound ADP is associated with F_1 as isolated and cannot be removed by ammonium sulfate precipitation, we believe there may also be a very tightly
bound ADP site designated Site 4 . We have not ruled out the possiblity that sites
3 and 4 are identical. Finally, under certain conditions ADP is a rather specific
nucleoside diphosphate inhibitor of NTP hydrolysis. The K_i for ADP inhibition of NTP
hydrolysis is much greater than the K_d of ADP binding to site 3. We therefore suggest
an ADP inhibitor site designated Site 5 . There may be additional nucleotide binding sites on F_1. Equilibrium dialysis assays using ATP-depleted F_1 detect, in addition to the 1 mol of AMP-PNPMg bound to site 2, one - two additional mol/mol enzyme.
(For references refer to Table 12).

To date it has not been established which of the five classes of nucleotide binding
sites are involved in ATP synthesis and which are involved in ATP-dependent functions.
The usual assumption that site 1, the Mg^{2+}-dependent NTP hydrolytic site, is involved
in ATP synthesis has not been proven. Moreover, the recent view (see text) that
ATP synthesis may occur by an alternating catalytic site mechanism at the level of
F_1 has not been proven (nor is it ruled out by nucleotide binding information summarized in the figure).

(B) Possible functional groups on OS-ATPase of rat liver associated with ATP-P_i exchange activity. The suggestion of specific functional groups is based on data presented in Table 13. Relative locations of suggested functional groups are based on
other studies with isolated F_1 or isolated OS-ATPase to be presented elsewhere. Parenthesis designate those functional groups whose location on F_1 or F_O is uncertain

184

Table 13. Effect of covalent labeling agents on the ATP-P_i exchange reaction of inner membrane vesicles[a]

Covalent labeling agent	Concentration	Suggested mode of action (always debatable)	% inhibition ATP-P_i exchange
N-ethylmaleimide	5 mM	Cysteine	65
p-chloromercuribenzoate	5 mM	Cysteine	99
1-Fluoro-2,4-dinitrobenzene	5 mM	Cysteine, Amino (or both)	79
Dicyclohexylcarbodimide	5 µg/mg	Carboxyl	100
NBT chloride[b]	5 mM	Tyrosine	100
Tetranitromethane	50 nmol/mg	Tyrosine	88
Ethoxyformic anhydride	5 mM	Histidine	92
Phenylglyoxal monohydrate	5 mM	Arginine	46
2,3-Butanedione	5 mM	Arginine	31

[a]Purified, freshly prepared Lubrol inner membrane vesicles (Chan et al., 1970) were incubated for 20 min on ice with the concentration of inhibitor indicated. 10 mg of inner membrane vesicles in 0.5 ml isolation medium (see above references), pH 7.4 were used. Final volume was 1.0 ml. After incubation centrifugation was carried out at 100,000 g at 0°-4^OC for 1 h. The sediment was resuspended in 0.5 ml isolation medium and assayed for ATP-P_i exchange activity.

[b]7-chloro-4-nitrobenzo-2-oxa-1,3-diazole.

Three of these five sites (sites 2, 3, and 4) can be generally classified as "tight" ($K_d < 2.5$ µM) whereas two other sites (site 1 and 5) can be generally catergorized as "less tight" or "loose". As emphasized in Table 12 and the legend to Figure 13, there may be additional nucleotide binding sites on rat liver F_1.

The "ATPMg site" and the "ADP inhibitor site" would appear at this time to represent regulatory sites, whereas the "NTP hydrolytic site", the "free ADP site", and the "very tightly bound ADP site" are likely candidates for sites involved in functional activities (Table 12 and Fig. 13). As emphasized in the preceding section, we must establish which of these sites are directly involved in ATP synthesis and which are directly involved in ATP-dependent functions. This remains a critically important problem to resolve by future experimentation.

Possible Candidates for Essential Amino Acid Residues

Table 13 and Figure 13B summarize recent work that we have initiated to identify amino acid residues on OS-ATPase of rat liver involved in ATP synthesis. As indicated in Table 13 several reagents (each of which has been suggested in the literature to have specificity for a certain type of functional group) were tested for their capacity to inhibit the ATP-P_i exchange reaction catalyzed by inner membrane vesicles. All reagents tested produced significant inhibition of the ATP-P_i exchange reaction. These results suggest that the mechanism of ATP synthesis catalyzed by OS-ATPase may be more complex than the average garden variety enzyme. Thus, there are indications from these data and from much other work in the literature (Penefsky, 1967; for reviews see Senior, 1977; and Koslov and Skulachev, 1977) that OS-ATPases may uti-

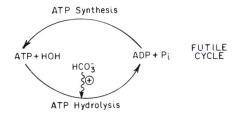

ATP Synthesis

ATP + HOH   ADP + P_i   FUTILE
CYCLE
HCO₃⁻ ⊕

ATP Hydrolysis

Fig. 14. Why there is a need to regulate the F_1-component of OS-ATPase. The rate of ATP synthesis in inner membrane vesicles of the rat liver is usually less than 0.3 μmol/min/mg (Table 8; Pedersen and Hullihen, 1978a,b), whereas the rate of ATP hydrolysis can be as high as 21-33 μmol/min/mg in the presence of bicarbonate (Table 2; Soper and Pedersen, 1976). Significantly, bicarbonate is present at high levels in the matrix space and may normally surround F_1-ATPase. If the site(s) on F_1 responsible for ATPase activity were not subject to regulation a futile cycling of newly synthesized ATP would occur

lize in their reaction mechanism for ATP synthesis as many as five amino acid residues [cysteine, histidine, glutamic acid (or aspartic acid), tyrosine, and arginine (or lysine)]. We also emphasized about ten years ago the important role of tightly bound Mg^{2+} in the catalytic mechanism (Pedersen and Schnaitman, 1969), a view that Racker (1977) now supports.

It has been of interest to try to map out the location on OS-ATPase of functional groups associated with $ATP-P_i$ exchange activity. This has been done by examining the effects of covalent labeling agents in isolated OS-ATPase, purified F_1-ATPase, and on OS-ATPase in the inner membrane (Soper and Pedersen, 1976; Table 13; and unpublished experiments). A tentative functional group map is presented in Figure 13B.

This is an exciting, new area that we have just begun to explore, and it may well shed light not only on the mechanism of ATP synthesis, but also on the question of whether ATP synthesis and ATP-dependent activities are catalyzed by the same site, separate sites, or overlapping sites (Fig. 12).

Regulation of the F_1-ATPase Component of OS-ATPase

Why There Is a Need for Regulation

For reasons that we do not fully understand rat-liver inner membrane vesicles have an enormous capacity for hydrolyzing ATP. Purified inner membrane vesicles prepared by the digitonin-lubrol method (Chan et al., 1970) have a specific ATPase activity at 25°C near 1 μmol ATP hydrolyzed min^{-1} mg^{-1}; when prepared by a digitonin-sonication technique such vesicles exhibit a specific ATPase activity at 25°C between 2.5 - 4.0 μmol ATP hydrolyzed min^{-1} mg^{-1} (Pedersen and Hullihen, 1978a,b); and when prepared by a modified digitonin-lubrol procedure and washed five times in KP_i the resultant vesicles exhibit specific ATPase activities at 37°C between 12 - 15 μmol ATP hydrolyzed min^{-1} mg^{-1} (Soper and Pedersen, 1976). In all cases, specific activities refer to values obtained in assays conducted in TrisCl buffer. Assays conducted in bicarbonate, an anion present in high concentrations in the mitochondrial matrix, result in about twofold higher values for rates of ATP hydroly-

186

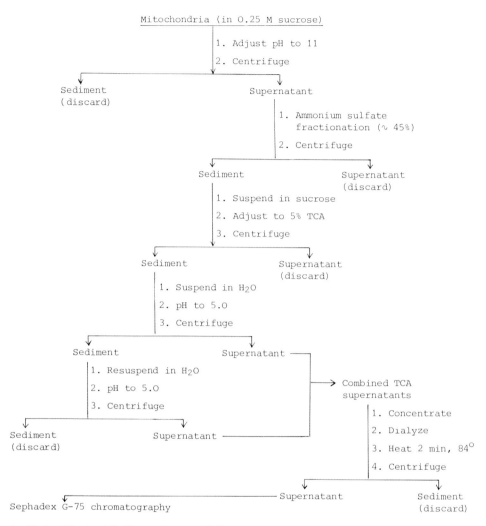

Mitochondria (in 0.25 M sucrose)

1. Adjust pH to 11
2. Centrifuge

Sediment   Supernatant
(discard)

1. Ammonium sulfate
   fractionation (∿ 45%)
2. Centrifuge

Sediment   Supernatant
                            (discard)

1. Suspend in sucrose
2. Adjust to 5% TCA
3. Centrifuge

Sediment   Supernatant
                      (discard)

1. Suspend in H_2O
2. pH to 5.0
3. Centrifuge

Sediment   Supernatant

1. Resuspend in H_2O
2. pH to 5.0   → Combined TCA
3. Centrifuge   supernatants

                                      1. Concentrate
Sediment   Supernatant   2. Dialyze
(discard)   3. Heat 2 min, 84°
                                      4. Centrifuge

                          Supernatant   Sediment
Sephadex G-75 chromatography   (discard)

1. Elute first with 20 mM KP$_i$, pH 6.7, to remove
   proteins contaminating the inhibitor. The inhibitor adheres to the column.

2. Elute next with 500 mM KP$_i$, pH 6.7

   ┌──────────┐ Specific activity = 4500-5500 units/mg
   │ INHIBITOR │ Gives single band on SDS gels
   └──────────┘

Fig. 15. Outline of scheme used in this laboratory to purify an ATPase inhibitor
peptide from rat liver mitochondria. (Cintrón and Pedersen, 1978). See text for a
definition of a "unit" of inhibitor activity

sis (Pedersen, 1976; Soper and Pedersen, 1976; for more detailed studies
of the effect of anions on rat liver ATPase see also Mitchell and Moyle,
1971; Lambeth and Lardy, 1971; Ebel and Lardy, 1975; Lardy et al.,
1975; Shuster et al., 1975; Moyle and Mitchell, 1975).

Table 14. Effect of the ATPase inhibitor peptide of rat liver mitochondria on ATP-dependent activities and ATP synthesis catalyzed by inner membrane vesicles[a]

Activity	TCA supernatant[b]
ATPase	+
ATP-dependent transhydrogenase	+
ATP-dependent reverse e$^-$ flow	+
ATP-P$_i$ exchange	+
ATP synthesis	−

[a]From Cintrón and Pedersen (1978).

[b] + = inhibition; - = no effect. All inhibitory effects observed were greater than 50%. The TCA supernatant fraction also inhibited the ATPase activity of purified F$_1$-ATPase of rat liver by greater than 50%.

If one considers that maximal rates of ATP synthesis catalyzed by inner membrane vesicles are usually less than 0.3 μmol min^{-1} mg^{-1} (Table 8; Pedersen and Hullihen, 1978), it becomes immediately clear that the F$_1$ component of OS-ATPase must be subject to some type of regulation. Otherwise, as indicated in Figure 14, a futile cycling of newly synthesized ATP may result. In intact mitochondria, where the F$_1$ component faces the bicarbonate-containing matrix, the problem of net ATP hydrolysis relative to net ATP synthesis may be compounded in the absence of effective ATPase inhibitors.

Peptide Inhibitors of ATP Hydrolysis

Figure 15 shows that an ATPase inhibitor peptide can be isolated from rat liver mitochondria (Cintrón and Pedersen, 1978). The purification scheme employed is not remarkably different from earlier schemes employed by Pullman and Monroy (1963), Horstman and Racker (1970), and Chan and Barbour (1976).[4] However, when the protein fraction remaining after the heat step (which exhibits one major band on SDS gel electrophoretic columns) is subject to gel filtration, there is a 20-25-fold enhancement of specific activity.

Table 14 emphasizes that the ATPase inhibitor peptide of rat liver, similar to AMP-PNP (Table 11), has the capacity to inhibit ATPase activity of membrane-bound F$_1$ and the capacity to inhibit ATP-dependent functions. However, the inhibitor is without effect on oxidative phosphorylation (see result in Table 14 with TCA supernatant fraction). In this regard the ATPase inhibitor of rat liver resembles the bovine heart ATPase peptide inhibitor of Pullman and Monroy (1963) (see also Asami et al., 1970).

---

[4]The rat-liver ATPase inhibitor preparation of Chan and Barbour (1976) has a specific activity of only 700 units/mg compared to 5000 units/mg for the Pullman and Monroy (1963) inhibitor and 5,000 units/mg for the Cintrón and Pedersen (1978) inhibitor. The Chan and Barbour (1976) preparation may be impure because these workers failed to use a chromatographic step. Unfortunately, they used a single "broad band" on SDS gels as the only criterion of purity. (Specific activity: One unit is that amount of protein inhibitor which results in a 50 % inhibition of 0.2 units of ATPase activity under conditions described elsewhere (Cintrón and Pedersen, 1978).)

188

Table 15. Effect of nucleotides and Mg^{2+} on ATP hydrolysis and ATP synthesis in rat liver inner membrane vesicles

Substance	ATP hydrolysis	ATP synthesis
ADP	Inhibits competitively[a]	Serves as substrate
ATP	Serves as substrate and as a negative effector[b]	Positive effector when added in reconstitution experiments[c]
Mg^{2+}	Inhibits when membranes are incubated with Mg^{2+} prior to assay[d]	Probably a substrate

[a]Catterall and Pedersen (1974); [b]Pedersen (1976); [c]Pedersen and Hullihen (1978a); [d]Moyle and Mitchell (1975).

As yet, we do not know whether the ATPase peptide inhibitor of rat liver is a "true" physiological regulator of OS-ATPase. Before we suggest too strongly that such inhibitors represent physiological regulators, we must first make sure that ATPase peptide inhibitors isolated from this and other laboratories have not arisen artifactually from the purification scheme employed, and we must ascertain whether such inhibitors can be shown to inhibit OS-ATPase at physiological pH. For reasons which remain puzzling, animal ATPase inhibitors, to exert their inhibitory effect, must be preincubated with F_1-ATPase (or membrane-bound ATPase) at pH 6.5 - 6.6 with ATP + Mg^{2+} present.

Nucleotide and Mg^{2+} Effects on ATP Synthesis and ATP Hydrolysis

As already emphasized in this report we believe that the F_1 component of OS-ATPase has at least 4 - 5 nucleotide binding sites, two of which, an "ATP Mg site" and an "ADP inhibitor site", may be regulatory sites (see above, Table 12, and Fig. 13). Nucleotide binding at these sites is thought to effect negatively the NTP hydrolytic site (Fig. 13). As indicated in Table 15, ATP (or AMP-PNP) has a positive effect on ATP synthesis when included in reconstitution experiments (F_1 + urea particles) (Pedersen and Hullihen, 1978a). Mg^{2+}, on the other hand, when added to purified or membrane-bound F_1-ATPase in the absence of ATP, results in a marked loss of ATPase activity (Catterall and Pedersen, 1972; Moyle and Mitchell, 1975).

Conclusions and Unanswered Questions

We are led to suggest on the basis of work summarized in this section that OS-ATPase of rat liver may be subject to a set of complex controls (Fig. 16). Nucleotides, in particular ATPMg and ADP, seem to be involved in one type of control mechanism designed to directly suppress ATP hydrolytic activity. At least one ATPase peptide inhibitor may be involved in a separate control process. However, we emphasize that as of this writing ATPase peptide inhibitors of animal cells have been shown to only weakly inhibit ATP hydrolysis catalyzed by F_1 or membrane-bound F_1 when added to the ATPase assay at physiological pH. Preincubation of the inhibitor with F_1 or membrane-bound F_1 at pH 6.5 - 6.6 in the presence of ATP + Mg^{2+} is required to observe optimal inhibition.

It seems safe to say that regulatory mechanisms involved in directing energy flux in mitochondria toward ATP synthesis or ATP-dependent functions are poorly understood and that most of the crucial experiments

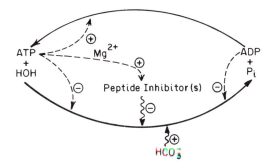

Fig. 16. Physiological effectors of ATP synthesis and ATP hydrolysis in rat liver mitochondria (see text for discussion). In particular, note that three different types of effector molecules are capable of suppressing ATPase activity

remain to be done. In particular, it will be of interest in future experiments to try to correlate changes in nucleotide binding, ATPase peptide inhibitor binding, phosphate potential, and membrane potential with changes in the functional and hydrolytic activities of OS-ATPase. Only with such data at hand will we be in a position to speak knowledgeably about the complex controls governing this complex enzyme.

Summary Statement and Direction of Future Work

The 1977 issue of the *Annual Review of Biochemistry* contains a series of articles by scientists who have been working vigorously for many years on the complex subject of oxidative phosphorylation (Boyer, 1977; Slater, 1977; Ernster, 1977; Racker, 1977; Mitchell, 1977). Unfortunately, rather than being able to agree on the basic principles underlying energy coupling and ATP synthesis each author expressed an independent view of the area. These articles dealt mainly with the subject of how energy from electron transport is coupled to ATP synthesis at the level of OS-ATPase. The main debate seemed to center around the four views summarized in Figure 17, i.e., *chemical, protonophoric, modified protonophoric*, and *electroconformational. Despite a great deal of discussion, it seemed clear that sufficient molecular and chemical information was not available to reach any major conclusions about energy coupling or ATP synthesis*. Moreover, the subject of ATP-dependent functions was almost completely neglected.

It is our view that the problem concerning the mechanism of action of OS-ATPases extends far, far beyond the views expressed in these articles. On the one hand, it is certainly important to establish at the molecular and chemical levels how energy derived from electron flow is coupled to OS-ATPases, and especially important to establish whether the electron transport chain interacts directly or indirectly with these enzyme complexes. However, this is only *one* aspect of the overall problem. A second problem is to establish *at a chemical level* how these enzymes use the energy (in whatever form) to dehydrate ADP + P_i at the level of OS-F_1. A third problem, as emphasized in Figure 12, is to establish how these enzymes also participate in ATP-dependent functions. Finally, a fourth problem is to identify those regulatory factors (be they chemical or electrochemical) which direct energy flux via OS-ATPases toward ATP synthesis or ATP-dependent functions.

190

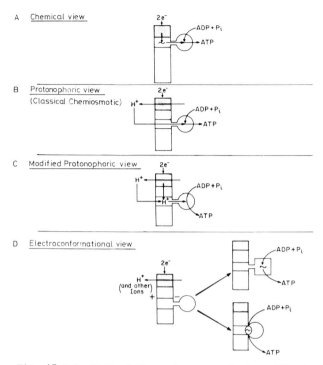

A Chemical view

B Protonophoric view
   (Classical Chemiosmotic)

C Modified Protonophoric view

D Electroconformational view

Fig. 17 A-D. Some of the major current views illustrating how energy from electron transport may be coupled to ATP synthesis. (A) Chemical View — According to this view, the electron transport chain generates a high energy chemical intermediate during electron flow which is chemically coupled to OS-ATPases in such a way as to energize P_i, ADP, or both molecules. Despite claims to the contrary, a strictly chemical view for ATP synthesis in *intact mitochondria* has not been ruled out (see text). (B) Protonophoric view — This view assumes that protons generated during electron flow (or equivalent protons) are directly involved in the dehydration of ADP and P_i at the level of OS-ATPase. (C) Modified Protonophoric view — Differs only from (B) above in that all protons are not necessarily ejected into the external space. Some may be released within the membrane, i.e., within membrane "pockets" or compartments". (D) Electroconformational view — This view states that an electrochemical gradient of protons (or protons + other ions) may drive ATP synthesis without the ions participating directly in the dehydration of ADP and P_i. The electrochemical gradient may drive a conformational change at the level of OS-ATPase, or it may simply "pull" F_1 partially into the hydrophobic phase of the membrane where dehydration of ADP and P_i may be facilitated.

The authors take no credit for the originality of these views. They have all appeared in the literature in one form or the other (see text). They are only summarized here to indicate that one of the problems involved in the elucidation of the mechanism of action of OS-ATPases is to establish how these enzymes are coupled to energy derived from electron flow. A second problem is to establish how these enzymes use the energy (in whatever form) to dehydrate ADP + P_i at the level of OS-F_1. A third problem (not illustrated here but illustrated in Fig. 12) is to establish how these enzymes also participate in ATP-dependent functions. Finally, a fourth problem is to identify those regulatory factors which direct energy flux via OS-ATPase toward ATP-synthesis or ATP-dependent functions. Future directions taken by this laboratory will entail efforts to help resolve all four problems.

Efforts in this laboratory have resulted in the isolation of the OS-ATPase complex of rat-liver mitochondria, the purification and crystallization of the F$_1$ component of this enzyme complex, and the isolation of an ATPase inhibitor peptide. Additional work has been successful in helping to define the complex subunit pattern of rat liver OS-ATPase, the number, type, and affinity of nucleotide binding sites on the enzyme complex, and the asymmetric unit of the F$_1$ component. Finally, experiments designed to probe the active site(s) of OS-ATPase of rat liver mitochondria are currently underway. We believe, therefore, that we have helped to set the stage to examine, at the molecular and chemical levels, the four important questions about OS-ATPase addressed above.

Throughout this lecture, in compliance with the organizer's suggestions, we have focused mainly on a review of work carried out in our own laboratories. New students, however, may wish to consult the following review articles for a more in-depth survey of the current state of knowledge of structural, functional, and regulatory properties of ATPases of mitochondria, chloroplasts, and bacteria (Penefsky, 1974a; Pedersen, 1975a; Simoni and Postma, 1975; Abrams, 1976; Nelson, 1976; Panet and Sanadi, 1976; Senior, 1977; Koslov and Skulachev, 1977; and Harris, 1978).

Acknowledgments. We are grateful to the National Cancer Institute (Grant CA 10951) for continued support of our work.

References

Abrams, A.: Structure and function of membrane-bound ATPase in bacteria. In The Enzymes of Biological Membranes. Martonosi, A. (ed.). New York: Plenum Publishing Corp., 1976, Vol. 3, pp. 57-73

Adolfsen, R., Moudrianakis, E.N.: Binding of adenine nucleotides to the purified 13S coupling factor of bacterial oxidative phosphorylation. Arch. Biochem. Biophys. 172, 425-433 (1976)

Amzel, L.M., Pedersen, P.L.: Adenosine triphosphatase from rat liver mitochondria, crystallization and x-ray diffraction studies of the F$_1$ component of the enzyme. J. Biol. Chem. 253, 2067-2069 (1978)

Andreoli, T.E., Lam, K-W., Sanadi, D.R.: Studies on oxidative phosphorylation. X. A coupling enzyme which activates reversed electron transfer. J. Biol. Chem. 240, 2644-2653 (1965)

Asami, K., Juntti, K., Ernster, L.: Possible regulatory function of mitchondrial ATPase inhibitor in respiratory chain-linked energy transfer. Biochim. Biophys. Acta 205, 307-311 (1970)

Bertina, R.M., Schrier, P.I., Slater, E.C.: The binding of aurovertin to mitochondria, and its effect on mitochondrial respiration. Biochim. Biophys. Acta 305, 503-518 (1978)

Boyer, P.D.: Coupling mechanisms in capture, transmission, and use of energy. Ann. Rev. Biochem. 46, 957-966 (1977)

Brooks, J.C., Senior, A.E.: Studies on mitochondrial oligomycin-insensitive ATPase. II. The relationship of the specific protein inhibitor to the ATPase. Arch. Biochem. Biophys. 147, 467-470 (1971)

Catterall, W.A., Coty, W.A., Pedersen, P.L.: Adenosine triphosphatase from rat liver mitochondria. III. Subunit composition. J. Biol. Chem. 248, 7427-7431 (1973)

Catterall, W.A., Pedersen, P.L.: Adenosine triphosphatase from rat liver mitochondria. I. Purification, homogeneity, and physical properties. J. Biol. Chem. 246, 4987-4997 (1971)

Catterall, W.A., Pedersen, P.L.: Adenosine triphosphatase from rat liver mitochondria. II. Interaction with adenosine diphosphate. J. Biol. Chem. 247, 7969-7976 (1972)

Catterall, W.A., Pedersen, P.L.: Structural and catalytic properties of mitochondrial adenosine triphosphatase. Biochem. Soc. Spec. Publ. 4, 63-88 (1974)

Chan, S.H.P., Barbour, R.L.: Purification and properties of ATPase inhibitor from rat liver mitochondria. Biochim. Biophys. Acta 430, 426-433 (1976)

Chan, T.L., Greenawalt, J.W., Pedersen, P.L.: Biochemical and ultrastructural properties of a mitochondrial inner membrane fraction deficient in outer membrane and matrix activities. J. Cell Biol. 45, 291-305 (1970)

Cintrón, N.M., Pedersen, P.L.: Isolation of an ATPase inhibitor fraction from rat liver mitochondria. Methods Enzymol., In Press (1978)

Cooper, C.: Effect of adenosine diphosphate on the exchange of oxygen between inorganic phosphate and water catalyzed by digitonin particles. Biochemistry 4, 335-342 (1965)

Cooper, C., Lehninger, A.L.: Oxidative phosphorylation by an enzyme complex from extracts of mitochondria. V. The adenosine triphosphate-phosphate exchange reaction. J. Biol. Chem. 224, 561-578 (1957)

Ebel, R.E., Lardy, H.A.: Stimulation of rat liver mitochondrial adenosine triphosphatase by anions. J. Biol. Chem. 250, 191-196 (1975)

Ernster, L.: Chemical and chemiosmotic aspects of electron transport-linked phosphorylation. Ann. Rev. Biochem. 46, 981-995 (1977)

Estabrook, R.W., Holowinsky, A.: Studies on the content and organization of the respiratory enzymes of mitochondria. J. Biophys. Biochem. Cytol. 9, 19-28 (1961)

Ferguson, S.J., Lloyd, W.J., Radda, G.K.: A method for determining the adenosine triphosphatase content of energy-transducing membranes. Biochem. J. 159, 347-353 (1976)

Gomez-Puyou, M.T.D., Gomez-Puyou, A.: A simple method of purification of a soluble oligomycin-insensitive mitochondrial ATPase. Arch. Biochem. Biophys. 182, 82-86 (1977)

Harris, D.A.: The interaction of coupling ATPases with nucleotides. Biochim. Biophys. Acta 463, 245-273 (1978)

Hinkle, P.C., McCarty, R.E.: How cells make ATP. Sci. Am. 238, 104-122 (1978)

Horstman, L.L., Racker, E.: Partial resolution of enzymes catalyzing oxidative phosphorylation. XXII. Interaction between mitochondrial adenosine triphosphatase inhibitor and mitochondrial adenosine triphosphatase. J. Biol. Chem. 245, 1336-1344 (1970)

Jones, D.H., Boyer, P.D.: The apparent absolute requirement of adenosine diphosphate for the inorganic phosphate $\rightleftharpoons$ water exchange of oxidative phosphorylation. J. Biol. Chem. 244, 5767-5772 (1969)

Kagawa, Y., Sone, N., Yoshida, M., Hirata, H., Okamoto, H.: Proton Translocating ATPase of thermophilic bacterium, morphology, subunits and chemical composition. J. Biochem. 80, 141-151 (1976)

Koslov, I.A., Skulachev, V.P.: H^+-adenosine triphosphatase and membrane energy coupling. Biochim. Biophys. Acta 463, 29-89 (1977)

Lambeth, D.O., Lardy, H.A.: Purification and properties of rat liver mitochondrial adenosine triphosphatase. Eur. J. Biochem. 22, 355-363 (1971)

Lardy, H.A., Schuster, S.M., Ebel, R.E.: Exploring sites on mitochondrial ATPase for catalysis, regulation, and inhibition. J. Supramol. Struct. 3, 214-221 (1975)

Leimbgruber, R.M., Senior, A.E.: Removal of "tightly bound" nucleotides from soluble mitochondrial adenosine triphosphatase (F_1). J. Biol. Chem. 251, 7103-7109 (1976)

Mitchell, P.: Chemiosmotic Coupling in Oxidative and Photosynthetic Phosphorylation. Glynn Research Laboratories, Bodmin, Cornwall, England, 1966, pp. 1-192

Mitchell, P.: Vectorial chemiosmotic processes. Ann. Rev. Biochem. 46, 996-1105 (1977)

Mitchell, P., Moyle, J.: Activation and inhibition of mitochondrial adenosine triphosphatase by various anions and other agents. Bioenergetics 2, 1-11 (1971)

Moudrianakis, E.N., Barnes, J.E.: Isolation of mitochondrial coupling factor F_1 in a new form, having low levels of intrinsic ATPase activity. Submitted for publication (1978)

Moudrianakis, E.N., Tiefert, M.A.: Synthesis of bound adenosine triphosphate from bound adenosine diphosphate by the purified coupling factor 1 of chloroplasts. Evidence for direct involvement of the coupling factor in this "adenylate-kinase-like" reaction. J. Biol. Chem. 251, 7796-7801 (1976)

Moyle, J., Mitchell, P.: Active/inactive state transitions of mitochondrial ATPase molecules influenced by Mg $^+$, anions, and aurovertin. FEBS Lett. 56, 55-61 (1975)

Muller, J.L.M., Rosing, J., Slater, E.C.: The binding of aurovertin to isolated F₁ (mitochondrial ATPase). Biochim. Biophys. Acta 462, 422-437 (1977)

Nelson, N.: Structure and function of chloroplast ATPase. Biochim. Biophys. Acta 456, 314-338 (1976)

Panet, R., Sanadi, D.R.: Soluble and membrane ATPase of mitochondria, chloroplasts and bacteria: Molecular structure, enzymatic properties, and functions. Curr. Top. in Membr. Transp. 8, 99-150 (1976)

Pedersen, P.L.: Mitochondrial adenosine triphosphatase. Bioenergetics 6, 243-275 (1975a)

Pedersen, P.L.: Interaction of homogeneous mitochondrial ATPase from rat liver with adenine nucleotides and inorganic phosphate. J. Supramol. Struct. 3, 222-230 (1975b)

Pedersen, P.L.: Adenosine triphosphatase from rat liver mitochondria: Separate sites involved in ATP hydrolysis and in the reversible, high affinity binding of ATP. Biochem. Biophys. Res. Commun. 64, 610-616 (1975c)

Pedersen, P.L.: ATP-dependent reaction catalyzed by inner membrane vesicles of rat liver mitochondria, kinetics, substrate specificity, and bicarbonate sensitivity. J. Biol. Chem. 251, 934-940 (1976)

Pedersen, P.L., Hullihen, J.: Adenosine triphosphatase of rat liver mitochondria, capacity of the homogeneous F₁ component of the enzyme to restore ATP synthesis in urea-treated membranes. J. Biol. Chem. 253, 2176-2183 (1978a)

Pedersen, P.L., Hullihen, J.: Resolution and reconstitution of ATP synthesis and ATP-dependent functions of liver mitochondria. Meth. Enzymol., In Press (1978)

Pedersen, P.L., Levine, H. (III), Cintrón, N.: Activation and inhibition of mitochondrial ATPase of rat liver mitochondria. In: Membrane proteins in transport and phosphorylation. Azzone, G.F. et al. (eds.). The Netherlands: North-Holland Publishing Co., 1974, pp. 43-54

Pedersen, P.L., Schnaitman, C.A.: The oligomycin-sensitive adenosine diphosphate-adenosine triphosphate exchange in an inner membrane-matrix fraction of rat liver mitochondria. J. Biol. Chem. 244, 5064-5073 (1969)

Penefsky, H.: Partial resolution of enzymes catalyzing oxidative phosphorylation. XVI. Chemical modifications of mitochondrial adenosine triphosphatase. J. Biol. Chem. 242, 5789-5795 (1967)

Penefsky, H.: Mitochondrial and Chloroplast ATPases. In: The Enzymes. Boyer, P.D., (ed.). New York: Academic Press, 1974a, Vol. 10, pp. 375-394

Penefsky, H.S.: Differential effects of adenylyl imidodiphosphate on adenosine triphosphate synthesis and the partial reactions of oxidative phosphorylation. J. Biol. Chem. 3579-3585 (1974b)

Pullman, M.E., Monroy, G.C.: A naturally occurring inhibitor of mitochondrial adenosine triphosphatase. J. Biol. Chem. 238, 3762-3769 (1963)

Pullman, M.E., Penefsky, H.S., Datta, A., Racker, E.: I. Purification and properties of soluble, dinitrophenol-stimulated adenosine triphosphatase. J. Biol. Chem. 235, 3322-3329 (1960)

Racker, E.: Mechanisms of energy transformations. Ann. Rev. Biochem. 46, 1006-1014 (1977)

Roy, H., Moudrianakis, E.: Synthesis and discharge of the coupling factor — adenosine diphosphate complex in spinach chloroplast lamellae. Proc. Natl. Acad. Sci. U.S.A. 68, 2720-2724 (1971)

Ryrie, I.J.: Reconstitution of ATP-³²Pᵢ exchange by phospholipid addition to the purified oligomycin-sensitive ATPase from yeast mitochondrial. Arch. Biochem. Biophys. 168, 704-711 (1975)

Sadler, M.H., Hunter, D.R., Haworth, R.A.: Isolation of an ATP-Pᵢ exchangease from lysolecithin-treated electron transport particles. Biochem. Biophys. Res. Commun. 59, 804-812 (1974)

Schnaitman, C.A., Greenawalt, J.W.: Enzymatic properties of the inner and outer membranes of rat liver mitochondria. J. Cell Biol. 38, 158-175 (1968)

Senior, A.E.: Mitochondrial adenosine triphosphatase: Location of sulfhydryl groups and disulfide bonds in the soluble enzyme from beef heart. Biochemistry 14, 660-664 (1975)

Senior, A.E.: The Mitochondrial ATPase. In: Membrane Proteins in Energy Transduction. Capaldi, R.A., Dekker, M. (eds.). New York: 1977, In Press

Senior, A.E., Brooks, J.C.: The subunit composition of the mitochondrial oligomycin-insensitive ATPase. FEBS Lett. 17, 327-329 (1971)

Serrano, R., Kanner, B.I., Racker, E.: Purification and properties of the proton-translocating adenosine triphosphatase complex of bovine heart mitochondria. J. Biol. Chem. 251, 2453-2461 (1976)

Shuster, S.M., Ebel, R.E., Lardy, H.A.: Kinetic studies on rat liver and beef heart mitochondrial ATPase. Evidence for nucleotide binding at separate regulatory and catalytic sites. J. Biol. Chem. 250, 7848-7853 (1975)

Shuster, S.M., Gertschen, R.J., Lardy, H.A.: Effect of Inosine 5'-(β,γ-imido) triphosphate and other nucleotides on beef heart mitochondrial ATPase. J. Biol. Chem. 251, 6705-6710 (1976)

Simoni, R.D., Postma, P.W.: Energy coupling to transport in cells and membrane vesicles. Ann. Rev. Biochem. 44, 523-554 (1975)

Slater, E.C.: Mechanism of oxidative phosphorylation. Ann. Rev. Biochem. 46, 1015-1026 (1977)

Soper, J.W., Pedersen, P.L.: Adenosine triphosphatase of rat liver mitochondria: Detergent solubilization of an oligomycin- and dicyclohexyl-carbodiimide sensitive form of the enzyme. Biochemistry 15, 2682-2690 (1976)

Spitsberg, V., Haworth, R.: The crystallization of beef heart mitochondrial adenosine triphosphatase. Biochim. Biophys. Acta 492, 237-240 (1977)

Stiggall, D.L., Galante, Y.M., Hatefi, Y.: Preparation and properties of an ATP-P$_i$ exchange complex (Complex V) from bovine heart mitochondria. J. Biol. Chem. 253, 956-964 (1978)

Tiefert, M.A., Roy, H., Moudrianakis, E.N.: Conversion of bound adenine nucleotides by the purified coupling factor of photophosphorylation. Biochemistry 16, 2404-2409 (1977)

Tzagoloff, A., Meager, P.: Assembly of the mitochondrial membrane system. VI. Mitochondrial synthesis of subunit proteins of the rutamycin-sensitive adenosine triphosphatase of yeast mitochondria. J. Biol. Chem. 246, 7328-7336 (1971)

Protonophoric Action of Coupling Factor ATPase – Its Reconstitution from Purified Subunits, Reconstitution Into ATP-Synthesizing Vesicles, and Image Reconstruction

Y. KAGAWA

Introduction

I would like to express our thanks to Germany briefly in German.

Zur wissenschaftlichen Entwicklung bis zum heutigen Stand in Japan hat die große deutsche Wissenschaft viel beigetragen. In unserem Gedächtnis sind zahlreiche deutsche Wissenschaftler behalten, die uns Japaner bisher viel gelehrt haben, unter anderen P. F. von Siebold, der 1823 in Japan eine medizinische Schule gegründet hat, und Prof. Leonor Michaelis, der 1922 in Japan Biochemie gelehrt hat, und viele andere. Auch heute hat man in Japan immer noch das größte Interesse für die deutsche Wissenschaft. Bei dieser Gelegenheit möchte ich den deutschen Wissenschaftlern für ihre Freundlichkeit uns gegenüber unseren herzlichsten Dank sagen.

The protonophoric action of H^+-translocating ATPase $|EC\ 3.6.1.3|$ (H^+-ATPase) is essential for oxidative and photophosphorylation. Most biological energy is derived from biomembranes consisting mainly of both H^+-ATPase and an electron transport system. The function of the electron transport system is to translocate H^+ across the membrane and thereby generate an electrochemical potential of H^+ ($\Delta\bar{\mu}H^+$), which consists of difference in pH ($Z\Delta pH$) and electrical potential ($\Delta\Psi$) across the membrane (Mitchell, 1976; Witt, 1976, Boyer et al., 1977). The $\Delta\bar{\mu}H^+$ is utilized for ATP synthesis by the H^+-ATPase in the membrane operating in reverse (Sone et al., 1977; Koslov and Skulachev, 1977; Kagawa, 1978; Witt et al., 1976). A cruce preparation of this ATPase was isolated for the first time from beef heart mitochondria (Kagawa and Racker, 1966b) and was shown to be composed of a catalytic portion (F_1) and a hydrophobic portion (F_O) that renders F_1 sensitive to energy transfer inhibitors (Kagawa and Racker, 1966a,b).

The translocating activity of H^+ can be measured only after the ATPase ($F_O \cdot F_1$) is reconstituted into lipid bilayers (Kagawa, 1972). However, further purification and reconstitution from its individual subunits have been difficult, since in mesophiles, the membrane system for these functions is highly unstable during such drastic procedures. For this reason, a stable ATPase was prepared from the thermophilic bacterium PS3, which was isolated from a Japanese hot spring. The use of the thermophiles enabled us to achieve the following results.

1. Crystallization of F_1 (Kagawa, 1976; Kagawa et al., 1976) and computerized image reconstruction of the F_1 molecule (Wakabayashi et al., 1977).

2. Complete reconstruction of F_1 from its five subunits ($\alpha,\beta,\gamma,\delta,\varepsilon$) (Yoshida et al., 1977b; Kagawa, 1976).

Abbreviations: CF_1, F_1 of chloroplasts F_1, coupling factor 1 (soluble ATPase); FCCP, carbonylcyanide p-trifluoromethoxyphenyl-hydrazone; F_O, oligomycin sensitivity-conferring factor; DCCD, N,N' dicyclohexylcarbodiimide; TF_1, thermophilic coupling factor 1.

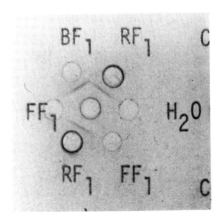

Fig. 1. Immunodiffusion profiles of F_1 from three different species. *Central well:* antiserum against F_1 of beef heart. BF_1, beef heart F_1; RF_1, rat liver F_1; FF_1, F_1 purified from a thermophilic bacterium of *Flavobacterium* species (now identified as *Thermus species*). H_2O, control without protein. The preparation of rabbit anti-serum was described in detail under the section on radioimmunoassac of F_1. (Kagawa, 1974)

3. Characterization of the channel function of F_o (Okamoto et al., 1977).

4. Identification of the gate function of the $\gamma\delta\epsilon$-complex of F_1 subunits (Yoshida et al., 1977a).

5. Determination of the molecular species of phospholipids of the thermophiles that form a stable liposome without unsaturation (Kagawa and Ariga, 1977).

6. Reconstitution of pure $F_o \cdot F_1$ into vesicles capable of H^+ translocation and exact determination of the $\Delta\bar{\mu}H^+$, ΔpH and $\Delta\Psi$ (Sone et al., 1976; Kagawa et al., 1977).

7. Net ATP synthesis driven by an artificially imposed $\Delta\bar{\mu}H^+$ in the vesicles reconstituted from purified $F_o \cdot F_1$ (Sone et al., 1977; Kagawa et al., 1977).

8. Physical studies on the F_1 and other proteins during energy transformation such as hydrogen-deuterium exchange (Ohta et al., 1978).

These results are summarized in a recent review (Kagawa, 1978).

Stable Biomembranes Suitable for the Study of H^+-ATPase

Selection Principle

1. Universal Distribution of H^+-ATPase Among Species

The isolation, reconstitution, and analysis of H^+-ATPase depended largely on the selection of suitable starting material. Since H^+-ATPase is so universal in both prokaryotic and eukaryotic cells and is essential for many kinds of functions (Harold, 1977), we have freedom to choose the starting material. In fact, F_1 portions of these H^+-ATPases are identical with 90-Å particles distributed over the surface of these membranes (Kagawa and Racker, 1966c), and they all have similar molecular weights of 3.8×10^5 (see discussions below), and can be resolved into five similar subunit polypeptides — three large (α, β and γ) and to small (δ and ϵ) (Pedersen, 1975, Knowles and Penefsky, 1972; Nelson, 1976). Moreover, F_1 from rat liver was found to be immunologically identical to F_1 of beef heart (Fig. 1, to be published). Thus the conclusions obtained from the F_1 of one species may be true for any F_1.

For mitochondrial components, the traditional methods of extraction
with cholate (Okunuki, 1966) and reconstitution with phospholipids
(Kakiuchi, 1927) have been improved upon in the studies of H^+-ATPase
(Kagawa, 1967, 1972, 1974). However, for further studies on the molecular properties of these H^+-ATPases, large quantities of stable, pure
preparations are essential, because extraction, dissociation, and reassembly of subunits and the other methods used in these studies are
drastic procedures involving great losses of material. For these reasons, thermostable, acid-stable, and alkali-stable rapidly growing
bacteria have been surveyed.

Thermophilic Versus Mesophilic Proteins

Most macromolecules in thermophilic bacteria are thermostable. These
proteins are also resistant to dissociation agents that, even at very
low concentrations, denature mesophilic counterparts. It was found
that the greater stability of the enzymes from thermophiles was due
mainly to an increased number of salt bridges (Arg-Glu, Asp-Arg, etc.).
For example, ferredoxins from several thermophilic *Clostridia* contain
more salt bridges that link residues near the N-terminus to others
near the C-terminus than those from the mesophilic *Clostridia* (Perutz
and Raidt, 1975).

Among the oligomeric proteins examined, the thermophilic tetramer D-
glyceraldehyde-3-phosphate dehydrogenase [EC 1.2.1.12] also contained
three additional salt bridges across the subunit interface (Arg 197 –
Asp 293, etc.), but its catalytic centers were essentially the same
as those of mesophilic counterparts (Biesecker et al., 1977). Usually,
a change in only one amino acid residue causes a temperature-sensitive
mutation, but some point mutations (Gly 211 to Asp, etc.) results in
a thermophilic A-subunit of tryptophan synthetase [EC 4.2.1.20] (Yutani
et al., 1977).

Thus it seems that in proteins from thermophiles, the extra energy of
stabilization is provided by a few extra salt bridges on the molecular
surface or subunit interface, without disturbance of the tertiary or
quarternary structure. In contrast to hydrogen and hydrophobic-bonding,
which are short-ranged, the electrostatic force of the salt bridges is
long range, and facilitates the reconstitution of a conformation of
proteins after complete denaturation.

Another advantage of thermophilic proteins is their very low content
of SH-groups, which will be discussed below.

Selection of Thermophilic Bacterium PS3

Several thermophilic aerobic bacteria were tested as sources of a
stable H^+-ATPase (Kagawa, 1976). All the strains tested were isolated
from Japanese hot springs and were kindly supplied by Dr. T. Oshima
of Mitsubishi Life Science Institute, Machida City, Tokyo. Strain PS3
was selected as starting material for further studies because it showed
high membrane ATPase activity (2.25 µmol Pi/min/mg protein, at 80°C),
high sensitivity to DCCD (85% at 80°C), and rapid growth: at 65°C, the
doubling time of PS3 was 35 min, as determined by measuring 600 nm absorption of a vigorously aerated medium. The medium contained 0.8%
polypeptone, 0.4% yeast extract, and 0.3% NaCl, pH 7.0.

PS3 was isolated from Mine hot spring in Shizuoka prefecture. It was
variable in gram staining and was a flagellated spore former, similar

to *Bacillus stearothermophilus* but more heat resistant. The content of the terminal oxidase (absorption maximum at 605 nm in reduced states) was only 0.6 nmol/mg protein, which was much less than the content of cytochrome *b* (4.0 nmol/mg protein) and *c* + *c₁* (2.6 nmol/mg protein) and flavin (3.1 nmol/mg protein); this low content may be explained by the rapid diffusion of O_2, despite its low concentration in hot water.

Acidophilic and Alkalophilic Bacteria

Since the $\Delta\bar{\mu}H^+$ of 3.5 pH units is theoretically required to synthesize ATP (Mitchell, 1966), it is interesting to examine H^+-ATPase in acidophilic and alkalophilic bacteria. *Bacillus acidocaldarius* TA6 and BA152 were isolated from Goshogake and Beppu acidic hot springs, respectively, and grown at 65°C at pH 3.0 in a medium containing $0.025N$ H_2SO_4, 0.5% polypeptone, 0.5% glucose, 0.2% yeast extract, and inorganic salts (Oshima et al., 1977). These cells were flagellated spore-forming rods.

Alkalophilic bacteria of *Bacillus* genus were kindly supplied from Dr. A. Ando of Tokyo University and Prof. Y. Nosoh of Tokyo Institute of Technology, and were isolated from the alkaline bath of a traditional factory of indigo staining. Both bacteria were grown at pH 10.5 in a medium containing $0.0375N$ NaOH, 1% polypeptone, 1% glucose, 0.15% yeast extract, and inorganic salts (Ohta et al., 1975).

The H^+-ATPases of these bacteria were neither acidophilic nor alkalophilic, and their activities were very low (less than 0.02 μmol Pi/min/mg protein) with optimal pH around 7.0. The intracellular pH of these bacteria was neutral (Oshima et al., 1977). The Pi-ATP exchange reaction of the inside-out particles of these bacteria was also very low, and their internal water phases should be extremely acidic or alkaline during the proton translocation by H^+-ATPase. The maintenance of such a large ΔpH across the membrane of these special bacteria may be explained by the coexisting cation transport systems and ω-cylohexyl-lipids and other lipids.

Oligomeric Structure of F_1

Stability of F_1

The H^+-ATPase ($F_o \cdot F_1$), F_1, and F_o were purified according to the method reported for mitochondria (Kagawa, 1974), chloroplasts (Nelson, 1976), and plasma membranes of PS3 (Yoshida et al., 1975; Sone et al., 1975).

Comparison of thermophilic and mesophilic membrane proteins is not usually possible, because mesophilic membrane proteins are difficult to purify. However, F_1 has been prepared from several mesophiles (Pullman et al., 1960; Lardy and Ferguson, 1969; Pedersen, 1975; Schäfer and Bäuerlein, 1977; Takeshige et al., 1976; Dose et al., 1977). To differentiate F_1 of beef heart mitochondria from that of the thermophilic bacterium PS3, the former is called F_1 and the latter, TF_1. F_1 from chloroplasts is called CF_1.

The stabilities of F_1 and TF_1 were compared in the presence of various concentrations of sodium dodecyl sulfate (SDS) (Fig. 2A), ethanol (Fig. 2B), urea (Fig. 2C), and a chaotropic agent (Fig. 2D, KSCN). The stability of TF_1 in the presence of the compounds shown in this figure,

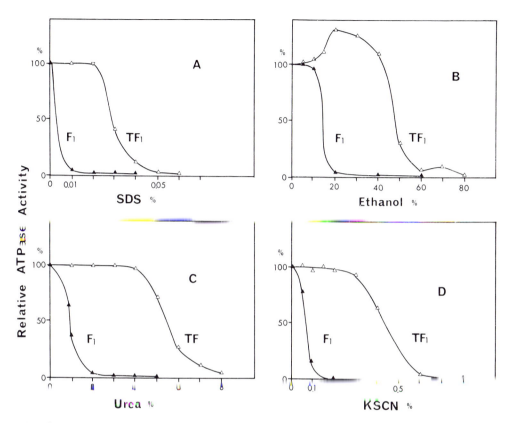

Fig. 2A-D. Effects of various dissociating agents on the ATPase activities of F_1 and TF_1. The activity in the absence of the agents is taken as 100%. (A) SDS (sodium dodecyl sulfate). (B) Ethanol. (C) Urea. (D) KSCN. F_1 and TF_1 in 50 mM Trischloride, pH 7.4, and 0.05 mM EDTA at a protein concentration of 0.1 mg per ml were treated with the indicated reagents at 30°C. After 30 min, 20 μl of the mixture was taken and ATPase was determined in a final volume of 0.5 ml as described previously (Kagawa et al., 1976). Since F_1 (from beef heart) was taken from a stock solution containing 2 mM ATP, 2 mM EDTA, 0.25 M sucrose, and 10 mM Tris-sulfate, pH 7.4, it was protected with 0.2 mM ATP during the inactivation process. The TF_1 preparation contained no ATP and there was no ATP added during the process

as well as other detergents, organic solvents and chaotropic agents, is useful in the purification and reconstitution of TF_1. The acido- and alkalo-philic F_1 were not as stable.

TF_1 is more heat resistant than F_1 and CF_1 and is not labile in the cold. The latter lability is characteristic of mesophilic F_1, whose oligomeric structure is easily dissociated even at room temperature, depending on the conditions of the medium. The secondary structure of TF_1 is also heat resistant, even without ATP (Fig. 3). The circular dichroism (CD) spectra of TF_1 were hardly affected by increasing the temperature from 2°C to 75°C, but the spectra changed at 81°C, and the trough decreased markedly at 96°C.

The reason for the greater stability of TF_1 than F_1 is unknown, but amino acid analysis showed that TF_1 had a higher content than F_1 of

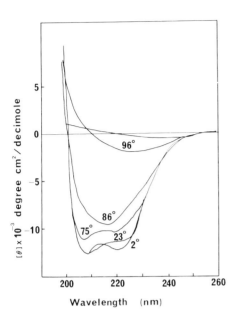

Fig. 3. Circular dichroism spectra of TF_1 at various temperatures. TF_1 (0.16 mg/ml) in 5.0 mM Tris-chloride, pH 8.0, and 0.03 mM EDTA was put in a quartz cell of 1.0-mm light path and the circular dichroism spectra at various temperatures were recorded at a scanning speed of 1 cm/min, 5-nm wavelength/cm, with a Jasco ORD/UV-S spectrophotometer (Tokyo, Japan). The percentage of α-helices and β-structures were calculated with a computer model YHP 2100A, Yokokawa-Hewlett-Packard (12 kilobits, Tokyo, Japan)

amino acid residues forming salt bridges (Arg 6.02 in TF_1, 5.49 in F_1; Glu 13.29 in TF_1, 12.54 in F_1) and a much lower content of SH residues (Cys 0.08 in TF_1, 0.35 in F_1) (values in parentheses are mol %) (Yoshida et al., 1975). The helical contents of TF_1 and F_1 at $25^{\circ}C$ were 31.5% and 34.5%, respectively.

Molecular Weight and Subunit Stoichiometry of F_1

The molecular structure of F_1 is still controversial, mainly because the F_1 oligomer readily dissociates. TF_1 does not dissociate appreciably and its molecular weight is 390,000 (value obtained in collaboration with Prof. N. Ui, to be published). The frequently cited molecular weight of CF_1 is 325,000, but in the same reference the value of 358,000 is described (Farron, 1970). The molecular weight of F_1 has been reported to be 284,000 - 360,000 (see review, Pedersen, 1975; Nelson, 1976). However, when dissociated components were removed by gel filtration, and further dissociation was retarded by the addition of methanol, the molecular weights of CF_1 and F_1 were also found to be 380,000 and 390,000, respectively (by Prof. N. Ui). Yeast F_1 was also shown to be 400,000 daltons (Takeshige et al., 1976) and *Micrococcus* F_1, 380,000 daltons (Dose et al., 1977).

The subunits of H^+-ATPase, which is composed of F_o and F_1, are summarized in Table 1. Of the eight subunits, five (α,β,γ,δ,ε) belong to F_1 (Fig. 4). There are many reports supporting a subunit stoichiometry of α:β:γ:δ:ε = 2:2:1-2:1-2:1-2 (Nelson, 1976; Racker, 1976; Baird and Hammes, 1976). However, our analysis (Kagawa et al., 1976) and those of others (Pedersen, 1975; Bragg and Hou, 1975; Dose et al., 1977) indicate a stoichiometry of α:β:γ:δ:ε = 3:3:1:1:1. Our analysis was based on the radioactivity incorporated in subunits of TF_1 of PS3 grown in a medium containing a universally labeled $|^{14}C|$amino acid mixture.

There are more than 12 SH-groups per mole of mesophilic F_1, but TF_1 contains only 3 (1 SH/α subunit). TF_1 also contains 22 tryptophan res-

Table 1. Subunits of H⁺-ATPase of the thermophilic bacterium PS3

No.	H⁺-ATPase ($F_o \cdot F_1$)		Molecular weight	Copies per $F_o \cdot F_1$	SH-group per subunit
	TF_1	F_o			
1	α	(−)	56,000	3	1
2	β	(−)	53,000	3	0
3	γ	(−)	32,000	1	0
4	(−)	(+)[a]	19,000	1	−
5	δ	(−)	15,500	1	0
6	(−)	(+)[b]	13,500	2 – 3	−
7	ε	(−)	11,000	1	0
8	(−)	(+)[c]	5,000	5 – 6	−

[a] TF_1-binding activity and H⁺-channel activity of F_o are not lost by removal of this subunit on a CM-cellulose column.

[b] TF_1-binding activity is detected in this subunit after removal of other subunits with Sephacryl-S-200 (1 x 105 cm) in 0.2% SDS.

[c] This subunit specifically binds [¹⁴C]DCCD, and this is the only subunit in H⁺-ATPase that is soluble in chloroform-methanol.

Removal of other subunits of F_o by proteolysis of F_o liposomes does not interfere with the H⁺-channel activity of this subunit.

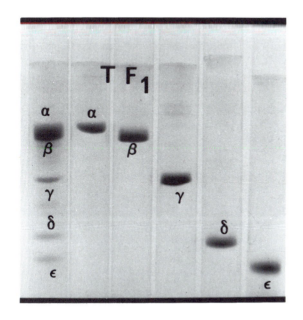

Fig. 4. Analysis of native TF_1 and its purified subunits by polyacrylamide gel elctrophoresis. Gels were loaded with 30 µg of TF_1 or subunits and electrophoresed in 10% polyacrylamide gel in the presence of 0.1% SDS. The gels were then stained with Amido black and destained with 5% acetic acid at 60°C

idues (4 in α, 2 in β, 1 in γ, 1 in δ, and 2 in ε, as shown by Mr. S. Ohta). From these physical and chemical analyses it was concluded that the oligomer structure of F_1 is the α3β3 type (Kagawa et al., 1976; Takeshiga et al., 1976; Dose et al., 1977).

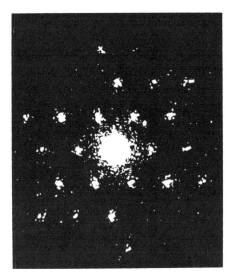

Fig. 5. Diffraction pattern of a two-dimensional crystal of TF_1 obtained by a parallel beam from a He-Ne-gas laser (NEC, Model GLG 2009, Tokyo, Japan). (Kagawa, Y., Wakabayashi, T., Yoshida, M., to be published)

Fig. 6. Optically filtered image of a two-dimensional crystal to TF_1. (Kagawa, Y., Wakabayashi, T., Yoshida, M., to be published)

Computerized Image Reconstruction of F_1

The elucidation of the structure of F_1 is essential for an understanding of the protonophoric action of H^+-ATPase. The detailed structure must be determined by crystallographic analysis. The only membrane ATPase that has ever been crystallized is F_1 (Spitsberg and Haworth, 1977) and TF_1 (Kagawa et al., 1976). Although the molecular weight of F_1 is too large to permit X-ray diffraction studies on the three-dimensional crystal, computerized image reconstruction from electron-microscopic figures of the two-dimensional crystal is promising. Two-dimensional paracrystals of F_1 have already been obtained, but they were unstable (Kagawa, 1974). Therefore, two-dimensional crystals of TF_1 were analyzed.

The diffraction pattern (Fig. 5) and optically filtered image (Fig. 6) of TF_1 showed a hexagonal structure. To obtain these figures, a parallel beam from a He-Ne laser (NEC, Tokyo, GLG 2009) was projected onto

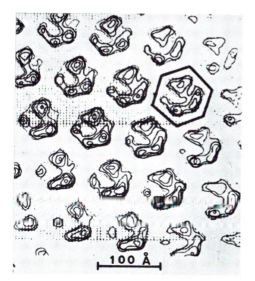

Fig. 7. A translationally filtered
image of TF_1 obtained by computerized
image reconstruction. (Wakabayashi et
al., 1977)

the electron micrograph and the diffraction pattern was observed at
the focal plane of the lens behind the micrograph. An opaque plate
with holes in it to permit selected beams to pass through, was placed
in the diffraction plane to achieve optical filtration, and a lens behind the diffraction plane produced the filtered image.

Numerical Fourier analysis was applied to images selected by the optical diffraction method. The optical density on the film was digitized
by a computer-linked microdensitometer equipped with a moving stage
scanner. The resulting digitized image was Fourier-transformed numeri-
cally. The translationally filtered image was produced by numerical
Fourier synthesis by combining only the Fourier components that were
approximately consistent with the translational symmetry of the crystal. The resulting digital image was then displayed as a line-printer
output, and contours were drawn around density peaks (Fig. 7). The image showed that TF_1 is hexagonal and has pseudo 6-fold and 3-fold symmetry, the spacing along the a axis and b axis being 90 Å. Since the
subunit composition of TF_1 is 3:3:1:1:1, the 6- or 3-fold symmetry may
not be perfect. Every TF_1 molecule has a low density region near its
center (Fig. 7), and it is tempting to conclude that H^+ flows through
this hole during oxidative phosphorylation (Wakabayashi et al., 1977).

It is interesting to compare these results with those on three-dimensional crystals of F_1. In contrast to the observation that F_1 is a
sphere of 90 Å diameter (Kagawa and Racker, 1966c), a unit cell with
the dimensions a = 158 ± 10 Å, b = 158 ± 10 Å, γ = 90° was seen in an
electron micrograph of F_1 (Spitsberg and Haworth, 1977). There were
approximately 3.5 molecules of F_1 per unit cell and at least one dimension of the F_1 molecule was calculated to be no more than about
60 Å. These discrepancies in structure between unstable F_1 and stable
TF_1 remain to be resolved.

Table 2. Reconstitution of ATPase from various combinations of subunits of TF$_1$

Combination numbers	α (41 μg)[a]	β (40 μg)	γ (8 μg)[b]	δ (4 μg)	ε (3 μg)	ATPase activity (μmol/min/mg) 40°C	60°C
1	+	+				0.15	0.08
2	+	+	+			1.20	1.90
3	+	+	+		+	1.62	2.48
4	+	+		+		1.21	0.35
5	+	+		+	+	1.18	0.40
6	+	+	+	+		2.14	2.10
7	+	+	+	+	+	2.33	2.39
8	+	+			+	0.24	0.08
9-12	βγ complex without α, plus or minus δε: variable						
13-31	pure subunits and other combinations: negligible						

[a]A portion of this subunit formed a dimer with an S-S bridge.

[b]This subunit was not soluble in the absence of urea, so that it was stored in 2 M urea before addition to other subunits.

Reconstitution of H$^+$-ATPase from Its Purified Subunits

Reconstitution of TF$_1$

As shown in Table 1, H$^+$-ATPase was dissociated into 8 polypeptides, 5 derived from F$_1$ and 3, from F$_0$. TF$_1$ was dissociated with 8 M guanidine hydrochloride solution, and the solvent of the denatured protein was replaced by a solution of 8 M urea-buffer, and applied to a column of CM-cellulose equilibrated with the same buffer. Subunits α, β, δ, and ε were not adsorbed, and subunit γ was then eluted with 0.06 M NaCl in the same buffer. The unadsorbed fraction was applied to a DEAE-cellulose column equilibrated with the same buffer. Subunit ε was not adsorbed and elution with a linear gradient of NaCl (0-0.08 M) NaCl in the same buffer resulted in elution of the δ, α, and β subunits in this order (Yoshida et al., 1977b). All the isolated subunits (Fig. 4) lost their secondary structure in the buffer containing 8 M urea.

Among ATPase that have been purified, only TF$_1$ has been reconstituted from purified individual subunits. There are 31 possible combinations of the isolated subunits for reconstituting ATPase (Table 2). Of these combinations, those containing the αβγ or αβδ complex were active. Assemblies containing αβγ had the same mobility as the native TF$_1$ molecule on gel electrophoresis, whereas those without the γ subunit moved more rapidly toward the anode. The ATPase activity of the αβδ complex was thermolabile and insensitive to NaN$_3$, whereas the activities of the αβγ complex and those containing αβγ were thermostable and sensitive to NaN$_3$, like the native TF$_1$ (Fig. 8). As reported earlier (Yoshida et al., 1977a), the βγ complex sometimes shows activity, but further confirmation is needed, since anti-α antibody inhibited the activity of TF$_1$. Purified β subunit bound ATP (Ohta, Kagawa, Yoshida, to be published).

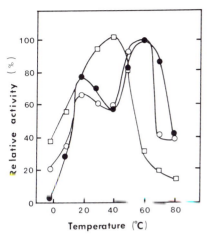

Fig. 8. Effect of temperature on the ATPase activities of complexes (α+β+δ) and (α+β+γ) and native TF$_1$. Solutions containing 2 mM MgSO$_4$, 20 mM Tris-maleate (pH 7.0), 0.74 mg/ml of α subunit, 0.58 mg/ml of β subunit, and either 0.154 mg/ml of γ subunit or 0.073 mg/ml of δ subunit were incubated at 30°C for 2 h. Ten-μl samples of the above solutions and of the TF$_1$ solution (0.15 mg/ml) were then assayed for ATPase for 20 min at various temperatures. Assay mixtures contained 50 mM Tris-maleate, pH 7.0, 5 mM MgSO$_4$, and 5 mM ATP. The maximal activity was taken as 100%. (Yoshida et al., 1977b)

□ α+β+δ; ○ α+β+γ; ● TF$_1$

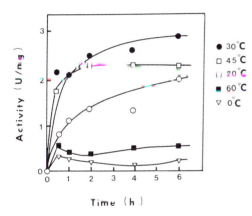

Fig. 9. Reconstitution of ATPase from α, β, and γ subunits at different temperatures. Solutions containing 80 μg/ml of α subunits, 73 μg/ml of β subunits, 21 μg/ml of γ subunits, 5 mM MgSO$_4$, and 50 mM Tris-sulfate at pH 8.0 were incubated at the temperatures indicated. At appropriate times 30-μl samples were removed for assay of ATPase activity for 15 min at 60°C. (Yoshida et al., 1977b)

Good reconstitution could be achieved by mixing the subunits at 20°C - 45°C, but not at 0°C or 60°C (Fig. 9). On transfer of a mixture from 0°C to 30°C, reconstitution was immediately initiated, whereas on transfer from 60°C to 30°C no further reconstitution occurred. These results are interesting from the standpoint of thermophily of the oligomers, as discussed previously. In fact the numbers of residues of Glu + Gln, Asp + Asn, and Arg per mole of each subunit of TF$_1$ are more than those in the subunits of F$_1$ (Yoshida and Kagawa, to be published). The results of ion exchange chromatography during isolation of the subunits show that subunit β is the most acidic and subunit γ is the most basic. Thus, these two subunits may form some salt bridges to make the complex thermostable.

Subunit β was essential (βγ complex shows some activity), indicating that the latent catalytic center of ATP hydrolysis must be located on subunit β. Subunit α+γ or α+δ must elicit function of subunit β. This conclusion is supported by observations with Fourier-transform infrared spectroscopy of ATP-β-subunit complex (Ohta, Yoshida, Kagawa, to be

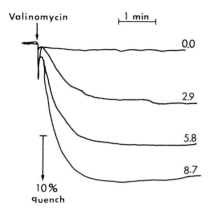

Fig. 10. Proton conduction driven by an artificially imposed membrane potential in F_O liposomes. K^+-loaded F_O liposomes were suspended at a final volume of 2.0 ml in a solution containing 5 μM 9-aminoacridine, 0.5 M sucrose, 2.5 mM $MgSO_4$, and 10 mM Tricine-sodium. The values on the *right side* of the curves indicate F_O mg/mg of phospholipids. Proton uptake was started by adding 20 ng of valinomycin and measured by fluorometry using a Hitachi fluorometer (model 204, Tokyo, Japan) at an excitation wavelength of 365 nm and emission wavelength of 451 nm at 25°C. (Okamoto et al., 1977)

published) and chemical labeling with ATP analogs (Koslov and Skulachev, 1977; Scheurich et al., 1977) bound to β.

Many hypotheses have been proposed to explain the role of SH-groups in F_1, because the SH-group of the γ subunit is labeled with N-ethylmaleimide in the energized condition (see review by Nelson, 1976). However, there is no SH-group in this subunit of TF_1. The presence of lipoic acid or pantotheine in any of the subunits of TF_1 was also denied.

F_1, CF_1, and several bacterial F_1s contain polysaccharides, but TF_1 contains a lesser amount of the polysaccharides (30 μg hexose/mg F_1 and 2 μg hexose/mg TF_1). These hydrophilic polysaccharides are characteristic of extrinsic membrane proteins, but in TF_1, larger amounts of ionic groups in the amino acid residues may compensate for the low polysaccharide content.

The H^+-Channel (F_o) of H^+-ATPase

Since crude F_O was extracted from mitochondria as a factor conferring F_1 with sensitivity to oligomycin (or other energy transfer inhibitors such as DCCD) (Kagawa and Racker, 1966a,b), F_O has been assumed to be a H^+-channel (Mitchell, 1967). F_O is usually obtained from H^+-ATPase by dissociating F_1 with urea. In the case of thermophilic H^+-ATPase, 8 M urea was used and the F_O thus obtained was composed mainly of three polypeptides (Table 1).

When F_O was incorporated into liposomes loaded with K^+, the addition of valinomycin caused a rapid uptake of H^+ through F_O, shown by measuring fluorescence quenching of 9-aminoacridine (Fig. 10). Both the velocity and extent of H^+ translocation were greatly enhanced by increasing the amount of F_O added (Fig. 10). Because the F_O liposome preparation contained liposomes without F_O, FCCP could induce further H^+ uptake by the preparation that had already attained equilibrium after the addition of valinomycin to establish inside negative membrane potential. The number of remaining unloaded liposomes can be estimated by the Poisson distribution

$$P = m\ e^{-ax} \tag{1}$$

where m is the total number of liposomes in the absence of F_O, a is the efficiency of incorporation of F_O into the liposomes, and x is the

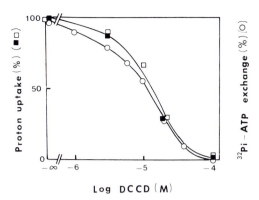

Fig. 11. Inhibition of ^{32}Pi-ATP exchange reaction and proton uptake by DCCD. Proton uptake was assayed as described for Figure 10 (*black square*) or by pH meter (*open square*). For ^{32}Pi-ATP exchange, F_O-vesicles were mixed with excess TF_1 (5.7 mg TF_1/mg F_O), washed once, and incubated for 10 min at 25°C with indicated concentrations of DCCD. Then the exchange activity was assayed (*open circles*). (Okamoto et al., 1977)

molar ratio of the F_O in the liposomes (about 10^9 daltons). If all the F_O is incorporated and the molar ratio is 1, 36.8% of the liposomes are unloaded. This P value can be seen in Figure 10, expressed as the extent of H$^+$ uptake of F_O liposomes. Any reaction in each F_O liposome is "quantumized" and "isolated". If three DCCD-binding subunits are required to form one oligomeric F_O (Altendorf et al., 1977) and if these subunits are dissociated during the reconstitution, the Γ-function of the subunit distribution in liposomes will reveal the number of the aggregation, 3. This system may be useful for analyzing the transport mechanism. For example, small amounts of a contaminating enzyme system, if it is essential for energy transformation, may be diluted and isolated from most of the H$^+$-ATPase.

The initial velocity of H$^+$-uptake into the F_O liposomes was found to be 6 H$^+$/s per F_O molecule at a membrane potential of 100 mV (Okamoto et al., 1977). The membrane potential could be adjusted by changing the concentration of K$^+$ inside and outside the liposome, and the value was calculated from the diffusion potential of Nernst (Sone et al., 1976). The velocity of H$^+$ translocation obeys Ohm's law. The velocity corresponds to a unit conductance of 9.5 x 10^{-18} mho/F_O at pH 8.0 and the maximum unit conductance is 1.6 x 10^{-16} mho measured by titration at different pH values. The observed effect of the external pH on passive H$^+$ uptake by F_O liposomes showed that H$^+$, not OH$^-$, was the actual ionic species conducted. The pH-velocity curve of H$^+$ conductance was identical to a titration curve for monoprotonic acid of pKa = 6.76. This quantitative coincidence suggests the presence of an H$^+$-binding site and a 1:1 stoichiometry of this binding. This observation is interesting since a stoichiometry of 2 H$^+$ to 3 H$^+$/ATP has been reported for H$^+$-ATPase (Mitchell, 1966; Brand and Lehninger, 1977).

The proteolipid, essentially identical to the DCCD-binding protein of F_O, was extracted from chloroplasts and was shown to conduct H$^+$ (Nelson et al., 1977). As shown in Figure 11, the H$^+$-translocating activity of F_O is specifically inhibited by DCCD and the inhibitory concentration of DCCD also inhibits ^{32}Pi-ATP exchange of F_O liposomes loaded with F_1. The 5400-dalton subunit shown in Table 1 corresponds to the DCCD-binding protein that was purified to homogeneity from PS3 by the method of Fillingame (1976). Treatment of F_O liposomes with trypsin did not impare H$^+$-channel activity, and the DCCD-binding protein alone remained intact. If the DCCD-binding protein alone can form an H$^+$-channel, what are the roles of the other subunits in F_O? The remaining PS3 subunits of 19,000 and 13,500 daltons may correspond to those of 19,000 and 11,000 daltons of *Neurospora crassa* (Kagawa, 1978). The subunit of 19,000

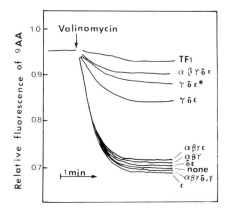

Fig. 12. Reconstitution of gates for H^+ on liposomes containing H^+ channels (F_O). The influx of H^+ into liposomes containing 17.5 µg of F_O was assayed by the self-quenching of fluorescence of 9-aminoacridine as described for Figure 10. The following amounts of subunit(s) were added to the F_O liposomes as indicated: α, 31 µg; β, 45 µg; γ, 15 µg; δ, 11 µg; ε, 11 µg. For the results marked as $\alpha\beta\gamma$, double amounts of α, β, and γ subunits were used. (Yoshida et al., 1977a)

may be identical to the OSCP (oligomycin) sensitivity-conferring protein), which was isolated from F_O with CM-cellulose chromatography and was 18,000 daltons and basic (MacLennan and Tzagoloff, 1968). The 13,500-subunit was isolated with Sephacryl S-200 equilibrated with 0.2% SDS and was shown to bind TF_1. These three subunits of F_O were found in precipitates of an anti-TF_1 + H^+-ATPase mixture (Sone et al., 1975). Thus, OSCP, TF_1-binding subunit, and DCCD-binding subunit may be essential components of F_O. The single-peptide structure of F_O is not consistent with the reported genetic analysis of F_O, which has two or more peptides (Gibson, 1977).

The H^+ Gate of H^+-ATPase

There are many gates, such as Na^+-, K^+-, and Ca^{2+}-gates, in biomembranes, but the gate in H^+-ATPase is the only one that has been reconstituted from purified subunits. Addition of F_1 to the F_1-depleted submitochondrial particles (Hinkle and Horstman, 1971) or of CF_1 to chloroplast membrane washed with EDTA (Berzborn and Schröer, 1976) is known to block their passive H^+-leakage, thus restoring phosphorylating activity. This suggests that there is an H^+ gate in the F_1 or CF_1.

The gating function of F_1 is distinct from its pumping function, since the ATPase activity of F_1 can be removed by chemical modification or antibody treatment without affecting the "structural" or gating role of F_1 (Racker, 1976; Berzborn, 1977). The gate in F_1 was opened when the $\Delta\bar{\mu}H^+$ across the membrane exceeded a certain value. In chloroplasts, for example, a sharp 1000-fold increase in H^+ conductance was observed when the ΔpH reached 3, and the characteristics of the flux-force were very similar to those of the potential barrier at the PN junction of *Zener-diode* (Schönfeld and Neumann, 1977). This voltage-dependent H^+ gate closing after passage of H^+, was demonstrated after CF_1 was modified with *N,N'-o*-phenyl diamine (Wagner and Junge, 1977).

The gating activity was examined by adding complexes of subunits of TF_1 to F_O liposomes loaded with K^+. As with the H^+-channel activity described above, valinomycin induced an inside negative $\Delta\Psi$, resulting in H^+ uptake through F_O. This activity was regulated by the addition of complexes reconstituted from the purified subunits of TF_1 (Fig. 12). TF_1 completely blocked this H^+ uptake, indicating that all the F_O molecules are arranged in the same direction to bind F_1, since the H^+-channel activity should be equal in both directions whereas F_O should be

Table 3. Reconstitution of H^+-ATPase by connecting the H^+-channel (F_O) to the energy transformer ($\alpha+\beta$) with the H^+-gate ($\gamma+\delta+\varepsilon$)

F_O liposomes Addition	ATPase activity bound to F_O liposomes (μmol P_i/min/mg F_O)				
	TF_1	$\alpha\beta\gamma\delta\varepsilon$	$\alpha\beta\gamma\delta$	$\alpha\beta\delta\varepsilon$	$\alpha\beta\gamma$
F_O	1.25 (38)[a]	1.14 (40)	0.27 (86)	0.31 (52)	0.14 (95)
$F_O + \delta$	1.21 (34)	1.21 (38)	0.32 (52)	1.00 (36)	0.34 (39)
$F_O + \varepsilon$	1.52 (38)	1.14 (37)	1.25 (37)	0.47 (66)	0.42 (55)
$F_O + \delta + \varepsilon$	1.44 (36)	1.33 (36)	1.44 (39)	1.17 (39)	1.21 (38)

[a] Numbers in parentheses are % of residual activity in the presence of DCCD.

The assembly solution added to an aliquot containing 1.5 μg of α subunits, 4.5 μg of β subunits, 1.5 μg of γ subunits, 0.54 μg of δ subunits, and 0.54 μg of ε subunits, as indicated. The resulting 20 mixtures containing F_O liposomes and subunits (or TF_1) were incubated for 20 min and then centrifuged at 140,000 g for 15 min. Then ATPase activity was measured after washing with 1% cholate (Yoshida et al., 1977a).

asymmetric. Among the assemblies of subunits added to F_o, those containing the $\gamma\delta\varepsilon$ complex inhibited the flux of H^+ (Fig. 12) (Yoshida et al., 1977a). These complexes inhibited the H^+ flow in both directions. Thus it is concluded that the $\gamma\delta\varepsilon$ complex is the gate of H^+ in H^+-ATPase (Yoshida et al., 1977a).

The organization of the gate was examined by studies on DCCD sensitivity conferred upon TF_1 by F_o, and the binding of TF_1 to F_o. Table 3 shows clearly that the δ and ε subunits are both required for the binding of the $\alpha\beta\gamma$ complex to F_o (Yoshida et al., 1977a). There is no preferable sequence of binding to F_o between δ and ε. Similar findings have been reported for other F_1 preparations. The assembly $\alpha\gamma\varepsilon$ of F_1 to $E.\ coli$ was obtained by freezing in salt solution (Vogel and Steinhard, 1976). Subunit was shown to be essential for binding to F_1-deficient particles (Futai et al., 1974). Thus it is concluded that the $\delta\varepsilon$ is the connecting bridge of $\alpha\beta\gamma$ complex and F_o (Yoshida et al., 1977a,b).

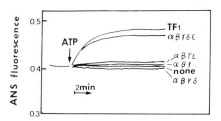

Fig. 13. Reconstitution of H^+-accumulating liposomes. Response of fluorescence of 1-anilinonaphthalene-8-sulfonate (ANS) upon addition of ATP with H^+-channel liposomes (F_O, 5.3 μg) pretreated with various subunit assemblies of TF_1 (each 12.5 μg). After the pretreatment for 15 min, the proteoliposome suspensions were placed in a fluorescence cuvette containing 2 ml of 50 mM Tris-sulfate (pH 8.0) and 2 mM MgSO4, 42°C, and then 20 μg of ANS was added. At the $arrow$, a mixture of 0.5 μmol of ATP and 0.25 μmol of MgSO4 (pH 8.0) was added. A Hitachi fluorometer was used at an excitation wavelength of 365 nm and emission wavelength of 480 nm. (Yoshida et al., 1977a)

Subunits Essential for H$^+$ Translocation

All five subunits are required in F$_o$ liposomes for synthesis of ATP driven by an H$^+$ gradient and for formation of $\Delta\bar{\mu}$H$^+$ driven by ATP hydrolysis (Fig. 13). In conclusion, $\alpha\beta$ is the energy transformer, $\gamma\delta\epsilon$ is the gate, and F$_o$ is the channel of the H$^+$ pump or H$^+$-ATPase. The components of H$^+$-ATPase are thus functionally established except for the role of three subunits in F$_o$.

Reconstitution of Liposomes Containing H$^+$-ATPase Capable of $\Delta\bar{\mu}$H$^+$
Formation and Net ATP Synthesis

The Molecular Species of Stable Phospholipids

In biomembranes of thermophiles, not only their proteins but also their phospholipids are physically and chemically stable. For this reason, liposomes capable of net ATP synthesis have been reconstituted from phospholipids of PS3 and H$^+$-ATPase (Sone et al., 1977).

The phospholipids were found to consist of 57.6% phosphatidylethanolamine, 19.2% cardiolipin, 12.9% phosphatidylglycerol, and 10.3% unidentified material (Kagawa et al., 1976). Gas chromatographic analysis of the methylesters of the phospholipid mixture indicated the following composition: 14:0-iso (0.3%), 15:0-iso (34.1%), 16:0-iso (10.8%), 16:0 (5.7%), 17:0-iso (34.0%), 17:0-anteiso (12.9%), and 18:0-iso (1.0%) (Sone et al., 1975). The fatty acyl groups of these phospholipids, unlike those of the usual phospholipids of mesophiles, were all saturated and were mixtures of even and old-numbered carbon chains.

The acetyldiglycerides derived from these phospholipids were subjected to a combination of mass chromatography, using chemical ionization with NH$_3$, and treatment of the phospholipids with phospholipase A$_2$ (EC 3.1.1.4). The intensities of all the fragments of all the molecular

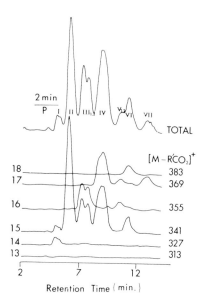

Fig. 14. Mass chromatography of acetyldiglycerides obtained from phosphatidylglycerol of thermophilic bacterium PS3. Peaks (*P* in the figure) I to VII correspond to acetyldiglycerdie containing acyl groups (*C numbers* are indicated on the *left side* of the figure) of C-15 + C-14 to C-17 + C-17. The intensities of the signals of $\left[\text{M-RCOO}\right]^+$ ions are plotted against the retention time according to the carbon number of the acyl group

Table 4. Molecular species of phospholipids of thermophilic bacterium PS3. (Values are relative amounts, % of each phosphatide)

| Molecular species | R_1 | C15 | C15 | C16 | C17 | C17 | C18 | C17 | C18 |
	R_2	C14	C15	C15	C15	C16	C15	C17	C16
Phosphatidyl-ethanolamine		---	24.9	20.4	31.5	5.9	13.2	2.9	---[a]
Phosphatidyl-glycerol		3.2	31.9	16.3	31.7	6.3	7.1	3.6	---[a]
Cardiolipin		3.9	22.5	18.5	28.5	8.8	10.1	6.2	1.0

[a] R_1, R_2 = C14, C14 species constituted about 1% of the total.
For details see Kagawa and Ariga (1977). The unsaturation in these acyl groups was not detected. Most of the acyl groups were branched (iso and anteiso).

species of acetyldiglyceride derivatives were stored and computed with a CC-MPSRAC 300 (Chimadwe Hyoky Tupan). The mass chromatograms of acetyldiglycerides obtained from phosphatidylglycerol are shown in Figure 14, and the results are summarized in Table 4 (Kagawa and Ariga, 1977). The main molecular species are 1,15-methylhexadecanoyl-2-13-methyltetradecanoyl and 1,2-di-13-methyltetradecanoyl-, sn-glycero-3-phosphate type.

In the membrane of acidophilic, thermophilic bacterium, *Bacillus acido-caldarius*, the major fatty acyl groups in the lipids are 11-cyclohexyl-undecanoyl and 13-cyclohexltridecanoyl groups (Oshima and Ariga, 1975). These acyl groups are less mobile in the membrane than those in the PS3 or mesophiles, thus maintaining the large H^+ gradient across the membrane.

In the purple membrane of *Halobacterium halobium*, highly stable lipids such as 2,3-di-O-phytanyl-sn-glycerol-1-phosphoryl-3'-sn-glycerol-1'-phosphate were the major components, and H^+-ATPase and the bacterio-rhodopsin were incorporated into the stable liposomes containing PS3 synthesized ATP on illumination (Kagawa et al., 1977).

Although oleoylphosphate has been claimed to be an essential inter-mediate in oxidative phosphorylation (Griffith et al., 1977), there is no oleic acid or any of its derivatives in the thermophilic bacte-rium PS3 (Kagawa and Ariga, 1977). Moreover, net synthesis of ATP from oleoylphosphate was not detected.

$\Delta\Psi$, ΔpH, and $\Delta\bar{\mu}H^+$ in the Liposomes Containing H^+-ATPase

If ATP is synthesized with H^+-ATPase driven by the flux of H^+ through F_o, the $\Delta\bar{\mu}H^+$ and H^+/P ratios should satisfy thermodynamic requirements, and the velocity of the H^+ flux should be consistent with the kinetics of photo- or oxidative phosphorylation (Boyer et al., 1977). The fa-miliar H^+/O ratio of 2 per site ($2H^+/2e^-$) was recently challenged, and a ratio of 3 or 4 was reported if an SH inhibitor was added to the in-tact mitochondria to block Pi transport (Brand et al., 1976). If the classical P/O ratio of 1 per site were correct, then the ratio of $3H^+/O$ per site of electron transport and the H^+/P ratio of 2 for the H^+-ATPase could not be correct, and in fact, ratios of 3 or 4 were ob-tained when an SH inhibitor was added to intact mitochondria (Brand and Lehninger, 1977). There are many controversies about this problem

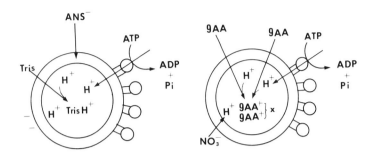

Fig. 15. H+ accumulation by H+-ATPase liposomes and measurements of ΔΨ by ANS fluorescence in the presence of permeable buffer (Tris) to remove ΔpH component of the
ΔμH+ (*left*) or ΔpH by 9AA fluorescence in the presence of permeable anion (NO₃⁻)
to remove the ΔΨ component of ΔμH+ (*right*; Sone et al., 1976)

(Boyer et al., 1977), but it must be pointed out that these experiments
have been performed with intact mitochondria and other crude membrane
systems. In intact mitochondria, H+ translocation accompanied by Pi
(Brand et al., 1976), electrogenic ATP-ADP exchange (Klingenberg and
Rottenberg, 1977), substrate translocations dependent on Pi and H+,
and leakage and transport of cations (Na+, K+, Mg^{2+}, Ca^{2+}, and H+) and
anions complicate the exact measurements of H+ and ΔμH+.

In stable H+-ATPase liposomes composed of only PS3 phospholipids and
H+-ATPase, and thus avoiding complicating side reactions, ΔμH+ was
found to be 253 mV (Sone et al., 1976). If the H+/ATP ratio is 2, a
ΔμH+ value of 210-250 mV is necessary for synthesis of ATP by the reverse reaction of H+-ATPase (Mitchell, 1966). The two components of
ΔμH+, ΔΨ and ΔpH, were measured separately using the fluorescent dyes
anilinonaphthalene sulfonate (ANS) and 9-amino-acridine (9AA) Sone et
al., 1976; Kagawa et al., 1977) as described in Figure 15. The measurement of ΔΨ is based on the linear relationship between the increment
of fluorescence (ΔF/F) upon addition of ATP and the diffusion potential
across the membrane. The diffusion potential of K+ mediated by valinomycin may be calculated by the following equation:

$$\Delta\Psi = RT \ln (K^+)_o / (K^+)_i \tag{2}$$

where $(K^+)_o$ and $(K^+)_i$ are the concentrations (more precisely, activities) of K+ outside and inside the liposomes, respectively. When the
external pH is below 9, ΔpH can be calculated by the following equation:

$$\Delta pH = \log Q/(1-Q) + \log/V \tag{3}$$

where Q is the fraction of the total fluorescence that is quenched by
added ATP and V is the volume of the osmotic compartment in liposomes
as a fraction of the total volume.

When a permeant buffer such as Tris is present, ΔpH becomes 0, whereas
when a permeant anion such as NO₃ is present, ΔΨ becomes 0, as shown
in Figures 15 and 16B. A compensatory rise in the other component of
the electrochemical potential was observed in each case (Fig. 16A).
The H+ accumulated inside the H+-ATPase liposomes were released by the
addition of FCCP (H+-carrier, Fig. 16C) and nigericin (H+-K+ antiporter, Fig. 16D). The H+-translocation of H+-ATPase liposomes was inhibited by the energy transfer inhibitors (Fig. 16E and F). Table 5

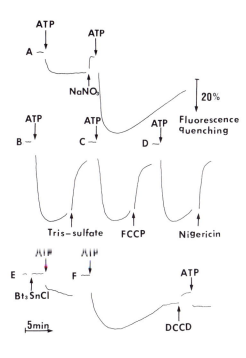

Fig. 16. H⁺ accumulation by H⁺-ATPase liposomes measured by fluorometry of 9AA and effects of various reagents. The reconstituted H⁺-ATPase liposomes (0.05 mg protein) were placed in a cuvette containing 2.0 ml of 20 mM Tricin-sodium, pH 7.7 at 45°C, 50 mM NaNO₃ (except *traces A and F*, 2 mm MgSO₄, and 4 µM 9AA. In *trace F*, NO₃⁻ was replaced by valinomycin (0.5 µg) and KCl (20 µmol). Additions were as follows: ATP (0.5 µmol containing 0.25 µmol of MgSO₄), NaNO₃ (0.10 mmol), Tris-sulfate (10 µmol), FCCP (2 nmol), nigericin (0.2 µg), tributyltin chloride (Bt₃SnCl, 20 nmol), DCCD (100 nmol). (Sone et al., 1976)

Table 5. $\Delta\Psi$, ΔpH, and $\Delta\bar{\mu}H^+$ formed in H⁺-ATPase liposomes following the addition of 0.25 mM ATP

Phospholipids (V values)	Medium	$\Delta F/F$ (%)	$\Delta\Psi$ (mV)	Q/1-Q	ΔpH (pH units)	$\Delta\bar{\mu}H^+$ (mV)
PS3 phospholipids (V = 0.83 µl/ml)	Tris	102	145	0	0	145
	Tricine	48	70	0.67	2.9	253
	Tricine + NO₃⁻	0	0	2.45	3.5	221
Soybean phospholipids (asolectin) (V = 0.74 µl/ml)	Tris	141	140	0	0	140
	Tricine	74	74	0.29	2.6	238
	Tricine + NO₃⁻	0	0	0.58	2.9	183

H⁺-ATPase (0.25 mg) was reconstituted into liposomes with either PS3 phospholipids or soybean phospholipids, and the relative volumes of the liposomes in the medium (V values in the table) were determined. The relative increase in the fluorescence of ANS ($\Delta F/F$), membrane potential ($\Delta\Psi$), the fraction of the total 9AA fluorescence quenched by the addition of ATP (Q), pH difference across the liposome membrane (ΔpH), and the electrochemical potential difference across the membrane ($\Delta\bar{\mu}H^+$) are described in the text. (Sone et al., 1976.)

summarizes the $\Delta\bar{\mu}H^+$ values in reconstituted H⁺-ATPase liposomes (Sone et al., 1976). The maximal $\Delta\bar{\mu}H^+$ was higher than the predicated value of 210 mV (Mitchell, 1966) that was compatible with the idea of a 2 H⁺/ATP ratio (Mitchell, 1966). When submitochondrial particles are used to determine the stoichiometry, there is no complicating transport phenomenon as in the case of intact mitochondria, and the ratio of 2 $H^+/2e^-$ was measured in all three oxidative phosphorylation sites (Hinkle and Horstman, 1971).

If we compare the values obtained with H^+-ATPase liposomes or submito-chondrial particles to those obtained with intact mitochondria (Brand and Lehninger, 1977), the discrepancy is evident, since a much higher H^+/ATP ratio than 3 is expected in the former, which have no transport system to consume the H^+ gradient. To confirm this, H^+-ATPase liposomes loaded with K^+ were added with valinomycin and ATP, to cancel $\Delta\bar{\mu}H^+$ formed (positive inside). To date, however, the H^+/ATP ratio measured with a pH meter has been nearly 2 (Kagawa, unpublished results). For measuring of $\Delta\Psi$, merocyanin dye gave faster and more accurate values than ANS in reconstituted liposomes, but the results were similar (Kagawa and Hinkle, unpublished results).

Net Synthesis of ATP by an Artificially Imposed $\Delta\bar{\mu}H^+$ in H^+-ATPase Liposomes

Net synthesis of ATP driven by an artificially imposed $\Delta\bar{\mu}H^+$ in a simple preparation of submitochondrial particles (Thayer and Hinkle, 1975) supported Mitchell's chemiosmotic hypothesis (Mitchell, 1976). However, these membranes contained many components, and it was difficult to conclude that only H^+-ATPase was responsible for this reaction. Net synthesis of ATP was also observed in reconstituted vesicles containing crude H^+-ATPase of beef heart mitochondria (Kagawa and Racker, 1971; Kagawa et al., 1973) in the presence of electron transport components called complexes, or bacteriorhodopsin (Racker, 1976). However, the H^+-ATPase used was not only crude but also unstable, and net ATP synthesis driven by "acid-base transition" in reconstituted H^+-ATPase liposomes has never been reported. The complexes of electron transport are very complicated preparations consisting of seven to 12 subunits, and the mechanism of H^+ translocation is still unknown.

The simplest ATP-synthesizing system is the reconstituted H^+-ATPase liposomes in which $\Delta\bar{\mu}H^+$ is imposed. These liposomes were first incubated in acidic malonate buffer at pH 5.5 with valinomycin. Then a base stage solution (glycylglycine buffer, final pH 8.33) containing KCl was rapidly injected. This instantaneous transition should create a $\Delta\bar{\mu}H^+$ composed of both ΔpH (2.83 units, acidic inside) and $\Delta\Psi$ (125 mV, positive inside) across the liposome membrane (Sone et al., 1977). Esterification of ^{32}Pi occurred at a velocity of 650 nmol/mg H^+-ATPase/ min. Considering the content of H^+-ATPase, this velocity should be faster than that coupled to substrate oxidation. The primary role of H^+-translocation in oxidative phosphorylation is thus substantiated (Kagawa et al., 1977).

The maximum level of Pi esterified was about 100 nmol/mg of H^+-ATPase in the reconstituted liposomes, while less than 2.5 nmol/mg of protein was synthesized by submitochondrial particles (Thayer and Hinkle, 1975) or bacterial membrane (Tsuchiya and Rosen, 1976). Taking the molecular weight of H^+-ATPase as 458,500 (Kagawa et al., 1976), it was calculated that 46 turnovers of the enzyme occurred with the optimal pulse of $\Delta\bar{\mu}H^+$. Figure 17 shows the effect of pH in the base stage (Fig. 17A) and acid stage (Fig. 17B). The decrease in the yield of ATP below pH 5.5 (Fig. 17B) and above pH 8.5 (Fig. 17A) was probably due to inactivation of the proteins. Since this acid-base treatment is a drastic procedure, liposomes made with phospholipids from PS3, which are stable, were active, but those made with soybean lipids, which contain many unsaturated fatty acyl groups, were not (Sone et al., 1977).

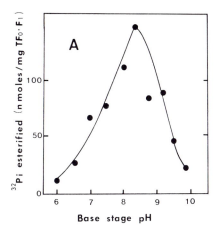
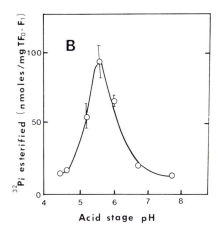

Fig. 17A and B. Net synthesis of ATP by an artificially imposed electrochemical gradient in H⁺-ATPase liposomes. (A) Dependence of ATP yield on the base-stage pH. The acid-stage pH was fixed at pH 5.5, with 10 μmol malonate. (B) Dependence of ATP yield on the acid-stage pH. The base-stage pH was fixed at pH 8.0. Experiments (A) and (B) were both carried out in the base-stage at 40°C for 5 min with reconstituted H⁺-ATPase liposomes (H⁺-ATPase, 34 μg in 2 mg of PS3 phospholipids) in a final volume of 2 ml in 0.15 M KCl, 2 mM ADP, 10 mM ³²Pi, 1 mM MgSO₄, 0.2 μg valinomycin, and 80 mM glycylglycine buffer with the indicated pH

The acid-base transition is the method for applying ΔpH or $\Delta \bar{\mu} H^+$ across the membrane. On the other hand, the application of $\Delta \Psi$ with the external electric field to the suspension of chloroplasts also produced ATP (Witt et al., 1976). This method is convenient for analyzing the rate of turnover, the extent of the electric field strength and $\Delta \Psi$, and the fraction of H⁺-ATPase that change their conformation. In the above-described H⁺-ATPase liposomes, application of the external electric field did not result in the formation of a large amount of ATP, since the diameter of the liposomes is less than 700 Å. Formation of $\Delta \Psi$ by loading K⁺ outside the vesicles and adding the valinomycin to them was also not sufficient to synthesize ATP. However, improvements of the conditions may solve the difficulty in the reconstituted liposomes.

Conformational Changes in H⁺-ATPase

The conformational change in the oligomeric H⁺-ATPase may transform energy from ATP to that of the transport of H⁺. The F_1 portion of H⁺-ATPase, though without an H⁺ channel (F_o), was reported to form a Volta potential of about 200 mV at an octane-water interface (Boguslavsky et al., 1975; Koszlov and Skulachev, 1977). The β subunit of F_1 was shown to be a catalytic site (see above) The simplest mechanism proposed by Mitchell was as follows (Mitchell, 1976): 2 H⁺ attack the O of PO_4^- in a complex with $ADPO^-$ and Mg^{2+} at the active site of F_1, from the F_o side, so that H_2O is released, leaving P^+O_3, while $ADPO^-$ makes a nucleophilic attack on the P^+ center, thereby producing ATP. This kind of protonation of phosphate does not take place in aqueous solution, and a theoretical prediction about an unknonw catalytic site of the H⁺-ATPase is difficult. However, a conformational change of F_1 is expected to convert the energy of $\Delta \bar{\mu} H^+$ or at least to smooth the energy profile of the chemical transition state.

In fact, light-induced, uncoupler-sensitive incorporation of ^{3}H in slow-exchanging-H (H of hydrogen bonds of helices, etc.) of CF$_1$ has been determined (Ryrie and Jagendorf, 1972). The conformational changes in TF$_1$ were studied with Fourier transform infrared (IR) spectroscopy (Ohta et al., 1978), since there is no time lag during the scanning of TF$_1$ while it exchanges its slow ^{1}H with ^{2}H in the solvent. A large change in pH during energy transformation accelerates the exchange reaction in peptide groups by the following equation for the rate constant k:

$$k = 50 \ (10^{-pH} + 10^{pH-6}) \ 10^{0.05 \ (\theta - 20)} \ \text{min}^{-1} \tag{4}$$

where θ is the temperature in $^\circ$C.

Many observations have been made on the role of bound nucleotides in relation to conformational change and ATP synthesis (Boyer et al., 1977). The rate of the ^{1}H-^{2}H exchange reaction of TF$_1$, determined by Fourier transform IR spectroscopy, decreased in the following order: free TF$_1$, TF$_1 \cdot$ADP, TF$_1 \cdot$ATP, and TF$_1 \cdot$AMPPNP. In contrast to TF$_1$, in ATP-consuming systems such as heavy meromyosin, the rate of the reaction was slower in the presence of ADP than in the presence of ATP.

Crystallization of TF$_1$ was possible in the presence and absence of AMPPNP or NaN$_3$, and the image reconstruction of these in detail may lead to further elucidation of conformational change.

Acknowledgments. The author thanks Drs. N. Sone, H. Hirata, and M. Yoshida of this laboratory, Dr. T. Oshima of Mitsubishi Life Science Institute, Dr. T. Wakabayashi of Tokyo University, Dr. T. Argia of Tokyo Metropolitan Clinical Institute, and Dr. Y. Nosoh of Tokyo Institute of Technology for their help during this work.

This research was supported by grants from the Ministry of Education of Japan, the Matsunaga Memorial Fund, and Toray Science Foundation. Part of this work was presented at the Symposium of Federation of Asian and Oceanian Biochemists, October 12, 1977.

References

Altendorf, K., Lukas, M., Kurth, R.: Charakterisierung des Dicyclohexylcarbodiimide-reaktiven Proteins des ATPase-Komplexes von *Escherichia coli*. Hoppe-Seyler's Z. Physiol. Chem. 358, 1385-1385 (1977)

Baird, B.A., Hammes, G.G.: Chemical cross-linking studies of chloroplast coupling factor 1. J. Biol. Chem. 251, 6953-6962 (1976)

Berzborn, R.J.: Rekonstitution der photosynthetischen ATP-Synthetase. Hoppe-Seyler's Z. Physiol. Chem. 358, 1384-1384 (1977)

Berzborn, R.J., Schröer, P.: Photophosphorylation: Mechanism of reconstitution by coupling factor 1 (CF$_1$). FEBS Lett. 70, 271-275 (1976)

Biesecker, G., Harris, J.I., Thierry, J.C., Walker, Wonacott, A.J.: Sequence and structure of D-glyceraldehyde 3-phosphate dehydrogenase from *Bacillus stearothermophilus*. Nature (London) 266, 328-333 (1977)

Boguslavsky, L.I., Kondrasin, A.A., Kozlov, I.A., Melelsky, S.T., Skulachev, V.V., Volokov, A.G.: Charge transfer between water and octane phases by soluble mitochondrial ATPase (F$_1$), bacteriorhodopsin and respiratory chain enzymes. FEBS Lett. 50, 223-226 (1975)

Boyer, P.D., Chance, B, Ernster, L., Mitchell, P., Racker, E., Slater, E.C.: Oxidative phosphorylation and photophosphorylation. Ann. Rev. Biochem. 46, 955-1026 (1977)

Bragg, P.D., Hou, C.: Subunit composition, function, and spatial arrangement in the Ca^{2+} and Mg^{2+}-activated adenosine triphosphatase of *Escherichia coli* and *Salmonella typhimrium*. Arch. Biochem. Biophys. 167, 311-321 (1975)

Brand, M.D., Lehninger, A.L.: H^+/ATP ratio during ATP hydrolysis by mitochondria: modification of the chemiosmotic theory. Proc. Natl. Acad. Sci. U.S.A. 74, 1955-1959 (1977)

Brand, M.D., Raynafarje, B., Lehninger, A.L.: Stoichiometric relationship between energy dependent proton ejection and electron transport in mitochondria. Proc. Natl. Acad. Sci. U.S.A. 73, 437-441 (1976)

Dose, K., Höckel, M., Hulla, F.W., Schmitt, M., Risi, S.: Charakterisierung und Funktion der F_1-ATPase (EC 3.6.1.3) aus *Micrococcus sp.* ATCC 398. Hoppe-Seyler's Z. Physiol. Chem. 358, 1385-1386 (1977)

Farron, F.: Isolation and properties of a chloroplast coupling factor and heat activated adenosine triphosphatase. Biochemistry 9, 3823-3828 (1970)

Fillingame, R.H.: Purification of the carbodiimide-reactive protein component of the ATP energy-transducing system of *Escherichia coli*. J. Biol. Chem. 251, 6630-6637 (1976)

Futai, M., Sternweis, P.C., Heppel, L.A.: Purification and properties of reconstitutively active and inactive adenosine triphosphatase from *Escherichia coli*. Proc. Natl. Acad. Sci. U.S.A. 71, 2725-2729 (1974)

Gibson, D.: Assembly and biochemical studies of the ATP synthase-coupling complex of *Escherichia coli*. First congress of the Federation of Asian and Oceanian Biochemists (FAOB) Abstract 1, 75-76 (1977)

Griffith, D.E., Hyams, R.L., Bertoli, E.: Studies of energy-linked reactions: dihydrolipoate- and oleate-dependent ATP synthesis in Yeast promitochondria. FEBS Lett. 74, 38-42 (1977)

Harold, F.M.: Ion currents and physiological functions in microorganisms. Ann. Rev. Microbiol. 31, 181-203 (1977)

Hinkle, P.C., Horstman, L.L.: Respiration-driven proton translocation in submitochondrial particles. J. Biol. Chem. 246, 6024-6028 (1971)

Kagawa, Y.: Preparation and properties of a factor conferring oligomycin sensitivity (F_O) and of oligomycin sensitive ATPase $(CF_O \cdot F_1)$ In: Methods in Enzymology. Estabrook, R.E., Pullman, M.E. (eds.). New York: Academic Press, 1967, Vol. 10, pp. 505-510

Kagawa, Y.: Reconstitution of oxidative phosphorylation. Biochim. Biophys. Acta 265, 297-338 (1972)

Kagawa, Y.: Dissociation and reassembly of the inner mitochondrial membrane. In: Methods in Membrane Biology. Korn, E.D. (ed.). New York: Plenum Press, 1974, Vol. 1, pp. 201-269

Kagawa, Y.: Transport activity of proteoliposomes reconstituted from crystalline ATPase or from solubilized alanine carrier. J. Cell. Physiol. 89, 569-574 (1976)

Kagawa, Y.: Reconstitution of the energy transformer, gate and channel — subunit reassembly, crystalline ATPase and ATP synthesis. Biochim. Biophys. Acta, in press (1978)

Kagawa, Y., Ariga, T.: Determination of molecular species of phospholipids of thermophilic bacterium PS3 using mass-chromatography. J. Biochem. (Tokyo) 81, 1161-1165 (1977)

Kagawa, Y., Kandrach, A., Racker, E.: Partial resolution of the enzymes catalyzing oxidative phosphorylation XXVI. Specificity of phospholipids required for energy transfer reactions. J. Biol. Chem. 248, 676-684 (1973)

Kagawa, Y., Ohno, K., Yoshida, M., Takeuchi, Y., Sone, N.: Proton translocation by ATPase and bacteriorhodopsin. Fed. Proc. 36, 1815-1818 (1977)

Kagawa, Y., Racker, E.: Partial resolution of the enzymes catalyzing oxidative phosphorylation, VIII. Properties of a factor conferring oligomycin sensitivity on mitochondrial adenosine triphosphatase. J. Biol. Chem. 241, 2461-2466 (1966a)

Kagawa, Y., Racker, E.: Partial resolution of the enzymes catalyzing oxidative phosphorylation, IX. Reconstitution of oligomycin-sensitive adenosine triphosphatase. J. Biol. Chem. 241, 2467-2474 (1966b)

Kagawa, Y., Racker, E.: Partial resolution of the enzymes catalyzing oxidative phosphorylation, X. Morphology and function in submitochondrial particles. J. Biol. Chem. 241, 2475-2482 (1966c)

Kagawa, Y., Racker, E.: Partial resolution of the enzymes catalyzing oxidative
  phosphorylation, XXV. Reconstitution of vesicles catalyzing ^{32}Pi-adenosine triphosphate exchange. J. Biol. Chem. 246, 5477-5487 (1971)
Kagawa, Y., Sone, N., Yoshida, M., Hirata, H., Okamoto, H.: Proton translocating
  ATPase of a thermophilic bacterium — Morphology, subunits and chemical composition. J. Biochem. (Tokyo) 80, 141-151 (1976)
Kakiuchi, S.: The significance of lipids in the oxygen consuming activity of tissues. J. Biochem. (Tokyo) 7, 263-269 (1927)
Klingenberg, M., Rottenberg, H.: Relation between the gradient of the ATP/ADP ratio
  and membrane potential across the mitochondrial membrane. Eur. J. Biochem. 73,
  125-130 (1977)
Knowles, A.F., Penefsky, H.S.: The subunit structure of beef heart mitochondrial
  adenosine triphosphatase. Physical and chemical properties of isolated subunits.
  J. Biol. Chem. 247, 6624-6630 (1972)
Kozlov, I.A., Skulachev, V.P.: H$^+$ Adenosine triphosphatase and membrane energy
  coupling. Biochim. Biophys. Acta 463, 29-89 (1977)
Lardy, H.A., Ferguson, S.M.: Oxidative phosphorylation. Ann. Rev. Biochem. 38,
  991-1034 (1969)
MacLennan, D.H., Tzagoloff, A.: Studies on the mitochondrial adenosine triphospha-
  tase system IV. Purification and characterization of the oligomycin sensitivity
  conferring protein. Biochemistry 7, 1603-1610 (1968)
Mitchell, P.: Chemiosmotic coupling in oxidative and photosynthetic phosphorylation.
  Biol. Rev. 41, 455-502 (1966)
Mitchell, P.: Active transport and ion accumulation. In: Comprehensive Biochemistry.
  Florkin, M., Stotz, E.M. (eds.). Amsterdam: Elsevier, 1967, Vol. 22, pp. 167-197
Mitchell, P.: Vectorial chemistry and molecular mechanics of chemiosmotic coupling:
  Power transmission by proticity. Biochem. Soc. Trans. 4, 399-430 (1976)
Nelson, N.: Structure and function of chloroplast ATPase. Biochim. Biophys. Acta
  456, 314-338 (1976)
Nelson, N., Eytan, E., Notsani, B-E., Sigrist, H., Sigrist-Nelson, K., Gitler, C.:
  Isolation of a chloroplast N,N'-dicyclohexylcarbodiimide binding proteolipid ac-
  tive in proton translocation. Proc. Natl. Acad. Sci. U.S.A. 74, 2375-2378 (1977)
Ohta, K., Kiyomiya, A., Koyama, N., Nosoh, Y.: The basis of the alkalophilic pro-
  perty of a species of *Bacillus*. J. Gen. Microbiol. 86, 259-266 (1975)
Ohta, S., Nakanishi, M., Tsuboi, M., Yoshida, M., Kagawa, Y.: Kinetics of hydrogen-
  deuterium exchange in ATPase from a thermophilic bacterium PS3. Biochem. Biophys.
  Res. Commun., 80, 929-935 (1978)
Okamoto, H., Sone, N., Hirata, H., Yoshida, M., Kagawa, Y.: Purified proton con-
  ductor in proton translocating adenosine triphosphatase of a thermophilic bacterium. J. Biol. Chem. 252, 6125-6131 (1977)
Okunuki, K.: Cytochrome and cytochrome oxidase. In: Comprehensive Biochemistry.
  Florkin, M., Stotz, E.H. (eds.). Amsterdam: Elsevier, 1966, Vol. 14, pp. 232-308
Oshima, M., Ariga, T.: ω-Cyclohexyl fatty acids in acidophilic thermophilic bacteria. Studies on their presence, structure, and biosynthesis using precursors
  labelled with stable isotopes and radioisotopes. J. Biol. Chem. 250, 6963-6968
  (1975)
Oshima, T., Arakawa, H., Baba, M.: Biochemical studies on an acidophilic, thermophilic bacterium, *Bacillus acidocardarius*. Isolation of bacteria, intracellular
  pH, and stabilities of biopolymers. J. Biochem. (Tokyo) 81, 1107-1113 (1977)
Pedersen, P.L.: Mitochondrial adenosine triphosphatase. Bioenergetics 6, 243-275
  (1975)
Perutz, M.E., Radit, H.: Stereochemical basis of heat stability in bacterial ferredoxins and haemoglobin A2. Nature (London) 255, 256-259 (1975)
Pullman, M.E., Penefsky, H.S., Datta, A., Racker, E.: Partial resolution of the enzymes catalyzing oxidative phosphorylation I. Purification and properties of soluble dinitrophenol stimulated adenosine triphosphatase. J. Biol. Chem. 235,
  3222-3229 (1960)
Racker, E.: A New Look at Mechanisms in Bioenergetics. New York: Academic Press,
  1976, pp. 1-197
Ryrie, I.J., Jagendorf, A.T.: Correlation between a conformational change in the
  coupling factor protein and the high energy state in chloroplast. J. Biol. Chem.
  247, 4453-4459 (1972)

Schäfer, G., Bäuerlein, E.: Charakterisierung von Proteinen energieübertragender Prozesse. Hoppe-Seyler's Z. Physiol. Chem. 358, 1379-1390 (1977)

Scheurich, P., Schäfer, H.J., Rathgeber, G., Dose, K.: 8-Azido-ATP, a photoaffinity label for F_1 ATPase from *Micrococcus luteus*. Hoppe-Seyler's Z. Physiol. Chem. 358, 1386-1386 (1977)

Schönfeld, M., Neumann, J.: Proton conductance of the thylacoid membrane: modulation by light. FEBS Lett. 73, 51-54 (1977)

Sone, N., Yoshida, M., Hirata, H., Kagawa, Y.: Purification and properties of a di-cyclohexyl-carbodiimide-sensitive adenosine triphosphatase from a thermophilic bacterium. J. Biol. Chem. 250, 7917-7923 (1975)

Sone, N., Yoshida, M., Hirata, H., Kagawa, Y.: Electrochemical potential of protons in vesicles reconstituted from purified, proton translocating adenosine triphosphatase. J. Membrane Biol. 30, 121-124 (1976)

Sone, N., Yoshida, M., Hirata, H., Kagawa, Y.: Adenosine triphosphate synthesis by an electrochemical proton gradient in vesicles reconstituted from purified adenosine triphosphatase and phospholipids of a thermophilic bacterium. J. Biol. Chem. 252, 2956-2960 (1977)

Spitsberg, V., Haworth, R.: The crystallization of beef heart mitochondrial adenosine triphosphatase. Biochim. Biophys. Acta 492, 237-240 (1977)

Takeshige, K., Hess, B., Bohm, M., Zimmermann-Telschem, H.: Mitochondrial adenosine triphosphatase from yeast, *Saccharomyces cervisiae*. Purification, subunit structure and kinetics. Hoppe-Seyler's Z. Physiol. Chem. 357, 1605-1622 (1976)

Thayer, W.P., Hinkle, P.C.: Kinetics of adenosine triphosphate synthesis in bovine heart submitochondrial particles. J. Biol. Chem. 250, 5336-5342 (1975)

Tsuchiya, T., Rosen, B.P.: Adenosine 5'-triphosphate synthesis energized by an artificially imposed membrane potential in membrane vesicles of *Escherichia coli*. J. Bacteriol. 127, 154-161 (1976)

Vogel, G., Steinhart, R.: ATPase of *Escherichia coli*: purification, dissociation, and reconstitution of the active complex from the isolated subunits. Biochemistry 15, 208-216 (1976)

Wakabayashi, T., Kubota, M., Yoshida, M., Kagawa, Y.: Structure of ATPase (coupling factor TF_1) from a thermophilic bacterium. J. Mol. Biol. 117, 515-519 (1977)

Wagner, R., Junge, W.: Gated proton conduction via the coupling factor of photo-phosphorylation modified by $N,N'-ortho$ phenyldiamine. Biochim. Biophys. Acta 462, 259-272 (1977)

Witt, H.T.: Biophysicalische Primärvorgänge in der Photosynthesemembran. Naturwissenschaften 63, 23-27 (1976)

Witt, H.T., Schlodder, E., Gräber, P.: Membrane bound ATP synthesis generated by an external electric field. FEBS Lett. 69, 272-276 (1976)

Yoshida, M., Okamoto, H., Sone, N., Hirata, H., Kagawa, Y.: Reconstitution of thermostable ATPase capable of energy coupling from its purified subunits. Proc. Natl. Acad. Sci. U.S.A. 74, 936-940 (1977a)

Yoshida, M., Sone, N., Hirata, H., Kagawa, Y.: A highly stable adenosine triphosphatase from a thermophilic bacterium. Purification, properties and reconstitution. J. Biol. Chem. 250, 7910-7916 (1975)

Yoshida, M., Sone, N., Hirata, H., Kagawa, Y.: Reconstitution of adenosine triphosphatase of thermophilic bacterium from purified individual subunits. J. Biol. Chem. 252, 3480-3485 (1977b)

Yutani, K., Ogasawara, K., Sugino, Y., Matsushiro, A.: Effect of a single amino acid substitution on stability of conformation of a protein. Nature (London) 267, 274-275 (1977)

Energy Transfer Inhibition in Photophosphorylation and Oxidative Phosphorylation by 3'-Esters of ADP

G. SCHÄFER, G. ONUR, and H. STROTMANN

A localization of adenine nucleotide binding sites on coupling factors can be achieved by use of covalently binding nucleotide analogs. 8-azido-(1-4), and 3'-arylazido-adenine nucleotides (5-7) have been employed as photolabels in F_1. The former analogs were found to be unsuitable in chloroplast coupling factor (8). The latter appear to be more promising probes since systematic investigations of nucleotide specificity of photophosphorylation and nucleotide binding have shown that reduction or methyl substitution in the C-3' position of ribose did not significantly affect the ADP molecule with regard to its interactions with membrane-bound CF_1 (9).

In order to ascertain an ADP-analogous behavior of 3'-arylazido esters of ADP in the chloroplast system, these compounds were employed in photophosphorylation and related reactions under conditions that exclude photoactivation of the azido group. These investigations yielded remarkable and unexpected results, which in addition to their photolabeling capabilities, make these compounds interesting probes in mechanistic studies of phosphoryl transfer in the terminal steps of energy conservation.

It has already been shown elsewhere (8) that 3'-arylazido-ADP effectively competed with ADP for the tight nucleotide binding site of CF_1, exhibiting a K_i of 6 µM, which is comparable to the affinity of ADP itself (K_d = 2.5 µM). Instead of being phosphorylated, 3'-arylazido esters of ADP are powerful inhibitors of ADP phosphorylation. Further modification of the arylazido moiety revealed that the azido group is not an essential constituent to show this property. Even analogs with aliphatic substituents in 3'-hydroxyl exert energy-transfer inhibition.

Comparative studies with submitochondrial particles lead to similar results, suggesting that the mechanism of inhibition by these ADP analogs is identical in coupling ATPases from different sources.

Experimental Results

Most of the experiments described here were performed with 3'-O-(4-N, 4-azido-2-nitrophenyl-aminobutyryl)-ADP (compound No. 1o in Fig. 4). In the following, the abbreviation aryl-N_3-ADP is used for this compound.

Table 1 demonstrates that this ADP analog is virtually not a substrate in photophosphrylation. The minute amounts of organic phosphate formation (3% - 4%) may result from ADP produced by partial hydrolysis of the 3'-ester bond. This reaction as well as a slow 3'-2'-transacylation (1o, 11) may occur under the conditions of the experiments (pH 8.0).

As shown in Figure 1, aryl-N_3-ADP strongly inhibits photophosphorylation of normal ADP and concomitantly decreases the rate of coupled

Table 1. Comparison of ADP and 3'aryl-N$_3$-ADP as phosphate acceptors in photophosphorylation. Incubation was performed in 25 mM tricine, pH 8.0, 50 mM NaCl, 5 mM MgCl$_2$, 5 mM ^{32}P-phosphate, 0.5 mM methylviologen, 10 mM glucose, and 13 units per ml of salt-free hexokinase at 20°C.
Chloroplast concentration was 35.7 µg chlorophyll/ml. Red light (filter RG 630, Schott) was used for energization

Phosphate acceptor	µmol P$_i$ incorporated/mg chlorophyll/h
None	2.5
ADP 100 µM	224.3
3'-aryl-N$_3$-ADP 5 µM	8.8
3'-aryl-N$_3$-ADP 25 µM	10.0
3'-aryl-N$_3$-ADP 200 µM	11.1

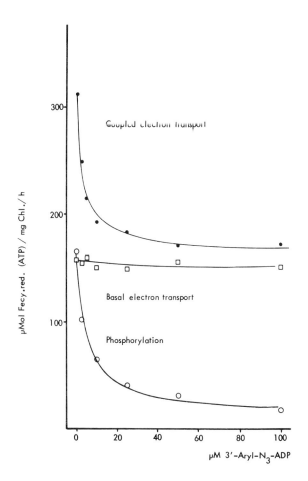

Fig. 1. Effect of 3'aryl-N$_3$-ADP on photophosphorylation and electron transport. Experimental conditions as given in Table 1, except that 100 µM ADP were present and 1 mM ferricyanide was used as electron acceptor instead of methylviologen. Ferricyanide reduction was followed at 420 nm. Details are reported in (8)

electron flow down to the level of basal electron transport. Basal electron transport (measured in the presence of ADP but absence of P$_i$) as well as uncoupled electron flow (8) are not affected at all, indi-

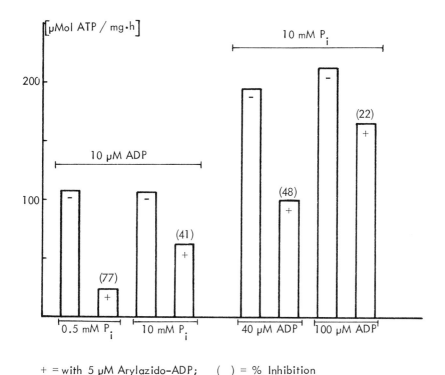

+ = with 5 µM Arylazido-ADP;   ( ) = % Inhibition

Fig. 2. Inhibition of ADP phosphorylation by 3'-aryl-N_3-ADP as influenced by concentration of ADP and of phosphate. Experimental conditions as given in Table 1

cating that aryl-N_3-ADP is a pure energy transfer inhibitor. Compared with most of the known energy transfer inhibitors in chloroplasts, its effectiveness is rather high. Under the conditions employed (presence of 100 µM ADP), the I_{50} value is about 6 µM (Fig. 1).

Figure 2 shows that the extent of inhibition at a fixed inhibitor concentration (5 µM) depends on the concentration of ADP in the assay. The type of inhibition, at least in part, is competitive with ADP, indicating that aryl-N_3-ADP occupies the catalytic ADP-binding site. However, inhibition kinetics are more complex since the inhibitory effect can also be diminished by increasing the phosphate concentration (Fig. 2).

The nucleotide structure of aryl-N_3-ADP is an essential prerequisite for showing the inhibitory effect: The nonesterified 3'-substituent N,4-azido-2-nitrophenyl-4-aminobutyric acid itself has no effect on phosphorylation (control phosphorylation: 141.1 µmol/mg Chl/h, phosphorylation in presence of 50 µM of this compound: 143.1 µmol/mg Chl/h). Figure 3 shows the effects of aryl-N_3-AMP, -ADP, and -ATP on photophosphorylation of ADP. While the AMP analog is completely inactive, aryl-N_3-ATP also exerts some inhibition. However, for half maximum inhibition a 20-fold higher concentration of aryl-N_3-ATP compared to aryl-N_3-ADP is required. Thus, under phosphorylating conditions the catalytic site of CF_1 discriminates between the diphosphate and the triphosphate analog with a remarkably high selectivity.

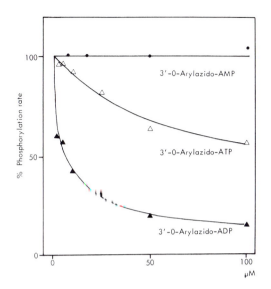

Fig. 3. Effects of 3'-aryl-N$_3$-AMP, -ADP, and -ATP on photophosphorylation. Experimental conditions were as given in Figure 1. 100% phosphorylation corresponds to incorporation of 191.7, 138.2, and 170.7 μmol P$_i$/mg chl/h

The azido group in the aryl substituent obviously is not essential for the inhibitory effect on photophosphorylation. Other derivatives were equally or even more effective (Fig. 4). Unfortunately the as yet most effective compound, which is derived from aminopropionic acid instead of aminobutyric acid, is difficult to deal with because of the instability of this 3'-ester against hydrolysis.

Even aliphatic esters of suitable chain length are found to be fairly good inhibitors whereas the 3'-methyl-ether as well as 3'-deoxy-ADP are substrates in photophosphorylation (E. Schlimme and H. Strotmann, unpubl.).

Figure 5 shows the effect of aryl-N$_3$-ADP on oxidative phosphorylation in beef heart submitochondrial particles (SMP). Oxidative phosphorylation is at least as effectively inhibited by this ADP analog as photophosphorylation, and similar results are obtained with other derivatives indicated in Figure 4. Inhibition in SMP by aryl-N$_3$-ADP is clearly competitive with respect to ADP (Fig. 5), demonstrating an interaction of the analog with the catalytic ADP-binding site of F$_1$. Thus, the geometries of these sites in CF$_1$ and F$_1$ as well as the mechanism of energy transfer in both systems seem to be similar.

Moreover, in SMP other coupled processes such as ATP-driven NAD-reduction by succinate are inhibited by aryl-N$_3$-ADP and related ADP analogs. Uncoupled ATP hydrolysis is practically not affected by concentrations of the analogs, which, in contrast, considerably inhibit NADH formation (Table 2). In fact, ATP hydrolysis by isolated F$_1$ as well as Ca^{2+}-dependent ATP hydrolysis by isolated CF$_1$ is rather insensitive to aryl-N$_3$-ADP (K$_i$ values about 550 μM (7) and 250 μM (K. Suhl and G. Schäfer, unpublished), respectively). Moreover, the corresponding ATP analogs themselves are not hydrolyzed by these ATPases.

ATP-driven reversed electron flow, like oxidative phosphorylation and photophosphorylation, involves the induction and utilization of an energized state that therefore appears to be a prerequisite for these inhibitors to interact with the enzmye. It should be emphasized that

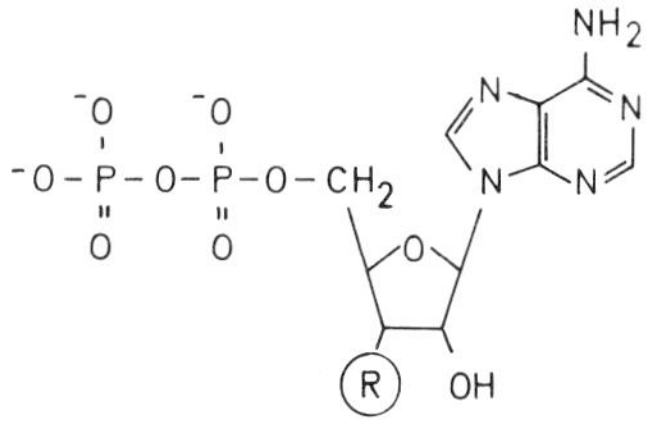

		SUBSTRATE K_m (µM)	INHIBITOR I_{50} (µM)
$-OH$	(1)	15	
$-H$	(2)	36	
$-O-CH_3$	(3)	29	
$-O-CO-CH_3$	(4)		85
$-O-CO-(CH_2)_4-CH_3$	(5)		20
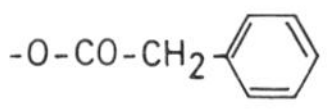 $-O-CO-CH_2-\phi$	(6)		9
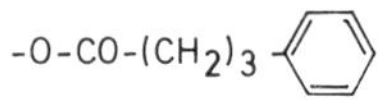 $-O-CO-(CH_2)_3-\phi$	(7)		10
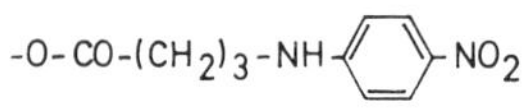 $-O-CO-(CH_2)_3-NH-\phi-NO_2$	(8)		2.5
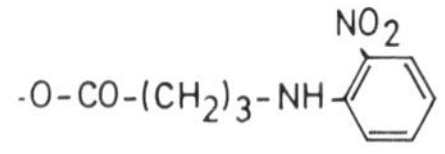 $-O-CO-(CH_2)_3-NH-\phi(NO_2)$	(9)		4.5
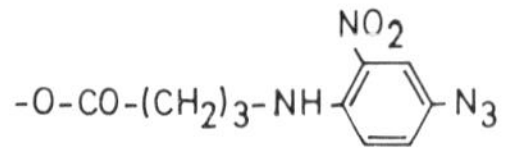 $-O-CO-(CH_2)_3-NH-\phi(NO_2)-N_3$	(10)		6

Fig. 4. 3'-modified ADP analogs as substrates or inhibitors of photophosphorylation

also in ATP-driven NAD reduction, aryl-N_3-ATP is much less effective compared with the ADP-analog (Table 3) and that aryl-N_3-AMP is without effect even at high concentrations (290 µM).

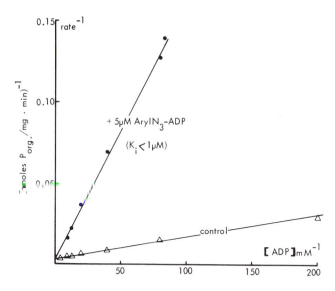

Fig. 5. Inhibition of oxidative phosphorylation by 3'-aryl-N$_3$-ADP in beef heart submitochondrial particles. Phosphorylating capacity was determined at 25°C in a medium containing 220 mM mannitol, 70 mM sucrose, 0.5 mM EDTA, 2.5 mM MgCl$_2$, 2.5 mM ^{32}P-phosphate, 10 mM Na-succinate, 0.5 mg/ml BSA. ATP was trapped by glucose/hexokinase (Fig. 1). Other conditions as indicated

Table 2. Effect of 3'-aryl-N$_3$-ADP on uncoupled ATPase and ATP-driven NAD reduction in beef heart SMP. ATP concentration was 500 µM in both systems. ATPase was measured by determination of phosphate liberation from ATP in the presence of 1 µM FCCP (12); energy-linked NAD reduction was measured spectrophotometrically (13), using an ATP-regenerating system

	ATPase		NADH production	
Conditions	Activity (µmol/min and mg)	Inhibition (%)	Activity (nmol/min and mg)	Inhibition (%)
Control	1.12	–	80.6	–
+ 50 µM aryl-N$_3$-ADP	1.03	8.04	44.6	44.7

Conclusions

1. 3'-O-esters of ADP are the first nucleotide analogs to become known that behave as highly specific and powerful inhibitors of energy transfer in photophosphorylation and oxidative phosphorylation.

2. In contrast to other energy transfer inhibitors, binding of these compounds can be precisely localized on the catalytic ADP-binding site of ATP synthetase.

226

Table 3. Effects of aryl-N_3-AMP, -ADP, and -ATP on ATP-driven NAD reduction
in beef heart SMP. Experimental conditions as given in Table 2

| Conditions | NADH production | |
	Activity (nmol/min and mg)	Inhibition (%)
Control	92.1	–
+ 40 µM aryl-N_3-ATP	86.7	5.8
+ 40 µM aryl-N_3-ADP	59.1	35.8
+ 290 µM aryl-N_3-AMP	96.7	0

3. The 3'-substituent by intra- or intermolecular steric hinderance
may either abolish phosphate binding to the enzyme, or phosphate activation, or the approach of phosphate to its acceptor, the β-phos-
phate group of the nucleotide. This may be the reason why 3'-esters
of ADP are not phosphorylated although they are bound to the enzyme
at the proper location.

4. The inhibitory effect of the ADP analogs on phosphorylation of ADP has
to be referred to their extremely high affinity to the catalytic ADP
site. It is even much higher than the affinity of the parent compound
ADP. From the competition experiments in oxidative phosphorylation as
well as in photophosphorylation, K_i values for aryl-N_3-ADP of about
1 µM are to be determined whereas the apparent K_m values in phosphorylation for ADP are at least one order of magnitude higher.

5. The reason for the high affinity of 3'-O-esters of ADP to the catalytic ADP-binding site is not completely understood. Considering the
structure of these analogs in comparison to ADP and also those ADP
analogs that are substrates in phosphorylation (3'-deoxy-, and 3'-O-
methyl-ADP), the 3'-esters contain a carboxy group that may interact
specifically with a group on the enzyme via hydrogen bonds. In addition, the 3'-substituents display a considerable hydrophobicity, although in some of them small polar groups are present. From the apparent increase of inhibitory activity with the size of the hydrophobic
side chain, it may be concluded that the substituent is able to occupy
a hydrophobic space at or close to the catalytic center. Both properties may contribute to the formation of a protein-ligand complex that
is more stable than the natural enzyme-ADP complex.

6. The fact that 3'-esters of ADP exert their inhibitory effect only
under conditions that involve an energized state of the membrane (phosphorylation, ATP-driven NAD reduction) points to the requirement for
an energy-dependent conformational transition of coupling factors for
the accessibility of the ADP-binding site to these compounds.

Acknowledgments. The skillful technical assistance of K. Edelmann, I. Luther, M.
Möller, and R. Magin is gratefully acknowledged. The work was supported by grants
from the Deutsche Forschungsgemeinschaft.

References

1. Haley, B.E., Hoffman, J.F.: Proc. Natl. Acad. Sci. USA 71, 3367-3371 (1974)
2. Wagenvoord, R.J., van der Kraan, I., Kemp, A.: Biochim. Biophys. Acta 460, 17-24 (1977)
3. Schäfer, H.J., Scheurich, P., Dose, K.: Hoppe-Seyler's Z. Physiol. Chem. 357, 278 (1976)
4. Scheurich, P., Schäfer, H.J., Dose, K.: Eur. J. Biochem., in press (1978)
5. Jeng, S.J., Guillory, R.J.: Supramol. Struct. 3, 448-468 (1974)
6. Russel, J., Jeng, S.J., Guillory, R.J.: Biochem. Biophys. Res. Commun. 70, 1225-1234 (1976)
7. Lunardi, J., Lauquin, G.J.M., Vignais, P.: FEBS Lett. 80, 317-323 (1977)
8. Schäfer, G., Onur, G., Edelmann, K., Bickel-Sandkötter, S., Strotmann, H.: FEBS Lett. 87, 318-322 (1978)
9. Strotmann, H., Bickel-Sandkötter, S., Edelmann, K., Schlimme, E., Boos, K.S., Lüstorff, J.: In: Structure and Function of Energy Transducting Membranes. Van Dam, K., van Gelder, B.F. (eds.). Amsterdam: Elsevier, 1977, pp. 307-317
10. Kalbitzer, H.H., Thorand, H.G.: J. Am. Chem. Soc. 85, 1997 (1963)
11. McLaughlin, C.S., Ingram, V.M.: Biochemistry 4, 1442-1448 (1965)
12. Summer, J.B.: Science 100, 413-418 (1944)
13. Ernster, L., Chuan-Pu Lee: In: Methods in Enzymology. Estabrook, R.W., Pullman, M.E. (eds.). New York: Academic Press, 1967, Vol. 10, pp. 738-744

Amino Acid Sequence of the Putative Protonophore of the Energy-Transducing ATPase Complex

W. SEBALD and E. WACHTER

Introduction

The energy transducing ATPase of mitochondria, chloroplasts, and bacteria is one of the most complex enzymes. It consists of at least ten different subunit polypeptides ranging in molecular weight from 60,000 to 8000 daltons. We have concentrated our efforts during the past few years on the smallest subunit of 8000 daltons, especially on its chemical characterization. This extremely hydrophobic polypeptide occurs in the complex as an oligomer, probably as a hexamer (Sebald et al., 1978) and is thus a major subunit, comprising about 10% of the total enzyme protein. Together with at least two further hydrophobic polypeptides it constitutes the membrane factor F_O (Sone et al., 1975; Sebald, 1977), which has been shown to exhibit the properties of a proton channel (Hinkle and Horstman, 1971; Okamoto et al., 1977).

This smallest ATPase subunit was detected about ten years ago by Beechey et al. (1967) by means of the ATPase inhibitor dicyclohexylcarbodiimide (DCCD). This hydrophobic carbodiimide was found to inhibit specifically and irreversibly the ATPase complex of beef heart mitochondria. By means of the radioactive compound, a DCCD-binding protein of low molecular weight was identified (Cattell et al., 1971). This protein was called a proteolipid due to its solubility in certain organic solvents. But it was never purified to homogeneity.

During these early studies, it was discussed that the DCCD-binding proteolipid plays an essential role in oxidative phosphorylation. Nevertheless, this protein lived a shadowy existence for several years, due mainly to the following two facts: (1) To organic chemists carbodiimides are known as highly reactive compounds (Khorana, 1953). Thus the specific inhibition of the ATPase complex as well as the specific binding to one protein only remained a puzzling result. (2) Due to the difficulties encountered in the purification and chemical analysis of hydrophobic proteins, the true nature of this proteolipid could not be established. The know-how for handling such proteins has been developed only in the last few years.

In the meantime, the DCCD-binding protein became one of the best characterized subunits of the ATPase complex. This applies to its function and especially to its structure. Recently, experiments have been reported which suggest that the isolated subunit when reconstituted into a lipid membrane exhibits protonophoric activity sensitive to DCCD or oligomycin (Nelson et al., 1977; Criddle et al., 1977). Thus the DCCD-binding protein may constitute the proton channel of the ATPase complex. We have studied two aspects of the amino acid sequence of this protein: (1) The DCCD-binding protein from mitochondria of different organisms as well as from bacteria and chloroplasts was analyzed in order to determine the general and typical features of the amino acid sequence and in order to see which residues have been conserved during evolution and therefore may be important for the function and structure of the putative protonophore. (2) The action of inhibitors was analyzed. The

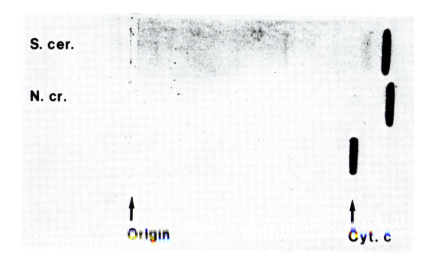

Fig. 1. SDS-gel electrophoresis of the DCCD-binding proteins from *Neurospora crassa* and *Saccharomyces cerevisiae*. The proteins were isolated as described by Sebald et al. (1978a), separated on 15% polyacrylamide gels and stained with coomassie blue. The cytochrome *c* standard was applied in a fivefold lower amount

DCCD-binding residue was identified, and amino acid exchanges could be established in oligomycin-resistant mutants from *Neurospora* and yeast. A brief summary of the results is presented in this communication.

Isolation and Amino Acid Composition of the DCCD-Binding Protein

The DCCD-binding protein was extracted from whole membrane by neutral chloroform/methanol. The protein from *Neurospora* and yeast mitochondria is thereby obtained in a pure form (Sebald and Jackl, 1975; Sebald et al., 1978a). Figure 1 shows an SDS-gel electrophoretic pattern of such preparations. The DCCD-binding proteins from *E. coli*, spinach chloroplasts (Wachter and Sebald, 1978b), and beef heart mitochondria (Graf and Sebald, 1978) are still heavily contaminated after extraction with chloroform/methanol. They can be finally purified by chromatography on DEAE- or CM-cellulose. This method was first described by Fillingame (1976) and by Altendorf (1977) for the isolation of the *E. coli* protein.

The amino acid composition of the DCCD-binding protein from five different sources is shown in Table 1. The proteins have a slightly different size — containing between 76 and 81 amino acids. All of them exhibit low polarity, especially the *E. coli* protein, which contains only 16% hydrophilic side chains. Generally, no tryptophan and histidine and only a few lysines and arginines are present. The large amount of the small amino acids glycine and alanine, which comprise about 25% of the total residues, is striking. All amino acid compositions, besides that of the chloroplast protein, were obtained by sequence analysis (Wachter and Sebald, 1978a; Wachter et al., 1978a; Wachter et al., 1978b). The sequences of the two mitochondrial proteins from *Neurospora* and yeast, as well as that of the *E. coli* protein, which was studied in cooperation with K. Altendorf, shall be described below.

230

Table 1. Amino acid composition of the DCCD-binding protein from mitochondria (*Neurospora crassa*, *Saccharomyces cerevisiae*, and beef heart), bacterial plasma membrane (*E. coli*) and chloroplasts (spinach). The composition of the chloroplast protein was determined by amino acid analysis after hydrolysis in 5.7N HCl at 105°C (Sebald, unpublished result). The composition of the other proteins is derived from sequence analysis. End groups were determined by the dansylation method. The formyl group was removed by incubating the protein for 4 h in 0.5 M methanolic HCl.

	Neuro-spora	Yeast	Beef heart	*E. coli*	Spinach chloroplasts
Lysine	2	2	2	1	1
Histidine	-	-	-	-	-
Arginine	2	1	1	2	2
Aspartic acid	4	3	3	5	2
Threonine	2	3	3	1	3
Serine	5	5	5	-	3
Glutamic acid	5	2	3	4	6
Proline	1	2	1	3	4
Glycine	11	10	12	10	11
Alanine	14	10	14	13	16
Valine	6	6	4	6	6
Methionine	4	3	3	8	2
Isoleucine	6	9	7	8	5
Leucine	11	12	9	12	11
Tyrosine	2	1	2	2	1
Phenylalanine	6	6	8	4	3
Cysteine	-	1	1	-	-
Tryptophan	-	-	-	-	-
Residues	81	76	78	79	76
End group	Tyr	F-Met	Asp	F-Met	F-Met
Polarity	24.7	21.1	21.8	16.5	22.4

Amino Acid Sequence of the DCCD-Binding Protein

Distribution of Hydrophilic and Hydrophobic Residues

One distinct feature of the amino acid sequence of all DCCD-binding proteins analyzed thus far is the concentration of hydrophilic, or hydrophobic, residues in certain stretches of the polypeptide chain (Fig. 2). This clustering is most pronounced in *E. coli*, but it is clearly seen also in the two mitochondrial proteins. In a short segment at the N-terminus, hydrophilic side chains (acidic, basic, and neutral) are accumulated. A long hydrophobic sequence of 25 amino acids follows. (*N. cr.* and *S. cer.*: positions 16 to 40; *E. coli*: positions 12 to 36.

	5	10	15	20	25
N. cr.	Tyr-Ser-Ser-Glu-Ile-Ala-Gln-Ala-Met-Val-Glu-Val-Ser-Lys-Asn-Leu-Gly-Met-Gly-Ser-Ala-Ala-Ile-Gly-Leu-
S. cer.	f-Met-Gln-Leu-*Val*-Leu-Ala-Ala-*Lys*-Tyr-Ile-*Gly*-Ala-*Gly*-Ile-Ser-Thr-*Ile-Gly-Leu*-
E. c.	f-Met-Glu-Asn-Leu-Asn-Met-Asp-Leu-Leu-Tyr-Met-Ala-Ala-Ala-Val-Met-Met-Gly-Leu-Ala-Ala-

	30	35	40	45	50
N. cr.	Thr-Gly-Ala-Gly-Ile-Gly-Ile-Gly-Leu-Val-Phe-Ala-Ala-Leu-Leu-Asn-Gly-Val-Ala-Arg-Asn-Pro-Ala-Leu-Arg-
S. cer.	Leu-*Gly-Ala-Gly-Ile-Gly-Ile*-Ala-Ile-*Val*-Phe-Ala-*Ala*-Leu-Ile-*Asn*-Gly-Val-Ser-*Arg-Asn-Pro*-Ser-Ile-Lys-
E. c.	Ile-*Gly-Ala-Ala-Ile-Gly-Ile-Gly-Ile*-Leu-Gly-Gly-Lys-Phe-Leu-Gln-Gly-Ala-Ala-*Arg*-Gln-*Pro*-Asp-Leu-Ile-

	55	60	65	70	75
N. cr.	Gly-Gln-Leu-Phe-Ser-Tyr-Ala-Ile-Leu-Gly-Phe-Ala-Phe-Val-Glu-Ala-Ile-Gly-Leu-Phe-Asp-Leu-Met-Val-Ala-
S. cer.	Asp-Thr-Val-*Phe*-Pro-Met-*Ala-Ile-Leu*-Gly-Phe-*Ala-Leu*-Ser-Glu-Ala-*Thr-Gly-Leu-Phe*-Cys-*Leu-Met-Val*-Ser-
E. c.	Pro-Leu-Leu-Arg-Thr-Gln-Phe-Phe-Ile-Val-Met-Gly-Leu-Val-Asp-*Ala*-Ile-Pro-Met-Ile-Ala-Val-Gly-Leu-Gly-

	80
N. cr.	Leu-Met-Ala-Lys-Phe-Thr
S. cer.	Phe-Leu-Leu-*Phe*-Gly-Val
E. c.	Leu-Tyr-Val-Met-*Phe*-Ala-Val-Ala

Fig. 2. Amino acid sequences of the DCCD-binding proteins from *Neurospora crassa* (*N. cr.*), *Saccharomyces cerevisiae* (*S. cer.*), and *E. coli* (*E. c.*). The sequences were determined by automated solid-phase Edman degradation of the whole polypeptide and of peptides obtained after cleavage with cyanogenbromide, N-bromosuccinimide, pepsin, and trypsin as described by Wachter and Sebald (1978a) and Wachter et al. (1978a). The numbering is according to the *Neurospora* polypeptide.

The numbering is according to the *Neurospora* sequence.) In the mitochondrial proteins a few serines and threonines are present in this segment. Most of the hydrophilic residues are concentrated in the middle of the polypeptide chain (*N. cr.* and *S. cer.*: positions 41 to 52;

E. coli: positions 37 to 56). Toward the end of the chain a second sequence of about 25 residues is found which is largely hydrophobic (*N. cr.* and *S. cer.*: positions 53 to 78; *E. coli*: positions 57 to 82). Just in the center, however, a single acidic side chain (position 65) interrupts the hydrophobic character. As will be shown below, this residue reacts with the inhibitor DCCD. In *E. coli* and yeast, the C-terminus is hydrophobic. In *Neurospora*, two hydrophilic residues are present.

This striking distribution of hydrophobic and hydrophilic amino acids suggests a certain arrangement of the polypeptide chain in the membrane. It can be visualized that the two hydrophobic segments traverse the membrane lipid bilayer, whereas the polar regions are either exposed to the water phase or are involved in specific contacts with the other subunits of the complex. This as a working hypothesis of course has to be verified, e.g., by surface labeling experiments with membrane non-permeating probes.

Homology Among Different Species
_____

A more specific picture emerges when the amino acid sequences from different species are compared. The two mitochondrial proteins from *Neurospora* and yeast (Fig. 2) exhibit a high degree of homology, even though the two microorganisms are as unrelated from an evolutionary point of view as are, e.g., *Neurospora* and higher animals (Fitch and Margoliash, 1967). Forty positions are occupied by identical amino acids, as indicated by the underlined residues. In the case of cytochrome *c*, a similarly high degree of homology has been found between these two organisms. The slow alteration of cytochrome *c* during evolution has been attributed to the fact that this protein is involved in specific interactions with other proteins, namely with oxidase and reductase (Ferguson-Miller et al., 1978). A similar constraint may be exerted on the DCCD-binding protein, since this polypeptide has to interact with other subunits of the ATPase complex.

In comparing the two mitochondrial polypeptides, the hydrophilic sequence at the N-terminus appears to be quite different. The same applies to the C-terminus. In contrast, a remarkable homology exists in the polar segment in the middle of the polypeptide chain. The similarity is even greater when isopolar substitutions are considered. Similar or identical hydrophilic residues occur in positions 14, 41, 45, 46, 50, 52, and 65 (*Neurospora* protein numbering). The most conserved parts of the sequence, however, occur in the two hydrophobic segments (positions 16 - 40 and 53 - 78). At first view, this finding is unexpected. It could indicate that these hydrophobic amino acids are not only in contact with the lipid phase, but also exert a more specific role in the ATPase complex.

As is also documented in Figure 2, the protein from *E. coli* ATPase is clearly homologous to the mitochondrial proteins. In comparison with *Neurospora* and yeast, 18 and 14 positions, respectively, are identical. The ten underlined positions are identical in all three sequences. This may be considered as final proof of the homology of the bacterial and mitochondrial ATPase complexes. This of course has already been inferred from the many similarities in function and subunit composition. It might be concluded that this highly complicated enzyme protein was invented by the cell very early during evolution.

The few hydrophilic positions that are identical or similar in all three polypeptides shall be emphasized again. The only conserved acidic residue is found in the center of the second hydrophobic sequence

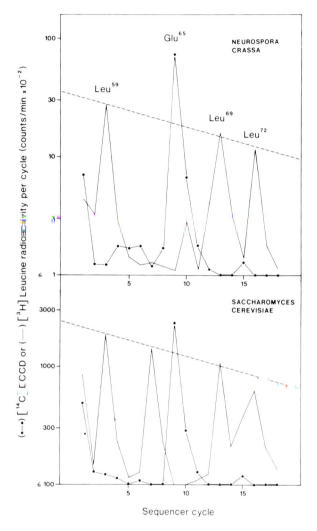

Fig. 3. Identification of the DCCD-binding residue in the DCCD-binding protein from *Neurospora crassa* and *Saccharomyces cerevisiae*. The (^{14}C)DCCD label of the DCCD-modified protein was recovered exclusively in a 17-residue fragment. This labeled fragment was coupled to γ-aminopropyl glass via a C-terminal homoserinelactone, and then submitted to 18 cycles of an automated Edman degradation. The protein was mass labeled with (^{3}H) leucine. The (^{14}C)DCCD radioactivity (● ●) and (^{3}H) leucine radioactivity (——) recovered at each cycle are plotted in a logarithmic scale. The *slope* of the *broken line* corresponds to a repetitive yield of about 95%

(position 65). In addition, one basic and two neutral positions (45, 41, and 46) are conserved. A further basic and neutral amino acid (positions 50 and 52) of the mitochondrial proteins can be correlated with *E. coli* (positions 54 and 56), when a four-residue insertion is postulated in the bacterial sequence. Such an insertion appears to be highly probable, since a phenylalanine in position 54 then also can be aligned.

In conclusion, there appears to be a minimal set of common features in all three proteins, including: (1) two hydrophobic sequences of about 25 residues, especially a glycine-rich segment (position 27 - 33); (2) one acidic residue in the center of the second hydrophobic sequence; (3) a central polar loop including two basic and three neutral hydrophilic residues and also a proline; (4) the polar character of the N-terminal sequence.

Action of Inhibitors

DCCD-Binding Residue

Further information on the specific role of individual amino acid residues may be obtained by inhibitor binding studies. Obviously, the
identification of the DCCD-binding site is of special interest in this
respect. The crucial experiment is shown in Figure 3. The DCCD-binding
protein was reacted with the ^{14}C-labeled inhibitor (Sebald et al.,
1978), and different fragments were isolated (Wachter and Sebald,
1978a,b). The bound (^{14}C)DCCD radioactivity could be traced back to
a 17-residue fragment, both in *Neurospora* and yeast. This labeled fragment was submitted to 18 cycles of an automated solid-phase Edman degradation. The continuous line in Figure 3 indicates the cleaved-off
leucine residues, which are measured by their (^{3}H)radioactivity, since
the protein was mass labeled with (^{3}H)leucine. The (^{14}C)DCCD radioactivity (closed circles) appears nearly exclusively at cycle 9, both
in *Neurospora* and yeast. This DCCD-binding residue corresponds to the
glutamic acid (position 65) present in the center of the second hydrophobic sequence (see Fig. 2). In *E. coli*, DCCD was found, by the same
method, to be bound to the aspartic acid in the identical position.

Thus, the only conserved acidic residue is further distinguished by
a specific reaction with the inhibitor DCCD. The location of this amino
acid in the center of a hydrophobic sequence may be correlated with
the observation that the ATPase complex is inhibited by hydrophobic
carbodiimides only, and not by water-soluble ones (Beechey et al.,
1967).

Amino Acid Exchanges Conferring Oligomycin-Resistance

Oligomycin is another inhibitor of the ATPase complex that acts like
DCCD on the membrane factor (Bulos and Racker, 1968). Thus, it was a
distinct possibility that the DCCD-binding protein also is involved
in the binding of oligomycin. Actually, amino acid exchanges could be
established in the DCCD-binding protein from oligomycin-resistant
mutants of *Neurospora* and yeast. Two different mutants from *Neurospora*
exhibited a phenylalanine-serine exchange in position 61 and a phenyl-
alanine-tyrosine exchange in position 70, respectively (Fig. 4) (Sebald-
Althaus et al., 1978). Oligomycin-resistant mutants from yeast have
been analyzed in cooperation with A. Tzagoloff (Sebald et al., 1978b).
Again, amino acid exchanges were detected: in one mutant a leucine-
phenylalanine substitution and in another mutant a cysteine-serine
substitution (Fig. 4). All amino acid exchanges can be explained by
a single base mutation. It may be mentioned, by the way, that the structural gene of the yeast protein is located on mitochondrial DNA (locus
OLI 1), whereas the corresponding *Neurospora* gene is located in the nu-
cleus (linkage group VII).

Remarkably, all mutations leading to oligomycin resistance were found
in the second hydrophobic segment (see Fig. 2), very near the DCCD-
binding residue. This location seems to be more than accidental, even
though only four exchanges have been identified thus far. Possibly,
this part of the polypeptide chain represents a inhibitor-binding domain. At the moment, however, it cannot be excluded that oligomycin
still binds to the resistant ATPase complex and that the mutations
interfere with the still unknown inhibitory mechanism of the antibiotic.

```
Neurospora crassa
                                    60              65                  70
Wild type            -Ala-Ile-Leu-Gly-Phe-Ala-Phe-Val-Glu-Ala-Ile-Gly-Leu-Phe-Asp-Leu-Met-
Mutant MSE-4         -Ala-Ile-Leu-Gly-Phe-Ala-Phe-Val-Glu-Ala-Ile-Gly-Leu-Thr-Asp-Leu-Met-
Mutant MSE-5         -Ala-Ile-Leu-Gly-Ser-Ala-Phe-Val-Glu-Ala-Ile-Gly-Leu-Phe-Asp-Leu-Met-

Saccharomyces cerevisiae
Wild type D273-10B   -Ala-Ile-Leu-Gly-Phe-Ala-Leu-Ser-Glu-Ala-Thr-Gly-Leu-Phe-Cys-Leu-Met-
Mutant D273-10B/A21  -Ala-Ile-Phe-Gly-Phe-Ala-Leu-Ser-Glu-Ala-Thr-Gly-Leu-Phe-Cys-Leu-Met-
Mutant D273-10B/A32  -Ala-Ile-Leu-Gly-Phe-Ala-Leu-Ser-Glu-Ala-Thr-Gly-Leu-Phe-Ser-Leu-Met-
```

Fig. 4. Alteration of the amino acid sequence of the DCCD-binding protein from oligomycin-resistant mutants from *Neurospora crassa* and *Saccharomyces cerevisiae*. All mutants were selected by their increased resistance to oligomycin in vivo. This resistance was retained in vitro at the level of the mitochondrial ATPase complex

Discussion

The present results shall be discussed in connection with a possible role of the DCCD-binding protein as a protonophore. In the last few years, several models have been proposed to explain on a molecular level the translocation of protons across biological membranes. One type of mechanism, considered, e.g., by Williams (1978) and by Nagle and Morowitz (1978), includes a net of hydrogen bonds provided by polar amino acid side chains of the protonophoric protein. It has been estimated that a minimal number of 10 to 20 of such residues is necessary to traverse the membrane. Another type of mechanism, developed by Boyer (1975), involves the migration of a charged amino acid side chain, or, the exposure of such a group to different sides of the membranes. Obviously, the latter mechanism depends on conformational changes of the protonophoric protein. Based on the assumption that the DCCD-binding protein alone constitutes the proton channel of the ATPase complex, then it appears to be difficult to construct a net of hydrogen bonds with the few essential hydrophilic residues present even in the hexameric form of the protein. Such considerations may favor a proton translocation via a charged group migration. At the moment, such a discussion remains highly speculative. Nevertheless, the sequence data provide a solid basis for further studies of this problem.

References

Altendorf, K.: FEBS Lett. 73, 271-275 (1977)

Beechey, R.B., Roberton, A.M., Holoway, C.T., Knight, I.G.: Biochemistry 6, 3867-3879 (1967)

Boyer, P.D.: FEBS Lett. 58, 1-6 (1975)

Bulos, B., Racker, E.: J. Biol. Chem. 243, 3891-3900 (1968)

Cattell, K.J., Lindop, C.R., Knight, I.G., Beechey, R.B.: Biochem. J. 125, 169-177 (1971)

Criddle, R.S., Packer, L., Shieh, P.: Proc. Natl. Acad. Sci., U.S.A. 74, 4306-4310 (1977)

Ferguson-Miller, S., Brautigan, D.L., Margoliash, E.: J. Biol. Chem. 253, 149-159 (1978)

Fillingame, R.H.: J. Biol. Chem. 251, 6630-6637 (1976)

Fitch, W.M., Margoliash, E.: Science 155, 279 (1967)

Graf, Th., Sebald, W.: FEBS Lett., submitted for publication

Hinkle, P.C., Horstman, L.L.: J. Biol. Chem. 246, 6024-6028 (1971)

Khorana, H.G.: Chem. Rev. 53, 145-166 (1953)

Nagle, J.F., Morowitz, H.J.: Proc. Natl. Acad. Sci. U.S.A. 75, 298-308 (1978)

Nelson, N., Eytan, E., Notsani, B., Sigrist, H., Sigrist-Nelson, K., Gitler, C.: Proc. Natl. Acad. Sci. U.S.A. 74, 2375-2378 (1977)

Okamoto, H., Sone, N., Hirata, H., Yoshida, K., Kagawa, Y.: J. Biol. Chem. 252, 6125-6131 (1977)

Sebald, W.: Biochim. Biophys. Acta 463, 1-27 (1977)

Sebald, W., Graf, Th., Lukins, H.B.: Eur. J. Biochem., submitted for publication (1978a)

Sebald, W., Jackl, G.: In: Electron Transfer Chains and Oxidative Phosphorylation. Quagliariello et al. (eds.). Amsterdam: North-Holland, 1975, pp. 193-198

Sebald, W., Wachter, E., Tzagoloff, A.: manuscript in preparation (1978b)

Sebald-Althaus, M., Sebald, W., Wachter, E.: manuscript in preparation (1978)

Sone, N., Yoshida, M., Hirata, H., Kagawa, Y.: J. Biol. Chem. 250, 7917-7923 (1975)

Wachter, E., Sebald, W.: Eur. J. Biochem., submitted for publication (1978a)

Wachter, E., Sebald, W.: manuscript in preparation (1978b)

Wachter, E., Altendorf, K., Sebald, W.: manuscript in preparation (1978a)

Wachter, E., Sebald, W., Graf, Th.: manuscript in preparation (1978b)

Williams, R.J.P.: FEBS Lett. 85, 9-19 (1978)

Chemical Mechanisms in Energy Coupling

E. BÄUERLEIN

As an introduction to the panel discussion I shall present a short review on chemical mechanisms for ATP synthesis by electron transport
phosphorylation, which have been proposed in chemical equations or are
on the tip of the tongue, but not expressed precisely in chemical terms.
The different mechanistic proposals will be compared, if possible, with
the related experience of inorganic and organic chemistry to give a
chemical basis for the discussion of the enzymatic process.

I shall begin with Mitchell's hypothesis in which there is a direct
coupling between the proton motive force, in form of the transmembrane
proton gradient, and the phosphorylation reaction, which apparently
occurs without the interaction of a protein.

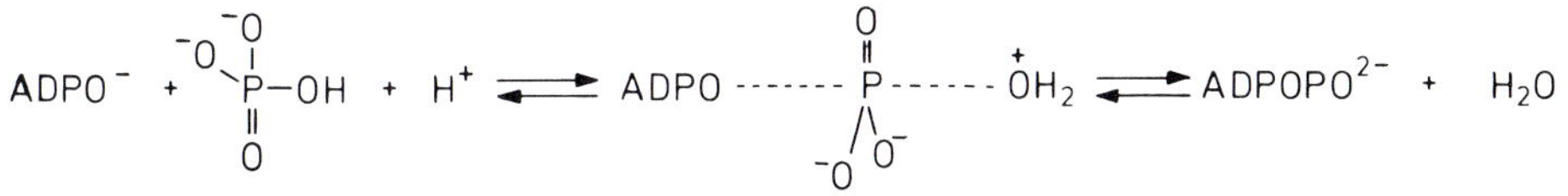

Fig. 1. S_{N_2}-mechanism for ATP synthesis (Mitchell, 1974)

This proposal (Mitchell, 1974) shows a nucleophilic substitution of
the second order (S_{N_2}), in which *water formation and phosphorylation* take
place in *one step* (Fig. 1). The difficulties inherent to this mechanism
are that an acid-base reaction within the molecule or between ADP-
anion and the postulated oxonium group ($-\overset{+}{O}H_2$) of the phosphate is much
more probable than nucleophilic attack at the posphorus center.

Recently Mitchell (1977) has proposed that *water formation precedes the
phosphorylation*, thus attempting to eliminate the difficulties of the
S_{N_2} mechanism. A dissociation of the P-O bond to a phosphorylium ion
and an oxygen two minus (O^{2-}) is proposed, and this oxygen two minus
is assumed to be translocated to protons of the F_O-side of the ATPase
complex; thus the ADP-anion is prevented from being protonated. Since
the translocation of oxygen two minus (O^{2-}) appears improbable, even
to Mitchell, protons are suggested to be translocated to phosphate
(this is formally equivalent to the O^{2-} translocation) in such a way
that the ADP anion is not protonized. To my mind this implies water
formation by concomitant formation of metaphosphate with the subsequent

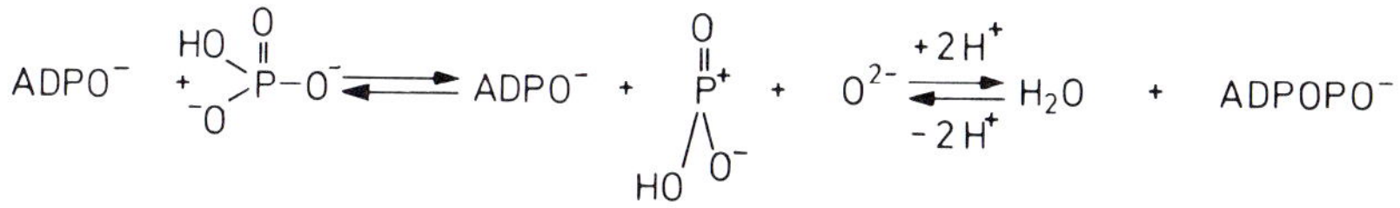

Fig. 2. O^2-group translocation after phosphorylium-oxyanion

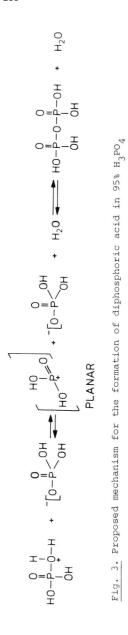

Fig. 3. Proposed mechanism for the formation of diphosphoric acid in 95% H_3PO_4

nucleophilic attack of the ADP-anion, this is identical with a S_{N_1}-mechanism.

This S_{N_1}-mechanism, which may be considered as the minimum chemistry for such a proton-driven phosphorylation reaction, is supported by the experience of inorganic chemistry, where diphosphoric acid is formed in at least 95% ortho-phosphoric acid (Huhti and Gartaganis, 1956; Keisch et al., 1958). The most probable mechanism is the protonization of the undissociated acid, thus forming a good leaving group.

The elimination of water probably yields planar metaphosphoric acid in the transition state, which is open to nucleophilic attack by the conjugated base, the monoanion of ortho-phosphoric acid. This model reaction is one of the essential features of William's (1961, 1978) hypothesis of ATP-formation by energized protons; but it is questionable whether such a superacid could be formed at the active center of an enzyme.

Finally, in the discussion of the minimum chemistry of the proton-driven ATP synthesis it should be borne in mind that Mitchell tries to verify the basic biochemical experience of the common intermediate which couples two reactions together — I cite Mitchell (1975), "in this case, the O^{2-} group of P_i and H_2O that is common to the proton translocation reaction and to the reversible ATP hydrolysis reaction in the F_oF_1 ATPase complex".

A nonphosphorylated high-energy compound was postulated in the original chemical hypothesis to be the common intermediate in electron transport phosphorylation, a hypothesis which has been recently reactivated by Griffiths (1977); he has reported that oleoyl-S-lipoate is this common intermediate and that lipoic acid and oleic acid are specific cofactors of oxidative phosphorylation.

ATP is formed analogous to substrate level phosphorylation and F_1 or F_oF_1 reacts as an oleoyl-thiokinase. Criddle (1977) reported recently that he was able to confirm this finding by forming ATP by incubating F_1 with the substrates ADP and oleoyl-phosphate.

It is obscure how the energy of electron transport is coupled to the synthesis of oleoyl-S-lipoate. Boyer (1977) proposed that the reduction of lipoic acid could be followed by the formation of the seven-ring thiolactone of the dihydrolipoic acid, which may be driven by energy-linked conformational events. Personally I prefer the analogy to chemical experience in that this thiolactone may be synthesized in acidic solution, i.e., by protons (Fig. 5), as the five- and six-ring thiolactones. However, if the thiolactone is cleaved by oleate (or phosphate), a mixed anhydride would be formed, and the reaction with another molecule of dihydrolipoic acid would then be necessary to obtain a thioester, if possible, the monooleoyl thioester of the dihydrolipoic acid.

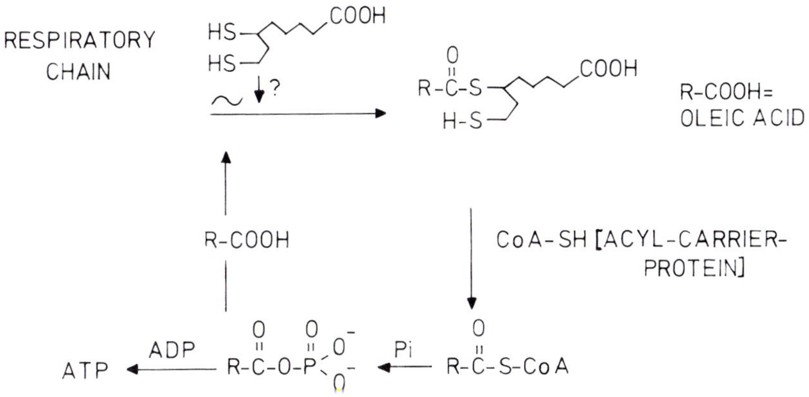

Fig. 4. The "oleoyl cycle" of oxidative phosphorylation (Griffiths, 1977)

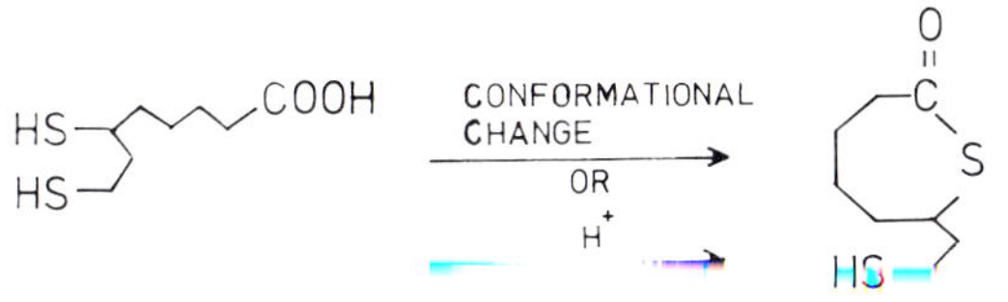

Fig. 5. Proposed formation of dihydrolipoic acid thiolactone

After pure inorganic and coenzyme-like chemistry, a third approach to ATP synthesis may be the involvement of functional groups of a protein, namely the ATP synthase, a proposal which includes once more a discussion of phosphorylated intermediates or corresponding transition states. In a recent working hypothesis which extended an earlier proposal of a protonized disulfide (Bäuerlein and Kiehl, 1976) I tried to combine the action of the two classical protons of the chemiosmotic hypothesis with protein chemistry, the activating step of which is assumed to be a proton-induced conformational change (Fig. 6).

With the first proton a disulfide bond is formed with the concomitant production of water; the disulfide group may be substituted by a thio-lactone, tyrosyllactone or carboxylimidazole group (Fig. 7).

The second proton induces a conformational changes which generates bond polarization of the disulfide, or one of the other three groups, thus facilitating the attack of the weak nucleophile phosphate anion. The *phosphorylation step*, which is thus *separated from the water formation* and the *direct interaction of the second proton*, has to be irreversible because of the strain of the attached disulfide. This postulated irreversibility reflects the findings of Penefsky (1974), Holland et al. (1974), Pedersen (1975), and Strotman et al. (1977) whose experiments with nucleotide analogs point to two active centers, or two different states

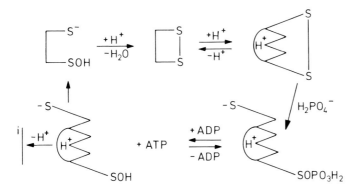

Fig. 6. A proton-driven ATP synthesis by chemical intermediates

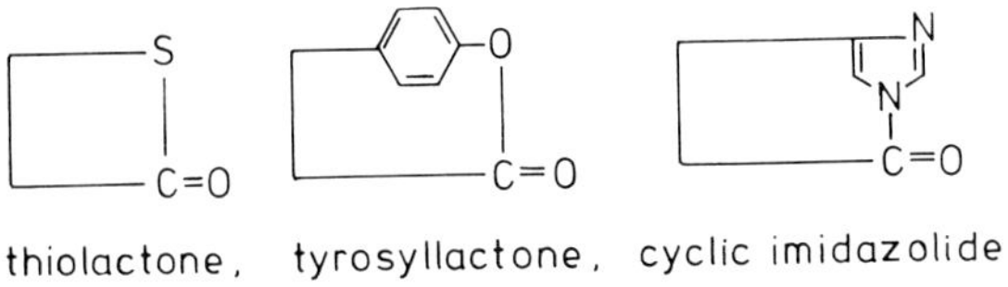

thiolactone, tyrosyllactone, cyclic imidazolide

Fig. 7. Other cyclic protein functions instead of disulfide

of the active center, one for ATP synthesis and another for the ATPase reaction. Another essential feature of this reaction scheme is that no phosphorylated protein can be isolated, because the proton-induced conformation would be released during any isolation procedure used, and the disulfide bond resynthesized.

A fourth approach to the mechanism of ATP synthesis are the extensive ^{18}O-exchange experiments of Paul D. Boyer, which aim directly at the chemical mechanism of the synthase. The present state of his conformational coupling hypothesis depends upon three crucial experimental observations:

1. The $P_i \rightleftarrows$ HOH exchange in tightly coupled mitochondria, which is absolutely dependent on the presence of ADP (Jones and Boyer, 1969).

2. The finding of Cross and Boyer (1973) that a membrane-bound mitochondrial protein, which at first appeared to be a phosphorylated intermediate of ATP synthesis, had bound tightly ATP, but not covalently the phosphoryl group.

3. The discovery of an oligomycin and DCCD-sensitive, uncoupler-insensitive $P_i \rightleftarrows$ HOH exchange accompanying the formation of P_i from ATP, an exchange which is apparently not dependent on a transmembrane proton gradient (Boyer et al., 1977).

Based on detailed studies of this exchange reaction, an alternating catalytic site mechanism was formulated, according to which an energy-requiring conformational change of F_1 promotes ADP and P_i binding in a manner capable of forming ATP at one catalytic site of the enzyme and simultaneously promoting the release of ATP from another catalytic

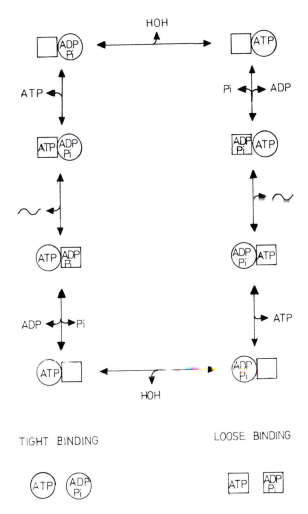

TIGHT BINDING   LOOSE BINDING

Fig. 8. Conformational coupling with subunit asymmetry (Boyer et al., 1977)

site (Boyer et al., 1977). Concerning the active center of ATP synthase, the reported P_i ⇄ HOH exchange may be regarded as resulting from reversal of the hydrolysis of bound ATP to bound ADP and P_i prior to the release of P_i to the medium. The capacity for reversible synthesis of ATP at the catalytic site has been suggested to depend upon the conformational state of the enzyme complex, which may be attained by energy input from oxidation or ATP cleavage. If one considers that the energy transduced through such conformational changes, serves *to release* ATP from *and to promote* productive binding of ADP and P_i to the enzyme, then if one reduces this proposal to the ATP-synthesizing step, it appears as if the productive binding of ADP and P_i would be a fixed transition state of a S_{N2} mechanism; the molecules ADP + P_i on the reaction coordinate could roll down to the reaction products ATP + H_2O only by energy input. Otherwise the β,γ pyrophosphate bond has to be formed before the release of the molecule, and it is obscure from where the energy of the pyrophosphate bond comes.

Concerning the binding of orthophosphate to F_1 the finding of Kasahara and Penefsky (1977) is of considerable interest, namely that the monoanion and not the dianion of phosphate is the binding species. Experiments with ATP and ADP as well as with analogs of these compounds suggested that the P_i binding site might be located at or near an adenine nucleotide site. Thus the hypothetical possibility is provided that the monoanion plus ADP could react to ATP plus H_2O; an event which is not possible with the dianion, because OH^- has to be expelled instead of H_2O. Finally, I want to mention the important experiments of Strotmann et al. (1977), which make uncertain a lot of evidence, which we felt was correct. Their systematic variation of nucleotide analogs makes it questionable whether the tightly bound nucleotides are really involved in the catalytic process of ATP synthesis.

With these sporadic remarks on binding studies, I shall finish my short review on phosphate chemistry, which may serve as a chemically orientated introduction to the discussion.

References

Bäuerlein, E., Kiehl, R.: FEBS Lett. 61, 68-71 (1976)

Boyer, P.D.: Ann. Rev. Biochem. 46, 965 footnote (1977)

Boyer, P.D., Gresser, M., Vinkler, Ch., Hackney, D., Choate, G.: In: Structure and Function of Energy-Transducing Membranes. van Dam, K., van Gelder, B.F. (eds.). Amsterdam: Elsevier Scientific Publishing Company, 1977, p. 261-274

Criddle, R.S.: presented at the Gordon Conference on Energy Coupling Mechanisms. Andover, N.H. Procter Academy, June 26 - July 1st (1977)

Cross, R.L., Boyer, P.D.: Biochem. Biophys. Res. Comm. 51, 59-66 (1973)

Griffiths, D.: Biochem. Soc. Trans. 5, 1283-1285 (1977)

Holland, P.C., La Belle, W.C., Lardy, H.A.: Biochemistry 13, 4549-4593 (1974)

Huhti, A.L., Gartaganis, P.A.: Canad. J. Chem. 34, 785-797 (1956)

Jones, D.H., Boyer, P.D.: J. Biol. Chem. 244, 5767-5772 (1969)

Kasahara, M., Penefsky, H.S.: In: Structure and Function of Energy-Transducing Membranes. van Dam, K., van Gelder, B.F. (eds.). Amsterdam: Elsevier Scientific Publishing Company, 1977, p. 295-305

Keisch, B., Kennedy, J.W., Wahl, A.C.: J. Amer. Chem. Soc. 80, 4778-4782 (1958)

Mitchell, P.: FEBS Lett. 43, 189-199 (1974)

Mitchell, P.: FEBS Lett. 50, 96 (1975)

Mitchell, P.: FEBS Lett. 85, 9-19 (1977)

Pedersen, P.L.: Biochem. Biophys. Res. Comm. 64, 610-616 (1975)

Penefsky, H.S.: J. Biol. Chem. 249, 3579-3585 (1974)

Strotman, H., Bickel-Sandkötter, S., Edelmann, K., Schlimme, E., Boos, K.S., Lüstdorff, J.: In: Structure and Function of Energy-Transducing Membranes. van Dam, K., van Gelder, B.F. (eds.). Amsterdam: Elsevier Scientific Publishing Company, 1977, p. 307-317

Williams, R.J.P.: Theoret. Biol. 1, 1-13 (1961)

Williams, R.J.P.: FEBS Lett. 85, 9-19 (1978)

On the Role and Interrelation of Nucleotide Binding Sites in Oxidative Phosphorylation

G. SCHÄFER

In connection with the problem of chemical mechanisms of oxidative phosphorylation the role of the different nucleotide binding sites in coupling factors (F1) has created substantial controversies which have not yet been solved and which may require a critical reevaluation.

Both membrane bound and isolated coupling ATPases have "tight", as well as other binding sites, which have been termed rapidly reversible, loose sites or catalytic sites. The kinetic behavior of the "tight" binding sites, however, differs by orders of magnitude in membrane-bound versus isolated ATPases, which exchange their tightly bound nucleotides extremely slowly (for review see (1)).

The number of tightly bound nucleotides is generally assumed to be 3 mol/mol ATPase; one additional site of ATPase is easily accessible.

Contradictory models of the function of these nucleotide binding sites have been proposed, ranging from the existence of entirely different sites for ATP hydrolysis and coupled ATP synthesis (2, 3, 4) to a multisite model involving all four binding sites in a cyclic revolving manner in the process of phosphorylation (5, 6). It has also been suggested that there may be one catalytic site together with "tight" bind ind sites, operating as regulatory sites (7 - 10).

A careful review of the available literature suggests that many discrepancies in understanding the role of these binding sites may result mainly from mutual application of findings obtained with completely different working models: i.e., intact, membrane-bound ATPases as in submitochondrial particles on the one hand, and solubilized ATPases (F1) from different sources on the other.

The most prominent difference between these preparations is that *only* membrane-bound coupling ATPases can be "energized"; hence, it is generally accepted that energization causes conformational changes in the enzyme (5, 11 - 13). In fact, only energized coupling ATPases can exchange their tightly bound nucleotides rapidly. Moreover, this exchange presumably is even faster than phosphorylation (14). This situation largely impedes the testing of models that require an *interconversion* of the four nucleotide binding sites during continuous phosphorylation.

Soluble ATPases thus lack this essential property of undergoing energization and the corresponding conformational changes. It is also uncertain whether other modifications of the enzyme occur upon removal from the membrane.

The function and properties of isolated, solubilized coupling factors may therefore more closely resemble the functions of the membrane-bound ATPase under uncoupled conditions.

Many of the arguments concerning the function of tight and catalytic sites have been derived from inhibitor studies or from comparisons of specificity to various nucleotide analogs (15, 16). There are ADP ana-

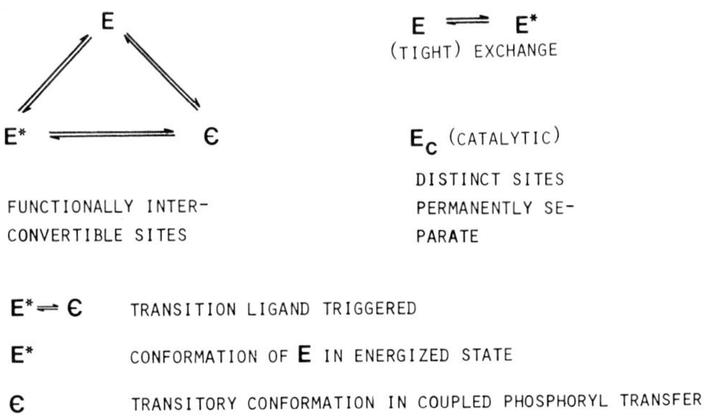

E

E ⇌ E*
(TIGHT) EXCHANGE

E* ⇌ Є

E_c (CATALYTIC)
DISTINCT SITES
PERMANENTLY SE-
PARATE

FUNCTIONALLY INTER-
CONVERTIBLE SITES

Fig. 1.

E* ⇌ Є   TRANSITION LIGAND TRIGGERED

E*   CONFORMATION OF E IN ENERGIZED STATE

Є   TRANSITORY CONFORMATION IN COUPLED PHOSPHORYL TRANSFER

logs, for example, that cannot be phosphorylated by submitochondrial particles (17), whereas the corresponding ATP analogs are fairly well hydrolyzed by ATPase. Other series of nucleotides show a certain specificity in energy-linked exchange at the tight binding sites, but their properties in coupled phosphorylation or with ATPase may be completely different (16).

It is dangerous to draw conclusions from these differences in specificity concerning a complete and permanent separation of catalytic and tight binding sites for the following reasons:

1. Binding studies are usually carried out under nonphosphorylating equilibrium conditions and in long-term experiments, leading to a final equilibrium level of exchange, whereas properties of nucleotides or inhibitors in phosphorylation are necessarily investigated in an open system characterized by irreversible fluxes. It may lead to erroneous conclusions, however, to compare *equilibrium levels* obtained with one system to *rates* observed with another system operating under entirely different conditions.

2. Properties of inhibitors or analogs in uncoupled systems or solubilized ATPases may not be comparable with their respective properties in coupled processes, which probably involve conformational transitions of the enzyme that are absent in the nonenergized state.

A more general model than the assumption of permanently distinct tight binding- and catalytic sites emerges from the view that both types of sites may be *functionally interconvertible* and that their momentary appearance is determined only by the state of the system: the effect of energization, phosphorylating, nonphosphorylating, or uncoupled conditions, and environmental factors such as ionic strength, divalent cations, and others.

To make this clearer, let us consider a single site of the membrane-bound coupling ATPase under the conditions demonstrated in Figure 1.

On the right side the situation is shown for fixed, separated functional sites: the tight site that exchanges nucleotides upon energization (E, E*) and the catalytic site (E_c). On the left, *interconvertible states* of the enzyme are symbolized by different lettering, E* designating the energized state. The following properties of the proposed model are postulated:

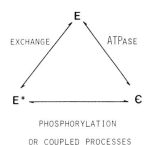

Fig. 2.

EXCHANGE   ATPASE

PHOSPHORYLATION
OR COUPLED PROCESSES

E*→Є   TRANSITION LIGAND TRIGGERED

E*   CONFORMATION IN ENERGIZED STATE

It requires three different functional states corresponding to three different conformational states E, E*, and Є.
It implies that any single nucleotide-binding site can undergo interconversion between these states under appropriate conditions.
The ligand affinity of each conformation may also be different.
Finally, the most important feature, these conformational transitions are *ligand
triggered*, which is a common characteristic of many enzymatic interactions. Thus,
for example, the presence of P_i and of ATP (or an equivalent analog) would be required to induce the proposed transition from E' to Є

This scheme easily accounts for various types of nucleotide interactions or inhibitor functions:

Whether or not a nucleotide analog is being phosphorylated and can
drive coupled processes depends only on the question of whether the
(enzyme-ligand)-complex can undergo the E* ⇌ Є transition or not. If
this is impossible, a nucleotide still may be exchangeable at the
tight sites or be hydrolyzed by ATPase.

This idea is also consistent with those proposed by other laboratories,
suggesting that, during phosphorylation, all nucleotide-binding sites
are equivalent and pass through the same conformational cycles (5, 6).
Accordingly, this would imply that the tight sites are directly involved in the phosphorylating or energy-transducing process, which as
yet has not been ruled out.

Figure 2 shows the different catalytic functions of membrane-bound or
isolated coupling ATPases and how they may be related to the single
transitions as proposed in the above scheme.

It should be emphasized finally that the ideas outlined by this scheme
should not be interpreted in terms of a single-site catalytic model.
As with other hypothetical dual-or multisite-models (5, 6, 16), the
problem of cooperation or of sequential interaction of single binding
sites for adenine-nucleotides requires the design of new experimental
approaches with much higher resolution than those presently available.
In other words, on/off rates must be determined at least within a milli-
second range.

Thus, the general conclusion is that one should be careful about interpreting the meaning of nucleotide-binding sites of coupling factors

and that, in light of the present state of knowledge, a clear-cut decision between models with "fixed-function" sites or "interconvertible function" sites cannot yet be made.

References

1. Harris D.A.: Biochim. Biophys. Acta 463, 245-273 (1978)
2. Holland, P.C., La Belle, W.C., Lardy, H.A.: Biochemistry 13, 4549-4573 (1978)
3. Pedersen, P.L.: Biochem. Biophys. Res. Commun. 64, 610-616 (1975)
4. Penefsky, H.S.: J. Biol. Chem. 249, 3579-3585 (1974)
5. Harris, D.A., Radda, G.K., Slater, E.C.: Biochim. Biophys. Acta 459, 560-572 (1977)
6. Kayalar, C., Rosing, J., Boyer, P.D.: J. Biol. Chem. 252, 2486-2491 (1977)
7. Hillborn, D.A., Hammes, G.G.: Biochemistry 12, 983-990 (1973)
8. Schuster, S.M., Ebel, R.E., Lardy, H.A.: J. Biol. Chem. 250, 7848-7853 (1975)
9. Schuster, S.M., Gerschen, R.J., Lardy, H.A.: J. Biol. Chem. 251, 6705-6710 (1976)
10. Kagawa, Y.: this volume, pp. 195-219 (1978)
11. Boyer, P.D.: In: Oxydases and Related Redoxsystems. King, T.E., Mason, H.S., Morrison, M. (eds.). New York: J. Wiley and Sons, 1965, pp. 994-1008
12. Boyer, P.D.: Proc. Natl. Acad. Sci. U.S.A. 70, 2837-2839 (1973)
13. Rosing, J., Kayalar, C., Boyer, P.D.: J. Biol. Chem. 252, 2478-2485 (1977)
14. Rosing, J., Smith, D.J., Kayalar, C., Boyer, P.D.: Biochem. Biophys. Res. Commun. 72, 1-8 (1976)
15. Lardy, H.A., Schuster, S.M., Ebel., R.E.: J. Supramol. Struct. 3, 214-221 (1975)
16. Harris, D.A., Gomez-Fernandez, J.C., Klungsøyr, L., Radda, G.K.: In: Structure and Function of Energy Transducing Membranes. Van Dam, K., Van Gelder, B.F. (eds.). Amsterdam: Elsevier, 1977, pp. 319-327
17. Barzu, O., Kiss, L., Bojan, O., Niac, G., Mantsch, H.H.: Biochem. Biophys. Res. Commun. 73, 894-902 (1976)

Photoaffinity Labeling of Coupling Factor ATPase from *Micrococcus luteus*

H.-J. SCHÄFER, P. SCHEURICH, G. RATHGEBER, and K. DOSE

Introduction

Photoaffinity labeling has found increasing application in characterizing the enzymic binding sites of substrates, effectors, or coenzymes
(Knowles, 1972). Photochemically labile reagents (photoaffinity labels)
appear to be superior to reagents with conventional functional groups
because the time and rate of reaction can be controlled relatively
easily by the illumination procedure. Most photoaffinity labels contain a photosensitive azido group. Upon irradiation azido derivatives
form highly reactive nitrenes that react rapidly with amino acid residues situated in or near the binding site of the enzyme. Different
azido derivatives of biological ligands have been synthesized for this
purpose.

In order to characterize nucleotide binding sites of F_1 ATPase from
Micrococcus luteus, we synthesized 8-azido-ATP (Haley and Hoffman, 1974;
Schäfer et al., 1976; Scheurich et al., 1978) and the fluorescent 8-
azido-1,N^6-etheno-ATP (Schäfer et al., 1978).

Results

8-Substituted ATP-analogs such as 8-bromo-ATP, 8-azido-ATP, and 8-
azido-1,N^6-etheno-ATP are hydrolyzed by F_1 ATPase from *M. luteus* in the
presence of divalent metal ions. The K_m value for hydrolysis of ATP/
Ca^{2+} is 0.2 mM, the corresponding K_m values for the 8-substituted analogs are four to ten times higher. This difference depends largely
on conformational differences between the nucleotides. ATP preferen-
tially attains the anti conformation, whereas the 8-substituted analogs
largely gain syn conformation, due to their bulky substituents in position 8 of the purine ring (Ikehara et al., 1972). Irradiation of F_1
ATPase with ultraviolet light in the presence of 8-azido-ATP or 8-
azido-1,N^6-etheno-ATP and divalent metal ions leads to a drastic reduction of enzymic activity. This inactivation is of the noncompetitive
type. It is probably related to the covalent and irreversible binding
of the analogs to the enzyme. After labeling the enzyme with 8-azido-
(^{14}C)ATP and subsequent conduction of SDS gel electrophoresis, most of
the radioactivity could be localized at the β subunit, as shown in
Figure 1.

When using the fluorescent 8-azido-1,N^6-etheno-ATP as a photoaffinity
label, again most of the fluorescence was found at the β subunit. Specific labeling by the above procedure is also indicated by the non-
specific labeling observed after replacing 8-azido-ATP by 8-azido-AMP.
In this experiment only a minute inhibition of enzymic activity was
observed in correlation with a minute overall incorporation of radioactivity. The amount of photoaffinity labeling and inactivation is also
decreased by adding ATP, ADP, or AMP-PNP prior to illumination in the

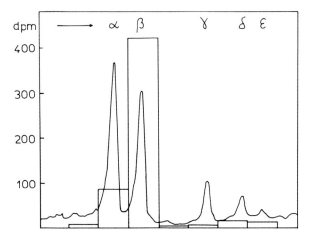

SDS gel electrophoresis densitogram of F_1 ATPase labeled by photolysis of 8-azido-(^{14}C)ATP (680 dpm/nmol). 300 µg protein were applied to the gel. The radio-activity of the slices is represented by the *bars*

Table 1

Label	Effector	dpm		Inactivation
		α	β	(%)
8-azido-ATP	Mg^{2+}	83	386	80
8-azido-ATP	Mg^{2+}, AMP	68	297	78
8-azido-ATP	Mg^{2+}, ADP	59	53	17
8-azido-ATP	Mg^{2+}, ATP	27	32	21
8-azido-ATP	Mg^{2+}, AMP-PNP	25	24	–
8-azido-AMP	Mg^{2+}	78	35	12

presence of 8-azido-ATP. Similiar results have been obtained with 8-azido-1,N^6-etheno-ATP (+ Me^{2+} ions) as label. AMP that is not specifically bound by F_1 ATPase causes no protection of the enzyme against the attack by the label. Table 1 summarizes the results obtained with 8-azido-(^{14}C)ATP.

If the number of modified 8-azido-(^{14}C)ATP residues bound to the β subunits of one ATPase molecule is plotted versus the irradiation-induced inhibition of the enzymic activity, extrapolation shows that 100% inactivation is reached after binding a single modified 8-azido-(^{14}C)ATP residue to one of the β subunits of one F_1 ATPase molecule.

Discussion

Our results are in qualitative agreement with those of Wagenvoord et al. (1977), who labeled the coupling factor ATPase from beef heart mitochondria with 8-azido-ATP. Contrary to our results the latter authors found complete inhibition after binding two modified 8-azido-ATP residues. Ferguson et al. (1975a,b, 1976), however, inactivated the enzyme by modifying one tyrosine residue at one β subunit with 4-chloro-7-nitro-benzofurazan, and Budker et al. (1977) inactivated the enzyme by labeling one of its β subunits with a single molecule of the ATP analog ATP γ-4-(N-2-chloroethyl-N-methylamino)benzylamidate.

Summary

1. 8-Azido-ATP and 8-Azido-1,N⁶-etheno-ATP are hydrolyzed by F_1 ATPase from *Micrococcus luteus*.

2. A nucleotide-binding site is located at the β subunit, which can be labeled both with 8-azido-ATP or 8-azido-1,N⁶-etheno-ATP. No labeling is obtained with 8-azido-AMP.

3. Photoaffinity labeling coincides with a decrease in enzymic activity.

4. Protection of the enzyme by ATP, ADP, and AMP-PNP decreases the binding of the photoaffinity label as well as the photoinactivation. AMP causes no protection.

5. Covalent binding of one modified 8-azido-ATP residue per F_1 ATPase is correlated with complete inhibition of its catalytic activity.

References

Budker, V.G., Kozlov, I.A., Kurbatov, V.A., Milgrom, Y.M.: FEBS Lett. 83, 11-14 (1977)
Ferguson, S.J., Lloyd, W.J., Lyons, M.H., Radda, G.K.: Eur. J. Biochem. 54, 117-126 (1975a)
Ferguson, S.J., Lloyd, W.J., Radda, G.K.: Eur. J. Biochem. 54, 127-133 (1975b)
Ferguson, S.J., Lloyd, W.J., Radda, G.K., Slater, E.C.: Biochim. Biophys. Acta 430, 189-193 (1976)
Haley, B.E., Hoffman, J.F.: Proc. Natl. Acad. Sci. U.S.A. 71, 3367-3371 (1974)
Ikehara, M., Uesugi, S., Yoshida, K.: Biochemistry 11, 830-836 (1972)
Knowles, J.R.: Acc. Chem. Res. 5, 155-160 (1972)
Schäfer, H.-J., Scheurich, P., Dose, K.: Hoppe-Seyler's Z. Physiol. Chem. 357, 278 (1976)
Schäfer, H.-J., Scheurich, P., Rathgeber, G., Dose, K.: Nucleic Acids Res. 5, 1345-1351 (1978)
Scheurich, P., Schäfer, H.-J., Dose, K.: Eur. J. Biochem. 08, 253-257 (1978)
Wagenvoord, R.J., van der Kraan, I., Kemp, A.: Biochim. Biophys. Acta 460, 17-24 (1977)

Long-Chain n-Alkylamines as Inhibitors
of Coupling and Uncoupling in Beef Heart Mitochondria

E. BÄUERLEIN and H. TRASCH

Following the working hypothesis that a protonized disulfide (Bäuerlein and Kiehl, 1976) or a conformationally strained disulfide (Bäuerlein, this Volume) may be involved in the phosphate-activating step of the electron transport phosphorylation, we used lipophilic trapping agents for thiol and sulfenyl groups (RS$^+$), which may be formed by the phosphorylytic cleagave of the disulfide. After the study of lipophilic maleimides (Kiehl and Bäuerlein, 1976, 1977), as well as lipophilic thioureas and thiouraciles (Bäuerlein and Kiehl, 1976), N-alkylthio-

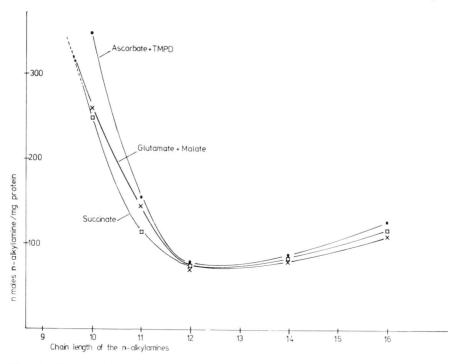

Fig. 1. The relation between the concentrations for the inhibition of coupled respiration (or uncoupling) and the chain length of n-alkylamines. x, the concentrations for complete inhibition of coupled respiration with glutamate + malate as substrates, and ● those for ascorbate + TMPF as substrates; □, the concentrations for complete uncoupling of state 4 respiration with succinate as substrate. For experimental details see legend of Figure 2

Abbreviations. TMPD = Tetramethylphenylendiamine; CCCP = Carbonylcyanide, m-Chlorophenylhydrazone; DNP = 2,4-Dinitrophenol; SF 6847 = p-Hydroxy-di-tert.butyl-benzyl-iden-malodinitrile.

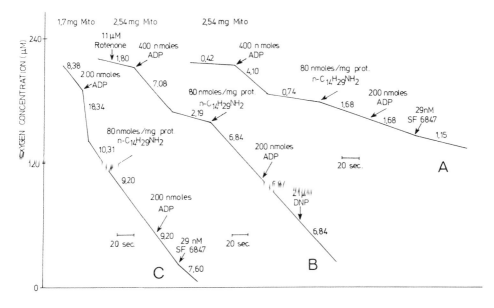

Fig. 2. Effects of n-tetradecylamine on the coupled respiration of beef heart mitochondria. *Lines* represent the output from an oxygen electrode. *Numbers on the lines*
are respiration rates, µmol of oxygen mg^{-1} protein h^{-1} at 25°C. Experiment A: Beef
heart mitochondria (2.54 mg) were added to a reaction mixture consisting of 2.4 ml
0.25 M sucrose containing 2.5 M glutamate, 2.5 mM D,L-malate, 5 mM malonate, 20 mM
KCl, 5 mM MgCl$_2$ 10 mM phosphate and 20 mM Tris-HCL, pH 7.3. Experiment B: Beef heart
mitochondria (2.54 mg) were added to a reaction mixture consisting of 2.4 ml 0.25 M
sucrose containing 10 mM succinate, 20 mM KCl, 5 mM MgCl$_2$, 10 mM phosphate and 20 mM
Tris-HCl, pH 7.3. Experiment C: Beef heart mitochondria (1.70 mg) were added to a
reaction mixture consisting of 2.4 ml 0.25 M sucrose containing 5 mM ascorbate, 0.25
mM TMPD, 20 mM KCl, 5 mM MgCl$_2$, 10 mM phosphate and 20 mM Tris-HCl, pH 7.3

imidazoles of increasing chain length,-of which the N-methyl compound
was reported to be the most effective sulfenyl group reagent as compared
with 6-n-propyl-2-thiouracile and other nucleophiles (Cunningham, 1964)-,
were shown to inhibit coupled respiration only if glutamate plus malate were the substrates. The finding was unexpected, that the sulfur-
free-N-alkylimidazoles were the better inhibitors (unpublished results).
Because N-alkyl-imidazoles were more basic than the related thioimida-
zoles, we concluded that the inhibition reaction may be ascribed to
the basic function of the lipophilic molecule. This encouraged us to
study n-alkylamines of increasing chain length, of which n-decyl compound
was reported by Skulashev et al. (1969) to be an uncoupler of oxidative
phosphorylation.

If the chain length was at least the n-undecyl group, we observed drastic changes in reactivity (Fig. 1). For example, n-tetradecylamine
prevented coupled respiration with 80 nmol/mg protein, if glutamate +
malate, β-hydroxybutyrate or ascorbate + TMPD were the substrates, and
these inhibitions could not be released by the uncouplers CCCP, DNP,
or SF 6847. With succinate state 4 respiration was stimulated to state
3 respiration (Fig. 2). With the corresponding hydrocarbon n-tetrade-
cane no effect could be detected up to 300 nmol/mg protein, if any of
the substrates was used. The inhibitory concentration of n-tetradecylamine together with equimolar amounts of picric acid uncoupled state 4
respiration; if, however, the mitochondria were incubated first with
the amine for 1 min, no uncoupling could be detected after the subsequent addition of picric acid.

It was concluded that the reported inhibition of coupling *and* uncoupling may be ascribed to the interaction of the basic amino group, which had to be linked at least to the n-undecyl group, with proton-translocating units of the electron transport chain. It is probable that the lipophilic amine is tightly bound both by the lipophilic alkyl chain and by the ionic linkage, formed by the acid-base reaction of an undissociated carboxyl group with the amine, in a channel or pump, by which proton transport, and concomitantly coupling and uncoupling, is prevented. Yet the minimal chain length of eleven carbons for the specific inhibition may be taken also as evidence for R.J.P. Williams's proposal of localized protons buried in the lipid of the membrane.

If one assumes that the proton translocation is directly coupled to the electron transport, the latter has to be inhibited concomitantly with the proton transport. We found, however, that with glutamate plus malate respiration was stimulated to 40% of state 3, if the concentration for the inhibition of coupled respiration was used (Fig. 2). Complete inhibition of electron transport was achieved with 300 nmol n-tetradecylamine per mg protein. Contrary to these findings, with ascorbate plus TMPD state 4 respiration was inhibited to 10%, if coupled respiration was prevented by the amine (Fig. 2). Maximally 35% inhibition of state 4 respiration could be obtained, if up to 500 nmol amine per mg protein were added. The study of the proton motive forces which is in progress may give us a more detailed information why three different effects were found with the three different substrates and if proton translocation and electron transport are coupled directly or not.

References

Bäuerlein, E., Kiehl, R.: FEBS Lett. 61, 68-71 (1976)
Cunningham, L.W.: Biochemistry 3, 1629-1634 (1964)
Kiehl, R., Bäuerlein, E.: FEBS Lett. 72, 24-28 (1976)
Kiehl, R., Bäuerlein, E.: FEBS Lett. 83, 311-315 (1977)
Skulashev, V.P., Jasaitis, A.A., Navickaite, V.V., Yagushinsky, L.S.: FEBS Symposium on Mitochondria Structure Function, 5th FEBS Meeting 1968, Vol. 17, p. 275-284 (1969)

Control and Dynamics of Energy Conservation

Control of Energy Flux in Biological Systems

D. F. WILSON, M. ERECINSKA, and I. SUSSMAN

Cells, in order to survive and grow, must maintain precise dynamic balance between the pathways which utilize ATP (biosynthetic pathways,
ion transport, etc.) and those which produce ATP (glycolysis, mitochondrial oxidative phosphorylation). In higher organisms most of the
ATP is supplied by mitochondrial oxidative phosphorylation. Complete
aerobic combustion of glucose to carbon dioxide and water yields approximately 17 times as much useful energy in the form of ATP as can
be obtained from the same glucose by anaerobic glycolysis alone. The
overwhelming importance of mitochondrial oxidative phosphorylation in
man becomes apparent when one realizes that a 69 kg man walking at 4
km/h in loose snow and carrying a 20 kg load is utilizing approximately 0.5 kg of ATP/minute! During a normal working day the ATP utilization of the same man may range from approximately 25 g ATP/min to
values in excess of 0.6 kg ATP/min. This high metabolic flux and large
dynamic range of control assures mitochondrial oxidative phosphorylation of a dominant role in cellular homeostasis.

Control site(s) of metabolic pathways are associated with reactions
that are strongly displaced from equilibrium (essentially irreversible).
This occurs because, when reactions are near equilibrium, the relationship of the products and reactants is determined by the state of the
system and not by the activity of the enzyme. In contrast, when reactions are essentially irreversible, the rate of product formation (flux
through the pathway) is strictly dependent on the enzyme activity.

Regulation of Mitochondrial Oxidative Phosphorylation

The behavior of mitochondrial oxidative phosphorylation indicates that
it may be analyzed in two parts: the reactions from the intramitochondrial NAD to cytochrome c (Eq. 1)

$$NADH + 2c^{3+} + 2ADP + 2Pi \rightleftharpoons NAD^+ + 2c^{2+} + 2ATP \tag{1}$$

and the reactions from cytochrome c to molecular oxygen (Eq. 2)

$$2c^{2+} + 1/2\ O_2 + 2H^+ + ADP + Pi \longrightarrow 2c^{3+} + H_2O + ATP \tag{2}$$

In reaction (1) the NADH and NAD^+ refer to the intramitochondrial pool
whereas ATP, ADP, and Pi are extramitochondrial (Wilson et al., 1974a;
Erecinska et al., 1974). The concentrations of the reactants in Eq. (1)
have been measured for suspensions of isolated mitochondria (Erecinska
et al., 1974) and for several different types of intact cells (Wilson
et al., 1974a,b; Erecinska et al., 1978b). These values have been used
to calculate the mass action ratio for the reaction in intact cells,
and the latter are presented in Table 1 along with the equilibrium constant for the reaction under cellular conditions. None of the measured
mass action ratios are significantly different from the equilibrium
constant, indicating that the reaction is near equilibrium. The use of

Table 1. A comparison of the measured mass action ratios and the equilibrium constant for oxidative phosphorylation from the NAD couple to cytochrome c (Eq. 1)

Type of cell	Experimental conditions	Mass action ratio (M^{-2})	Reference
Liver cells	No added substrate	5×10^7	Wilson et al., 1974a
Liver cells	Lactate + ethanol	3×10^7	Wilson et al., 1974a
Perfused liver	Endogenous	6×10^6	Wilson et al., 1974b
Ascites tumor cells	Endogenous substrate	1×10^7	Wilson et al., 1974b
Cultured Kidney cells	Endogenous substrates	5×10^7	Wilson et al., 1977b
Tetrahymena pyriformis	Endogenous substrates	4×10^7	Erecinska et al., 1978b
Paracoccus denitrificans	Glucose	5×10^7	Erecinska et al., 1978a
Equilibrium constant	1 mM Mg^{2+}	4.4×10^7	
Pigeon heart mitochondria	Glutamate + malate	7×10^6	Erecinska et al., 1974
Equilibrium constant	O Mg^{2+}	3×10^6	

The measured mass action ratios were taken from the indicated references while the equilibrium constants were calculated assuming a ΔG^O, of -7.6 kcal/mol in the presence of 1 mM Mg^{2+} (cellular conditions) and -8.4 kcal/mol at low Mg^{2+} (condition for isolated mitochondria). Also used in the calculation of the equilibrium constant were the half-reduction potential for the NAD couple (-0.320V) and for cytochrome c (0.235V).

the total cellular [ATP], [ADP], and [Pi] instead of the cytosolic values raises a question as to the possible errors due to compartmentalization.

Three lines of evidence suggest that the error introduced is small:

1. Near equilibrium was also observed for liver cells incubated in the presence of adenosine (Wilson et al., 1974b), a procedure that increases the ATP and ADP concentrations in the cytosolic compartment (Lund et al., 1975), making the mitochondrial compartment an insignificant fraction of the measured total cellular values.

2. Fractionation of liver cells into mitochondrial and cytosolic fractions (Elbers et al., 1974; Siess and Wieland, 1976; Zuurendonk and Tager, 1974) gives cytosolic [ATP]/[ADP][Pi] values only slightly (2-3-fold) greater than the total cell values used for Table 1.

3. In *Paracoccus denitrificans*, a prokaryotic (single compartment) organism with a respiratory chain very similar to that of mammalian mitochondria, reaction (1) is also near equilibrium.

If Eq. (1) is near equilibrium, then the regulation of mitochondrial respiration must occur in the reaction expressed in Eq. (2) where the overall reaction is strongly displaced from equilibrium (irreversible). The equilibrium constant for reaction (1) may be solved for the reduction of cytochrome c as a function of $[NAD^+]/[NADH]$ and $[ATP]/[ADP][Pi]$. This can then be substituted into the rate expression for reaction (2). The resulting rate expression should be applicable to the entire oxidative phosphorylation from the NAD couple to molecular oxygen.

A model has been developed for the cytochrome c oxidase-oxygen reaction (Eq. 2), which quantitatively describes the dependence of the respiratory rate on the reduction of cytochrome c, [ATP], [ADP], [Pi], and ox-

Table 2. The effect of glycerol on concentrations of metabolites in isolated liver cells

Metabolite or metabolite ratio	Control	10 mM glycerol		
		2 min	5 min	15 min
ATP	2.02 ± 0.27	1.66 ± 0.33	1.29 ± 0.35	0.95 ± 0.042
ADP	0.48 ± 0.15	0.76 ± 0.30	0.79 ± 0.27	0.75 ± 0.33
AMP	0.45 ± 0.07	0.47 ± 0.07	0.42 ± 0.11	0.42 ± 0.06
Pi	3.38 ± 0.04	1.39 ± 0.13	1.01 ± 0.03	0.80 ± 0.01
ATP/ADP	4.02 ± 1.05	2.60 ± 0.15	1.99 ± 0.14	1.57 ± 0.25
ATP/ADP x Pi (M^{-1})	1046 ± 329	1203 ± 165	1380 ± 204	1277 ± 381
Energy charge	0.76 ± 0.01	0.68 ± 0.01	0.65 ± 0.04	0.58 ± 0.02
$\frac{[cyt\ c\ ox]}{[cyt\ c\ red]}$	5.12 ± 2.0	4.74 ± 20.0	4.50 ± 2.0	4.46 ± 1.6
$\frac{[3\text{-}OH\text{-}butyrate]}{[acetoacetate]}$	1.41 ± 0.4	1.40 ± 0.35	1.45 ± 0.52	1.62 ± 0.48
O_2 uptake (μmol/min/g)	1.57 ± 0.4	1.64 ± 0.56	1.41 ± 0.47	1.20 ± 0.43

Experimental results are taken from Erecinska et al., 1977. Concentrations of ATP, ADP, and AMP and Pi are expressed in μmol/g wet weight. Results from four independent experiments ± S.D.

ygen (Wilson et al., 1977a). This rate expression, taken together with the equilibrium constant for reaction (1), gives an accurate description of mitochondrial respiration in vivo as a function of three variables: the intramitochondrial $[NAD^+]/[NADH]$, the cytosolic $[ATP]/[ADP]$ $[Pi]$, and the concentration of cytochrome oxidase (or cytochrome c). The details of the model have been presented elsewhere (Wilson et al., 1977a; Erecinska et al., 1978b) and will not be further discussed here.

It is implicit in this analysis that the concentration of reduced cytochrome c and thus the respiratory rate is dependent on $[ATP]/[ADP][Pi]$ and $[NAD^+]/[NADH]$ as well as on the concentration of molecular oxygen, which enters directly into the rate equation (Owen and Wilson, 1974; Wilson et al., 1977a). Thus it is predicted that cytosolic $[Pi]$ is an important determinant of cellular homeostasis.

Experimental support has been obtained (Erecinska et al., 1977) for this role of $[Pi]$ in intact cells by two experimental approaches, one of which is illustrated in Table 2. Suspensions of cells prepared from rat liver were incubated in the presence of glycerol, a substrate that leads to substantial conversion of inorganic phosphate to organic phosphates with a resultant decrease in the intracellular $[Pi]$. As shown in the Table, the fall in intracellular $[Pi]$ is accompanied by a decrease in $[ATP]/[ADP]$ so that the $[ATP]/[ADP][Pi]$ (and the respiratory rate) remain essentially constant.

A more complete description of the behavior predicted by this model is given in Figure 1 as a family of curves representing plots of the intramitochondrial $[NAD^+]/[NADH]$ against the cytosolic $[ATP][ADP][Pi]$, assuming saturating oxygen concentrations. Each of the family of curves is for a constant respiratory rate (expressed as the turnover number for cytochrome c). The measured parameters for the various cell lines

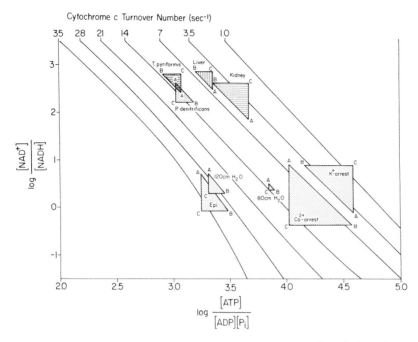

Fig. 1. The relationships of the intramitochondrial $[NAD^+]/[NADH]$, cytosolic $[ATP]/[ADP][Pi]$, and respiratory rate for mitochondria in vivo and in vitro. The *solid lines* represent the relationships of these parameters as calculated from the model of Wilson et al. (1977). Experimental results are plotted as three points: A $[ATP]/[ADP][Pi]$ is plotted versus the turnover number for cytochrome c; B Intramitochondrial $[NAD^+]/[NADH]$ is plotted versus the turnover number for cytochrome c; C $[ATP]/[ADP][Pi]$ is plotted versus the intramitochondrial $[NAD^+]/[NADH]$. The experimental points for perfused heart, isolated liver cells, and *T. pyriformis* are taken from Erecinska et al. (1978b), those for cultured kidney cells from Wilson et al. (1977b), and those for *P. denitrificans* from Erecinska et al. (1978a). *Shaded areas* are used to indicate more clearly the range of values for each type of cell or experimental condition

are plotted as points A, B, and C on the figure, where the points represent the three possible combinations of measured values taken two at a time. A perfect fit to the model would be seen as a superposition of the three points in each case. As may be seen, the fit of the cellular data to the model ranges from good to excellent.

This graphic presentation emphasizes that a cell, in order to regulate its respiration, has at its disposal three variables: the concentration of the respiratory chain; the intramitochondrial $[NAD^+]/[NADH]$, and the cytosolic $[ATP]/[ADP][Pi]$. The concentration of respiratory chain components remains constant for short-term experiments of less than a few hours with any particular cell. Thus cells must regulate their rate of ATP synthesis using the two parameters, either decreasing $[NAD^+]/[NADH]$ or $[ATP]/[ADP][Pi]$ or both in order to increase the rate of ATP synthesis or increasing these parameters to decrease the rate of ATP synthesis.

Interrelationships of Glycolysis and Respiration

Cells utilizing glucose as a substrate derive ATP from both glycolysis and respiration. When the cells are capable of sustaining most or all of their ATP requirements by either pathway alone, but under aerobic conditions use both, two classical phenomena may be observed, depending on the cell line and experimental conditions: (1) The *Pasteur effect* which is defined as inhibition of glycolysis (lactate production) that occurs upon introduction of oxygen to suspensions of anaerobic cells metabolizing glucose, and (2) the *Crabtree effect*, which is defined as inhibition of respiration that occurs upon addition to glucose of aerobic suspensions of cells capable of high glycolytic activity (Crabtree, 1929). It is this latter phenomenon that we will discuss in more detail. Many investigators have examined the Crabtree effect and made suggestions concerning its metabolic origin. Early work by Johnson (1941) and Lynen (1941) implicated inorganic phosphate, and it was suggested that both glycolysis and respiration were regulated by intracellular [Pi]. Wu and Racker (1959) later developed this idea and suggested that since [Pi] is an important regulator of phosphofructokinase, the decrease in [Pi] which follows glucose addition gives rise to inhibition of this key enzyme of glycolysis. In contrast Warburg (1959) and Chance and Hess (1959) placed primary emphasis on oxidative phosphorylation. Warburg suggested that cells that exhibit the Crabtree effect have defective respiratory chains and are incapable of adequate respiratory activity, while Chance and Hess (1959) postulated that the addition of glucose leads to an initial increase in cytosolic ADP, which is rephosphorylated in the mitochondria and sequestered there, making it unavailable for use in glycolysis.

These interactions of respiration and glycolysis are expressions of common elements of their respective regulatory mechanisms. Respiration behaves according to the relations discussed previously, while glycolysis is regulated primarily through the activation and inhibition of hexokinase and phosphofructokinase. We will discuss the regulation of the total flux of glycolysis without attempting to analyze the details of the regulatory mechanisms of each enzyme.

The Crabtree Effect in Ascites Tumor Cells

Ascites tumor cells incubated in Krebs-Henseleit medium, which is essentially free of phosphate, respond to glucose addition with a short, approximately 45 s, burst of respiration to approximately 150% of the preglucose rate (Fig. 2A). Measurements of glucose-6-phosphate (G-6-P) and fructose-1,6-diphosphate (F-1,6-P) show (Fig. 2B) a rapid accumulation of these compounds after glucose addition with a corresponding decrease in the intracellular inorganic phosphate from 3.8 mM to 1 mM. Essentially all of the decrease in inorganic phosphate can be accounted for by the increase in phosphate bound to G-6-P and F-1,6-P, indicating that other phosphorylated intermediates that accumulate are in much lower concentrations. Glycolysis, as measured by lactate production, occurs at a high rate for approximately 45 s and then decreases to a level where there is no net lactate production when the respiratory rate is minimum.

The total ATP production by the cell can be approximately calculated by assuming that 6 ATP are produced by oxidative phosphorylation for each molecule of oxygen consumed and that glycolysis produces 1 ATP

260

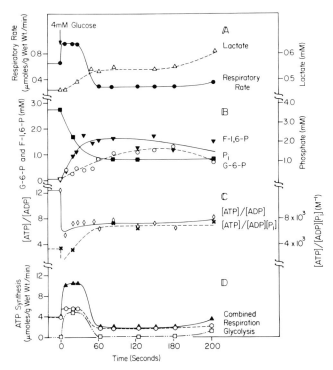

Fig. 2. The Crabtree effect in suspensions of sarcoma 180 ascites cells. The ascites cells were suspended at approximately 21.4 ± 0.8 mg wet weight/ml in Krebs-Henseleit medium containing 10 mM Tris-Cl instead of bicarbonate (pH 7.4) and no added phosphate. The measured metabolites are expressed as their calculated intracellular concentrations with the exception of lactate, which is given as the concentration in the total suspension. The rates of ATP synthesis were calculated assuming 6 ATP/O_2 for respiration and 1 ATP/lactate for glycolysis. The *curve labeled combined* was obtained by adding the rates of ATP synthesis by respiration and glycolysis. The presented data points are the average of three experiments

for each lactate synthesized from glucose. We have calculated the ATP synthesized by glycolysis based on the net lactate synthesis because when glucose is fully oxidized by the citric acid cycle (and oxidative phosphorylation), the ATP production by glycolysis is insignificant (1/17) compared to that by oxidative phosphorylation. During the burst of respiration and lactate production the total rate of ATP production (Fig. 2D) increases from about 4 μmol/g wet weight to about 11.7 μmol/g wet weight with glycolysis (lactate production) contributing nearly 50% of the total. The rate of ATP synthesis and utilization during the burst of respiration is equal to the preglucose rate plus the rate of ATP utilization for production of G-6-P and F-1,6-P, indicating that essentially all of the increased ATP utilization is due to the activity of hexokinase and phosphofructokinase. The respiratory rate is initially stimulated by the decrease in [ATP]/[ADP][Pi] but it remains high until the [ATP]/[ADP][Pi] rises above the levels observed prior to glucose addition, consistent with the increased reducing substrate concentration lowering the intramitochondrial [NAD^+]/[NADH].

Inhibition of respiration (Crabtree effect) begins approximately 60 s after glucose addition, with respiration being inhibited to one-half

of the preglucose level. However, the decrease in respiration is not
due to production of ATP by glycolysis, since glycolysis is also inhibited to the point where there is no net lactate production. Thus
at that point the total cellular rate of ATP production (and utilization) is decreased. This decrease almost certainly results from an inhibition of the ATP-utilizing reactions (such as cell growth) by low
intracellular $[Pi]$, because only inhibition of the ATP-utilizing reactions can explain the elevated $[ATP]/[ADP][Pi]$. Primary inhibition of
the ATP-producing reactions would lower this value.

It is useful to note that the decrease in both respiration and glycolysis occurs at a time when the $[ATP]/[ADP]$ is essentially constant.
Assuming that adenylate kinase is near equilibrium (Beis and Newsholme,
1975), this means that the adenylate charge (see, e.g., Atkinson, 1971)
is also constant and thus cannot be a primary regulator of glycolysis
or respiration (see also Erecinska et al., 1977). The nearly simulta-
neous inhibition of both respiration and the glycolytic flux suggests
that both pathways are regulated to provide ATP at the required free
energy of hydrolysis ($[ATP]/[ADP][Pi]$). The two pathways have some degree of independence because respiration depends on the intramitochondrial $[NAD^+]/[NADH]$, whereas glycolysis is coupled to the cytosolic
$[NAD^+]/[NADH]$. In addition glycolysis is subject to other regulatory
factors in order to allow the cell to use noncarbohydrate metabolites,
etc.

Experimental support for the role of low intracellular phosphate in
the inhibition of ATP utilization by ascites cells has been substantiated in two ways (see also Wu and Racker, 1959): (1) When the cells
are suspended in a medium with high phosphate (10 mM), the intracellular phosphate is less depleted following glucose addition, and little
or no inhibition of cellular ATP production is observed. The Crabtree
effect is less pronounced: Respiration is inhibited only 20% or less
with glycolysis (lactate production) providing the remaining ATP requirement. (2) When the cells are suspended in a medium containing
1 mM phosphate or less, the cellular ATP production is inhibited as
shown in Figure 2, but at longer times the G-6-P and F-1,6-P levels
decrease and intracellular $[Pi]$ rises. With increase in intracellular
$[Pi]$ and presumably other metabolic adjustments, the respiratory rate
and lactate production increase until the rate of cellular ATP production returns to a level equal to or slightly higher than that observed
prior to glucose addition.

The inhibition of respiration by addition of glucose to media with 1 mM
extracellular $[Pi]$ may be somewhat simplistically considered as occurring in two phases: (1) an early phase with low intracellular $[Pi]$ in
which the cellular ATP utilization is inhibited. This phase is characterized by a high $[ATP]/[ADP][Pi]$, relatively strongly inhibited respiration, and very low rates of lactate production. (2) A late phase
following recovery of the intracellular $[Pi]$, characterized by near
normal cellular ATP utilization and only slightly elevated $[ATP]/[ADP]$
$[Pi]$ — The respiration is somewhat inhibited while glycolysis (lactate
production) produces ATP at a rate that compensates for the decrease in
ATP synthesis by respiration.

The Crabtree effect is greater in tumor cells than, for example, liver
cells, partly because the activities of the enzymes involved in mitochondrial respiration and glycolysis are present in different amounts.
The cytochrome a content of liver cells is about 0.1 μmol/g wet weight
as compared to about 0.05 μmol/g wet weight in the sarcoma 180 ascites
tumor cells, while the glycolytic enzymes are more active in the ascites
tumor cells. The differences may reflect in part their respective growth

262

conditions. Ascites cells grow in the intraperitoneal cavity, where there is a relatively low concentration of oxygen but adequate supplies of glucose, whereas liver cells have adequate supplies of oxygen.

Summary

Respiration and glycolysis share a common regulatory parameter in the cytosolic $[ATP]/[ADP][Pi]$ but interact with different $[NAD^+]/[NADH]$ pools. Thus although both pathways are homeostatically regulated toward maintaining the cytosolic $[ATP]/[ADP][Pi]$, the balance between the two pathways is dependent on other regulatory parameters involved in establishing the reducing power ($[NAD^+]/[NADH]$) in each compartment and the carbon flow (oxidation of fats vs. carbohydrates etc.). Addition of glucose to suspensions of sarcoma 180 cells initially causes an increased ATP utilization for phosphorylation of glucose to G-6-P and F-1,6-P, a challenge met by a combination of increased glycolysis (lactate production) and respiration. As the concentrations of glycolytic intermediates rise, they inhibit hexokinase and phosphofructokinase, lowering the rate of ATP utilization with the resulting rise in $[ATP]/[ADP][Pi]$ and inhibition of *both* glycolysis and respiration. In these particular cells suspended in media containing 0 - 1 mM Pi, the increase in organic phosphates (G-6-P and F-1,6-P) depletes the intracellular $[Pi]$ and thus inhibits ATP utilization (cell-growth?) to below the control value so that respiration (and glycolysis) are strongly diminished. At longer times, intracellular $[Pi]$ and ATP utilization rise again, and both glycolysis and respiration are activated so that they share ATP production. The final "balance" of glycolysis and respiration and thus the long-term Crabtree effect are dependent on the relative total activities of the pathways and to a lesser extent on the regulatory properties of glycolysis in the cell.

References

Atkinson, D.E.: In: Metabolic Pathways. Vogel, H.J. (ed.). New York: Academic Press, 1971, Vol. V, pp. 1-21

Beis, I., Newsholme, E.A.: Biochem. J. 152, 23-32 (1975)

Chance, B., Hess, B.: Science 129, 700-708 (1959)

Crabtree, H.G.: Biochem. J. 23, 536-545 (1929)

Elbers, R., Heldt, H.W., Schmucker, P., Soboll, S., Wiese, H.: Hoppe-Seyler's Z. Physiol. Chem. 355, 378-393 (1974)

Erecinska, M., Kula, T., Wilson, D.F.: FEBS Lett. 87, 139-144 (1978a)

Erecinska, M., Stubbs, M., Miyata, Y., Ditre, C.M., Wilson, D.F.: Biochem. Biophys. Acta 462, 20-35 (1977)

Erecinska, M., Veech, R.L., Wilson, D.F.: Arch. Biochem. Biophys. 160, 412-421 (1974)

Erecinska, M., Wilson, D.F., Nishiki, K.: Am. J. Physiol. 234, c82-c89 (1978b)

Johnson, M.J.: Science 94, 200-202 (1941)

Lund, P., Cornell, N.W., Krebs, H.A.: Biochem. J. 152, 593-599 (1975)

Lynen, F.: Justus Liebigs Ann. Chem. 546, 120-141 (1941)

Owen, C.S., Wilson, D.F.: Arch. Biochem. Biophys. 161, 581-591 (1974)

Siess, E.A., Wieland, O.H.: Biochem. J. 156, 91-102 (1976)

Warburg, O.: Science 123, 309-314 (1956)

Wilson, D.F., Erecinska, M., Drown, C., Silver, I.A.: Am. J. Physiol. 233, c135-140 (1977b)

Wilson, D.F., Owen, C.S., Holian, A.: Arch. Biochem. Biophys. 182, 749-762 (1977a)

Wilson, D.F., Stubbs, M., Oshino, N., Erecinska, M.: Biochemistry 13, 5305-5311
  (1974a)
Wilson, D.F., Stubbs, M., Veech, R.L., Erecinska, M., Krebs, H.A.: Biochem. J. 140,
  57-64 (1974b)
Zuurendonk, P.F., Tager, J.M.: Biochem. Biophys. Acta 333, 393-399 (1974)

The Efficiency of Oxidative Phosphorylation

J. W. STUCKI

Introduction

In many experimental studies with incubated mitochondria it appeared
that at state 4, i.e., in the presence of oxygen and oxidizable substrates but in the absence of added ADP, oxidative phosphorylation be-
haved almost like a system at thermodynamic equilibrium. This conclusion was based exclusively on the comparison of the measured ΔG values
of the driving reaction (oxidation) and the driven reaction (phosphorylation) with their known equilibrium values. From this it was then
assumed that the efficiency of mitochondrial energy transductions, notably oxidative phosphorylation, can be treated within the framework
of classical equilibrium thermodynamics. Consequently, oxidative phosphorylation at state 4 appeared to operate with an efficiency of almost
100%. Furthermore, it was tacitly assumed that this thermodynamic treatment of oxidative phosphorylation can be extended to mitochondria incubated in the presence of ADP, i.e., under state 3 conditions. Consequently oxidative phosphorylation was assumed to operate with nearly
100% efficiency also at state 3.

Several authors have cast doubt on this straightforward extension of
the thermodynamic treatment of mitochondria at state 4 to mitochondria
at state 3 (1 - 4). In fact, this treatment was found to be in sharp
contrast with experimental observations with incubated mitochondria
(1). In light of these results, the tacit assumption that the efficiency of oxidative phosphorylation is independent of the state of
energetization of the mitochondria appears to be wrong.

It is well known that during state 4 the mitochondrial respiration
does not come to a complete rest. To be more specific, a basal rate
of oxygen uptake, the so-called state 4 respiration, is observed. Therefore, the entropy production of the mitochondria does not vanish at
state 4 and hence, by definition, the system is not in thermodynamic
equilibrium. Several studies showed that state 4 respiration is due to
energy-dissipating processes such as the recycling of Ca^{2+} and H^+ ions
across the inner membrane (4 - 7), to extramitochondrial ATPases (8),
and to the transhydrogenase reaction (1). These reactions prevent mitochondria from reaching the state of equilibrium, although they may
closely approach it. Stated differently, the system at state 4 is not
time independent as it would be if state 4 corresponded to the equilibrium state.

Such systems with nonvanishing flows close to equilibrium can be successfully treated within the framework of linear nonequilibrium thermodynamics (9, 10). Kedem and Caplan have proposed a simple model for
thermodynamic energy converters (11) that was also applied to mitochondrial energy transductions (2, 3). Several experimental studies
revealed a satisfactory qualitative agreement between the predictions
of this simple scheme and the measurements in intact mitochondria (4,
7).

In this paper we develop a theory for the efficiency of oxidative phosphorylation within the framework of linear nonequilibrium thermodynamics that relies heavily on the original scheme of Kedem and Caplan.
Section 2 summarizes some basic facts and definitions of nonequilibrium thermodynamics and presents the adopted scheme. Section 3 summarizes some results obtained by Kedem and Caplan that are pertinent to
our analysis. In addition, we present a useful formula for obtaining
the phenomenologic stoichiometry of oxidative phosphorylation from flow
measurements only. In Section 4 we discuss possible definitions of efficiency and its measures. In Section 5 we enquire about the optimal
operating conditions of oxidative phosphorylation. It will be shown
that there are two distinct degrees of coupling of oxidative phosphorylation: one allowing a maximal rate of ATP synthesis and the other
allowing maximal output power of ATP synthesis. Furthermore, we briefly discuss the effects of a fluctuating environment on the efficiency
of oxidative phosphorylation. Section 6 deals with the entropy production of oxidative phosphorylation. This treatment allows us to specify
under which conditions oxidative phosphorylation opperates at maximal
efficiency in a cellular environment. Furthermore, we present a simple
geometric interpretation of the entropy production of the mitochondria.
Section 7 concludes with some remarks about possible applications and
extensions of this theory.

A Linear Phenomenologic Description of Oxidative Phosphorylation

The aim of this study is to obtain some general information about the
efficiency of mitochondrial energy transductions, especially oxidative
phosphorylation. Therefore at the outset we consider a general scheme
with coupled chemical reactions and transport processes. This scheme
will then be simplified as far as possible in order to yield the required information about the efficiency of oxidative phosphorylation.
The second law of thermodynamics requires that the rate of entropy production be nonnegative

$$\frac{dS}{dt} = \dot{S} = \sum_{i=1}^{n} J_i X_i \geq 0 \tag{1}$$

where equality applies only at the state of thermodynamic equilibrium.
J_i and X_i are the generalized flows and conjugate forces, respectively,
of the ith process in the system (see, e.g., 10). This inequality al
lows that individual terms under the summation sign may be negative;
it is only the sum of all terms that must remain nonnegative. A negative sign of the ith term has the physical meaning that the process is running against its own natural driving force X_i (uphill) and that this
process builds up or maintains a potential. A typical example of such
a potential in mitochondria is the phosphate potential. For the ith
process running against its natural driving force, we have $X_i < 0$, $J_i >
0$. Hence it is through the negative sign of X_i that the ith term under
the summation sign in Eq. (1) acquires a negative sign. Such processes
are nonspontaneous and can only be realized when they are driven by a
spontaneous process that is running in the direction of its own driving
force (downhill). For the driving process we have $X_i > 0$ and $J_i > 0$.
Hence, the spontaneous processes are represented by positive terms in
the summation. The inequality (1) implies that there must be at least
one spontaneous process in the system.

In general the flows J_i are complicated nonlinear functions of the
forces X_i, which precludes an exact analytical treatment of Eq. (1).

It is, however, natural to expand the nonlinear relation between flows and forces into a Taylor series about the equilibrium state. Thus for the ith flow, one obtains

$$J_i = J(o) + \sum_{j=1}^{n} \left(\frac{\partial J_i}{\partial X_j}\right)_{eq} X_j + \text{higher order terms} \tag{2}$$

where, by definition, the term J_i (O) vanishes at equilibrium. Neglecting higher order terms in Eq. (2) yields the *linear relations*

$$J_i = \sum_{j=1}^{n} L_{ij} X_j \tag{3}$$

where

$$L_{ij} = \left(\frac{\partial J_i}{\partial X_j}\right)_{eq} \tag{4}$$

The partial derivatives evaluated at the state of equilibrium are the so-called *phenomenologic coefficients* L_{ij} of the system (9, 10). In what follows we will assume the L_{ij} to be time independent. This assumption is independent of the linearity assumption (3) above. The time independence of the L_{ij} may be violated if the departure from equilibrium entails nonlinear kinetics such as allosteric activation mechanisms, since the whole kinetics of the system is encapsulated in the L_{ij}. In cases when one ignores the kinetics of the system it is not possible to calculate these quantities. However, it is important to note that this approach still allows a treatment of the problem within the framework of a purely phenomenologic theory. In many cases it is possible to measure the coefficients L_{ij} experimentally. At this point it must be stressed that Eq. (3) is only first order approximation to the nonlinear problem. Therefore linear nonequilibrium thermodynamics is not an exact theory but only an approximative description of the real situation. In far-from-equilibrium situations, the neglected higher order terms in Eq. (2) may become so important that the linearity of Eq. (3) is no longer fulfilled. The validity of Eq. (3) has therefore to be carefully checked by suitable experiments. By inserting Eq. (3) into Eq. (1), one obtains the important inequality

$$\dot{S} = \sum_{i=1}^{n} \sum_{j=1}^{n} X_i L_{ij} X_j \geq 0 \tag{5}$$

which imposes restrictions on the possible values of the coefficients L_{ij}. In fact Eq. (5) is a nonnegative definite quadratic form and the matrix of the coefficients (L_{ij}) must therefore be nonnegative definite. This sign definitness of (L_{ij}) can be expressed in terms of Sylvester's criterion for the nonnegative definitness of a matrix: All successive principal minors of (L_{ij}) must be nonnegative (see, e.g., 12). With this, the restrictions imposed on the L_{ij} can be expressed in the form of the inequalities

$$\det (L_{11}) = L_{11} \geq 0$$

$$\det \begin{pmatrix} L_{11} & L_{12} \\ L_{21} & L_{22} \end{pmatrix} = L_{11} L_{22} - L_{12} L_{21} \geq 0 \tag{6}$$

Near equilibrium the system exhibits temporal symmetry (13). This property imposes further restrictions on the matrix (L_{ij}). In fact, temporal symmetry of the system near equilibrium implies that (L_{ij}) is symmetric, i.e.,

$$L_{ij} = L_{ji} \qquad (7)$$

These are the celebrated Onsager symmetry relations. At certain critical distances from equilibrium this temporal symmetry may get lost and the system might exhibit symmetry-breaking instabilities and give rise to the spectacular appearance of the so-called dissipative structures (13, 14). It is important to note that these phenomena cannot be described by linear nonequilibrium thermodynamics and necessitate the introduction of higher order terms in Eq. (3). For the following we will, however, restrict our analysis to the linear regime of nonequilibrium thermodynamics as defined by Eq. (3), since our main interest is to study the efficiency of oxidative phosphorylation in nonoscillating mitochondria.

The general scheme (3) shall now be reduced as far as possible in order to simplify the calculations. In mitochondria one deals typically with a situation where one energy-yielding reaction (oxidation of reducing equivalents in the respiratory chain) drives several energy-utilizing chemical reactions (oxidative phosphorylation, transhydrogenation) and energy-requiring transport processes (H^+, Ca^{2+}, and anion transport) across the inner membrane. Since our main interest is the efficiency of oxidative phosphorylation, we will truncate the full problem to a scheme where phosphorylation is driven by respiration and ignore all other chemical and transport processes. With this, J_1 and J_2 designate the flow of ATP synthesis and oxygen uptake, respectively, and the force X_1 designates the phosphate potential. In the following we will denote the difference in redox potentials between electron-donating and -accepting redox couples shortly by redox potential X_2. Inequality Eq. (1) becomes therefore

$$\dot{S} = J_1 X_1 + J_2 X_2 \geqslant 0 \qquad (8)$$

By making use of Eq. (7), the set of Eq. (3) reduces to

$$J_1 = L_{11} X_1 + L_{12} X_2 \qquad (9a)$$

$$J_2 = L_{12} X_1 + L_{22} X_2 \qquad (9b)$$

The validity of the linear approximation (3), the symmetry relations (7), and the truncation of the problem to the representation of two processes only requires some comments.

It is well known that the linearity assumption holds over a wide range of forces when only transport processes occur, while this range is severely restricted when chemical reactions occur in the system as well (13). However, several experimental and theoretical studies have demonstrated that linearity between flows and forces as well as time independence of the L_{ij} is observed in the case of oxidative phosphorylation in mitochondria within the range of the forces of practical interest (15). In fact, mitochondria were found to operate rather close to equilibrium, which may be the reason for the experimentally observed linearity (16, 17). Furthermore, the phenomenon of reversed electron transfer indicates that electron flow in the respiratory chain (with the exception of the last step) and phosphorylation are readily reversible. Therefore deviations from the symmetry relations (7) are probably minimal in experimental situations of practical interest. At this point it must again be stressed that a deviation from the linear relationship between flows and forces does not imply a breakdown of the validity of the nonequilibrium thermodynamics theory as is often assumed. As stated above, higher order approximations of Eq. (2) can be introduced to solve the problem to any required degree of precision.

Several treatments of oxidative phosphorylation within the framework
of linear nonequilibrium thermodynamics were aimed at furnishing new
criteria to support one or the other of the hypotheses of the mechanism of oxidative phosphorylation (2, 3, 18). In these studies it was
therefore necessary to introduce H^+ transport in addition to the processes considered in our relations (8) and (9). Since our main concern
is the efficiency of oxidative phosphorylation rather than its mechanism, we can ignore these other processes, and we can treat our problem without loss of generality, within the framework of a purely phenomenologic theory. The only restriction imposed on this approach is
that we can only treat the situations in which all flows and forces
not explicitly considered in Eq. (9) are at a steady state (see also
18).

Kedem and Caplan introduced a fundamental dimensionless parameter that
gives a measure on the interactions among the different processes going
on in the system: the *degree of coupling* (11), which is a normalized measure for the cross coupling L_{12}

$$q = \frac{L_{12}}{\sqrt{L_{11} L_{22}}} \qquad (10)$$

In our case this degree of coupling has to be interpreted as an overall
degree of coupling of oxidative phosphorylation with respect to respiration, which lumps together all individual q_{ij} of the reactions in
the full system that are linked to oxidative phosphorylation and respiration (see 18).

Inequality (6) yields immediately that

$$0 < |q| \leqslant 1 \qquad (11)$$

Since we are interested in the situation in which oxidation drives
phosphorylation, we have to consider only positive values of q.

To end this section, we may represent the system under consideration
graphically as a black box in which respiration drives phosphorylation
by a mechanism with unknown kinetics (Fig. 1). The only quantities of
interest at the exterior that can be measured experimentally are the
redox potential X_2 applied at the input pins and the developed phosphate potential X_1 at the output pins and, in addition, the flow of
oxygen J_2 and the rate of ATP synthesis J_1. The capacity of the mitochondria to store ATP, oxygen, and respiratory substrates can always
be rendered negligible with respect to the surrounding incubation medium by choosing suitable experimental conditions. Therefore it is sufficient to measure all these quantities outside the box. These measurements will then allow one to calculate the fundamentally important
parameter, the degree of coupling q.

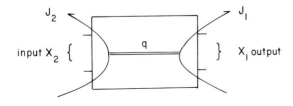

Fig. 1. Input-output relationship of oxidative phosphorylation in isolated mitochondria.
For the meaning of symbols see
text

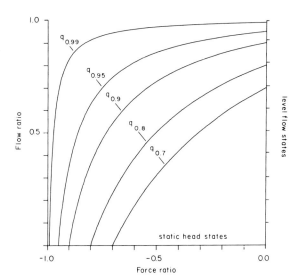

Fig. 2. Dependence of flow
ratio on force ratio. Plot
of Eq. (13) for the values
of the degree of coupling q
indicated in the Figure.
z = 1

Flow Ratios and the True Stoichiometry of Oxidative Phosphorylation

In this section we shall briefly recall some important relations between flows and forces as derived by Kedem and Caplan (11). These relations will then be used in our further analyses. By dividing Eq. (9a)
by (9b), dividing all terms in the numerator and the denominator by
$L_{22}X_2$, making use of Eq. (10), and introducing the new dimensionless
parameter, the *phenomenologic stoichiometry*

$$z = \sqrt{\frac{L_{11}}{L_{22}}} \qquad (12)$$

one finally obtains (11)

$$\frac{J_1}{zJ_2} = \frac{z\frac{X_1}{X_2} + q}{qz\frac{X_1}{X_2} + 1} \qquad (13)$$

Note that z does not generally represent the true mechanistic stoichiometry of the system, as is often assumed (15). However, for values of
q close to unity, z coincides with the mechanistic stoichiometry. Function (13) reflects the dependence of the *flux ratio* J_1/J_2 on the *force
ratio* X_1/X_2 and the degree of coupling q. In Figure 2 graphs of this
function in the region where oxidation drives phosphorylation, i.e.,
$X_1 \leqslant 0$, are depicted for several values of q. J_1/J_2 is the measured
P/O ratio of the mitochondria, and X_1/X_2 is the ratio of the developed
phosphate potential and the applied redox potential. Figure 2 displays
two important features of Eq. (13) that are of immediate practical interest. For values of q close to 1, the flow ratio vanishes at values
of the force ratio close to the theoretical equilibrium value of the
fully coupled reactions $zX_1/X_2 = 1$. Furthermore, for values of q close
to unity the flow ratio is strongly dependent on the force ratio only
at high values of the phosphate potential.

The states where either the flow ratio or the force ratio is zero have
a physical meaning of practical interest. At the point where the flow
ratio vanishes, the net rate of ATP synthesis is zero, and therefore
the mitochondria are not performing any net work under these conditions.
However, unless $q = 1$, the rate of oxygen uptake is not zero at this
state. Hence energy is still produced and expended by the mitochondria.
This means that the emphasis of the machinery of oxidative phosphorylation is no longer a net production of ATP but a maintenance of a maximal phosphate potential. Chance and Williams have introduced the term
state 4 to characterize this situation (19). In the terminology of Kedem
and Caplan this state has been called *static head* in order to avoid confusion with equilibrium (11). By setting $J_1 = 0$ in Eq. (13) the magnitude of the developed maximal phosphate potential at static head can
be calculated

$$(X_1)_{sh} = -qX_2/Z \qquad (14)$$

Since in any realistic situation $q < 1$, one understands why the phosphate potential measured at state 4 never exactly reaches the theoretical equilibrium value but can only approach this value up to a factor
of q/z.

By realizing that $q < 1$ summarizes formally all energy dissipating processes in mitochondria such as the recycling of H^+ and Ca^{2+} ions across
the inner membrane, etc., one arrives at a more intuitive interpretation of Eq. (14). In the presence of energy leaks the phosphate potential cannot attain the overall equilibrium value because the system
cannot further increase this potential once $(X_1)_{sh}$ is reached; rather,
it now uses all available energy to compensate the leaks and to maintain that potential. This situation is reminiscent of the impossibility
of cooling a refrigerator to absolute zero with realistic and hence
imperfect insulation.

Later on it will be shown that the static head is the natural steady
state of isolated mitochondria to which this system will evolve if X_1
is not constrained to a value other than $(X_1)_{sh}$ by the experiments.
This state is characterized by a minimum of entropy production. The
respiratory rate at state 4 $(J_2)_{sh}$ can readily be calculated from Eqs.
(9b) and (14) in terms of q, the "conductance" of the respiratory chain
L_{22} and the redox potential X_2 of the substrate

$$(J_2)_{sh} = L_{22} X_2 (1 - q^2) \qquad (15)$$

or in terms of the phosphate potential at state 4

$$(J_2)_{sh} = L_{22} z (X_1)_{sh} \left(q - \frac{1}{q} \right) \qquad (16)$$

Remembering that the total power of the input at static head is $(J_2)_{sh}$
X_2 and equals the entropy production at static head, one can calculate
the energy expenditure during the maintainance of a state 4 phosphate
potential

$$(\dot{S})_{sh} = (J_2 X_2)_{sh} = L_{22} X_2^2 (1 - q^2) \qquad (17)$$

or alternatively

$$(\dot{S})_{sh} = (J_2 X_2)_{sh} = L_{22} z^2 (X_1)_{sh}^2 \left(\frac{1}{q^2} - 1 \right) \qquad (18)$$

These are the important quadratic laws of Kedem and Caplan (11), which show that the energy waste at static head is quadratically dependent on the magnitude of the static head potential.

The situation at static head can physically be interpreted as an open-circuited cell. In constrast, the point where the force ratio vanishes has the physical meaning of a short-circuited cell. The term *level flow* was introduced by Kedem and Caplan to characterize this situation (11). There is no biochemical correlate to this term akin to the notion state 4. Again, as in the case of static head, no net work is performed by the mitochondria at this state since no phosphate potential is developed, i.e., $(X_1)_{1f} = 0$. From Eq. (13) one gets for the flux ratio at level flow

$$\left(\frac{J_1}{J_2}\right)_{1f} = qz \tag{19}$$

which is the P/O ratio of the mitochondria at a zero phosphate potential. It is important to note that during state 3 respiration, the phosphate potential is not zero. Therefore level flow should not be confused with state 3. In the classical respiratory jump experiment, after addition of ADP, the phosphate potential is constantly changing until static head is reached. Similarly, in experiments in which excess hexokinase is added (1), the phosphate potential is also not zero but is limited by the equilibrium of the hexokinase reaction. However, it should be possible to closely approach the level flow state by addition of a large excess of ATPase as has been described by Davis et al. (20) Nevertheless, the weak dependence of the P/O ratio on the force ratio near level flow (Fig. 2) for high values of q justifies the approximation of the flow ratio at level flow by the P/O ratio measured at state 3. In bioenergetics the P/O ratio at state 3 has been used as an estimate for the true stoichiometry of oxidative phosphorylation or, alternatively, as an indicator of the "quality" of the mitochondria, i.e., the degree of coupling q. In fact, Eq. (19) reveals that the P/O ratio at level flow is only dependent on these two quantities. Again it is apparent from this relation that unless $q = 1$ the measured P/O ratio at level flow is always smaller than the phenomenologic stoichiometry by a factor of q. This already indicates that the phenomenologic stoichiometry of oxidative phosphorylation can be determined from Eq. (19) once the degree of coupling is known. In this case it is not even necessary to have high quality mitochondria for this measurement. In fact, poorly coupled preparations can also be used as long as $q > 0$.

However, we will note the relations analogous to Eqs. (15-18) for the dependence of oxygen uptake at level flow on the applied redox potential

$$(J_2)_{1f} = L_{22}X_2 \tag{20}$$

and on the velocity of the flow $(J_1)_{1f}$

$$(J_2)_{1f} = \frac{(J_1)_{1f}}{qz} \tag{21}$$

For the entropy production at level flow one gets in terms of X_2

$$(\dot{S})_{1f} = (J_2X_2)_{1f} = L_{22}X_2{}^2 \tag{23}$$

and in terms of $(J_1)_{1f}$

$$(\dot{S})_{1f} = (J_2X_2)_{1f} = \frac{(J_1)^2{}_{1f}}{q^2z^2L_{22}} \tag{24}$$

These are again quadratic laws relating the square of the applied redox potential or the square of the flow $(J_1)_{1f}$ to the total energy expenditure of the system. For the sake of simplicity we will now assume
X_2 to be held constant at the outside of the mitochondria in the following discussion.

From Eqs. (15) and (20) one obtains the useful relation

$$q^2 = 1 - \frac{(J_2)_{sh}}{(J_2)_{1f}} \tag{25}$$

The ratio of oxygen uptake rates at minimal phosphate potentials and
at state 4 has been called the *respiratory control ratio*. This ratio can
be determined experimentally and is a good approximation of the exact
relation

$$r = \frac{(J_2)_{1f}}{(J_2)_{sh}} \tag{26}$$

In bioenergetics the respiratory control ratio has been used to measure
the quality of the mitochondria, i.e., the degree of coupling. This
procedure was introduced on intuitive grounds, and Eq. (25) shows that
the respiratory control ratio is indeed only a function of q. We are
now in the position to calculate exactly the phenomenologic stoichiometry of oxidative phosphorylation from the knowledge of the P/O ratio
at level flow and the respiratory control ratio only. By using Eqs. (19),
(25), and (26) we obtain

$$z = \frac{\sqrt{r} \left(\frac{P}{O}\right)_{1f}}{\sqrt{r - 1}} \tag{27}$$

As noted above, this formula is exact only in the case where the P/O
ratio and $(J_2)_{1f}$ are measured at level flow and $(J_2)_{sh}$ is measured at
state 4. This latter measurement imposes no practical problems. As
discussed above, the error introduced by performing the two other measurements not exactly at level flow is small if q is close to unity.
Hence Eq. (27) remains a good approximation of the phenomenologic
stoichiometry under these conditions.

Rottenberg recently derived additional formulae to Eq. (27) for exact
determination of the phenomenologic stoichiometry which allow relaxation of the requirement of performing flow measurements at level flow
(15). However, the practical application of his procedure necessitates
measurement of the phosphate potential and the redox potential at two
different states. These measurements are often subject to more experimental error than the determination of oxygen uptake rates only. In
summary, it may be mentioned that the commonly used practice of round-
ing off the maximal measured P/O ratio to the next higher integer number in order to determine z and of taking the respiratory control ratio
as an indicator of the degree of coupling appears justified, in light
of Eqs. (19) and (25), only for values of q close to unity.

Efficiency of Oxidative Phosphorylation and Its Measurement

Before starting the discussion about efficiency of oxidative phosphorylation we must first decide how to define this quantity. The choice of

an appropriate measure for the efficiency is still a controversial matter in bioenergetics. Efficiency has implicitly been assumed to be given by the so-called ΔΔG, which is

$$\Delta\Delta G_{oxphos} = \Delta G_{ox} - \Delta G_{phos} \tag{28}$$

Numerous studies have shown that at state 4 this value is close to zero (see, e.g., 16, 17). This indicates that phosphorylation is almost in equilibrium with respiration. From this it was then concluded that the efficiency of oxidative phosphorylation is almost 100%. Furthermore, it was tacitly assumed that efficiency is also nearly 100% at state 3, i.e., far from static head. This assumption has given rise to some conflicting results about the energy yield of oxidative phosphorylation. Remembering that the affinity is $X_i = -\Delta G_i$, Eq. (28) expressed in terms of the force ratio yields our Eq. (14) at static head. Since at this state the net production of ATP vanishes, one feels intuitively that Eq. (28) is not an adequate definition of the efficiency of oxidative phosphorylation. In fact, it is a measure of the distance of the phosphate potential from its theoretical equilibrium value with respect to the applied redox potential.

Another alternative is to take the flow ratio rather than the force ratio as a measure of the efficiency of oxidative phosphorylation (1, 4, 7). From Figure 2 we notice that the flux ratio indeed vanishes at state 4, where the net ATP production is zero as well. Therefore, the flow ratio seems to be a much better measure of efficiency than is the force ratio. However, this definition gives rise to troubles at the other end of the force ratio scale, namely, at level flow. At this state no phosphate potential is developed, and therefore it makes no sense to speak about efficiency in this case. Nevertheless, at level flow the flow ratio just attains its maximum value. From this we conclude that near static head the flow ratio could be used as a measure of efficiency and likewise the force ratio near level flow. Obviously a satisfactory definition of efficiency is a combination of these two ratios, as has been introduced by Kedem and Caplan (11).

$$\eta = -\frac{J_1 X_1}{J_2 X_2} = \frac{output\ power}{input\ power} \tag{29}$$

i.e., the product of the force ratio and the flow ratio. This definition seems to be the best available choice at the moment. In addition, it has the advantage of physical transparence because it relates the work delivered by the system to the input applied to this system. As long as we deal with efficiency of oxidative phosphorylation in isolated mitochondria, there is no a priori reason to look for still another definition. In what follows, we will always assume efficiency to be defined by Eq. (29).

By making use of Eqs. (9a, 9b), (10), and (12) one can derive an expression that relates efficiency to the force ratio (11)

$$\eta = -\frac{z\frac{X_1}{X_2} + q}{q + \frac{X_2}{zX_1}} \tag{30}$$

This function is plotted in Figure 3 for some values of q within the range of forces where oxidation drives phosphorylation. From this graph it is apparent that efficiency is optimal at a state intermediary between static head and level flow. Setting the derivative of Eq. (30)

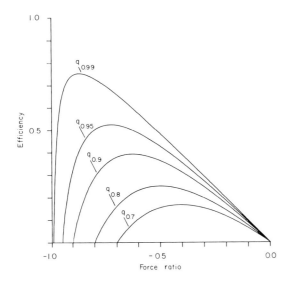

Fig. 3. Dependence of efficiency
on the force ratio. Plot of Eq.
(30) for the values of the degree
of coupling q indicated in the
Figure. z = 1

with respect to zX_1/X_2 to zero, solving the resulting quadratic equation, and inserting the result in (30), one obtains an expression for the optimal efficiency *that is only dependent on q*

$$\eta_{opt} = \frac{q^2}{(1 + \sqrt{1-q^2})^2} \qquad (31)$$

This is a highly nontrivial result. It is convenient to represent Eq. (31) in a parametric form. By choosing the parameter angle α as

$$\alpha = \arcsin q \qquad (32)$$

Eq. (31) assumes the compact form

$$\eta_{opt} = tg^2\left(\frac{\alpha}{2}\right) \qquad (33)$$

We will make frequent use of this important relation in the subsequent sections.

Let us now calculate the error introduced by approximating the efficiency of oxidative phosphorylation by the measured P/O ratio. From Eqs. (13) and (29) one obtains for the absolute error

$$E_{abs} = \frac{\left(z\frac{X_1}{X_2} + q\right)\left(1 + z\frac{X_1}{X_2}\right)}{qz\frac{X_1}{X_2} + 1} \qquad (34)$$

This error is plotted in Figure 4 for various values of q. For q = 1 the dependence of E_{abs} on the force ratio is linear. At static head we have $(E_{abs})_{sh} = 0$ while at level flow $(E_{abs})_{lf} = qz$. For the relative error we obtain

$$E_{rel} = -\left(\frac{X_2}{zX_1} + 1\right) \qquad (35)$$

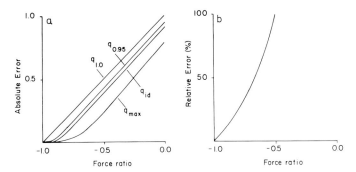

Fig. 4a and b. Error in flow ratio approximation of efficiency. Plot of Eq. (34) in *left panel* (4a) and of Eq. (35) in *right panel* (4b) $z = 1$. q as indicated in the figure. For q_{id} and q_{max} see text Section 5

The relative error is hyperbolically dependent on the force ratio as shown in Figure 4b. Note that this error is not defined at level flow. At static head it is 0 and at level flow $(E_{rel})_{lf} \to \infty$.

From these considerations we conclude that the P/O ratio is a tolerable measure of the efficiency of oxidative phosphorylation only near state 4 for $q > 0.9$. At any state different from static head within the driving region of oxidative phosphorylation, the P/O ratio is always an over-estimate of the efficiency. Therefore, the P/O ratio can at best be used as an upper boundary estimate of the efficiency.

A useful formula that allows an exact calculation of the efficiency from the flow ratio is the conjugate expression to Eq. (30)

$$\eta = - \frac{q - \dfrac{J_1}{zJ_2}}{q - \dfrac{zJ_2}{J_1}} \tag{36}$$

This expression, in addition to the flow ratio, contains q and z. As discussed above, these two quantities can be obtained from flow measurements with the help of Eqs. (25) and (27). Care should be exercised when using this procedure for the calculation of η in the far-from-static-head regime, i.e., near level flow, since the precision of this η estimate depends heavily on the precision of the P/O measurement in that range.

Optimal Operating Conditions for Oxidative Phosphorylation

In this section we consider the question of the optimal operating environment for mitochondria with respect to oxidative phosphorylation and the related question of what properties the mitochondria should have in order to phosphorylate ADP with the best possible efficiency. To be more specific, we look for the ideal degree of coupling of the phosphorylating machinery.

From Eq. (30) and Figure 3 it is apparent that mitochondria operate with the highest efficiency when the phosphate potential is clamped to the value of the force ratio corresponding to η_{opt} for a given q. This force ratio can be obtained by inserting Eq. (33) into Eq. (30) and solving for the force ratio. Taking the redox potential X_2 to be constant as assumed above one obtains

$$(X_1)_{opt} = - \sqrt{\eta_{opt}}\ \frac{X_2}{Z} = - \text{tg}\left(\frac{\alpha}{2}\right) \frac{X_2}{Z} \tag{37}$$

Similarly we can calculate the optimal flow ratio from Eqs. (33) and (36)

$$\left(\frac{J_1}{J_2}\right)_{opt} = \text{tg}\left(\frac{\alpha}{2}\right) z \tag{38}$$

This relation can also be expressed in terms of the flow ratio at level flow

$$\left(\frac{J_1}{J_2}\right)_{opt} = \left(\frac{J_1}{J_2}\right)_{lf} \frac{1}{1 + \cos \alpha} \tag{39}$$

It is easy to see that

$$\frac{1}{2}\left(\frac{J_1}{J_2}\right)_{lf} < \left(\frac{J_1}{J_2}\right)_{opt} < \left(\frac{J_1}{J_2}\right)_{lf} \tag{40}$$

Stated in words this inequality says that unless q = 1 a *maximal P/O ratio is incompatible with optimal efficiency*. In fact, the optimal P/O ratio is always bound within the limits given in inequality (40). This result is of immediate practical interest. It tells that "low" P/O ratios measured in vivo do not imply a poor coupling of the phosphorylating machinery but, on the contrary, may indicate that oxidative phosphorylation is operating under optimal conditions.

Similarly to the above, one obtains for the optimal rate of ATP synthesis

$$(J_1)_{opt} = (J_1)_{lf} \frac{\cos \alpha}{1 + \cos \alpha} \tag{41}$$

from which we obtain the lower and upper bounds

$$0 < (J_1)_{opt} < \frac{1}{2}(J_1)_{lf} \tag{42}$$

This result shows that a *maximal rate of net ATP synthesis is incompatible with optimal efficiency*. This situation is reminiscent of the well-known practical experience that one should not drive a car as fast as possible if one wishes to obtain maximal mileage from a given amount of fuel. Furthermore, inequality (42) shows that the net rate of ATP synthesis tends to zero as q tends to unity. This already indicates that degrees of coupling close to 1 lead to unfavorable outputs of the phosphorylating machinery. This is in contrast to what intuition would predict. For the respiratory rate at optimal efficiency of ATP synthesis we obtain

$$(J_2)_{opt} = (J_2)_{lf} \cos \alpha \tag{43}$$

which yields the bounds

$$0 < (J_2)_{opt} < (J_2)_{lf} \tag{44}$$

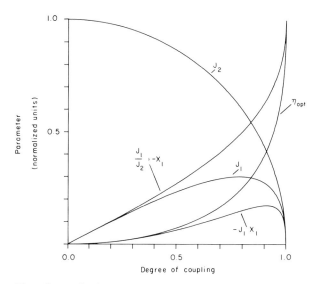

Fig. 5. Optimal parameters as a function of the degree of coupling. Plot of the optimal parameters η_{opt} (Eq. 33), $(X_1)_{opt}$ (Eq. 37), $(J_1/J_2)_{opt}$ (Eq. 38), $(J_1)_{opt}$ (Eq. 45), $(J_2)_{opt}$ (Eq. 46) and $(J_1/X_1)_{opt}$ (Eq. 51) versus q. All parameters normalized by $z = L_{22} = X_2 = 1$

Again, $(J_2)_{opt}$ vanishes as q tends to unity.

These calculations show that there must be a unique degree of coupling that allows the mitochondria to produce ATP at the fastest possible rate at optimal efficiency. To calculate this degree of coupling we consider the relations between $(J_1)_{opt}$, $(J_2)_{opt}$, and the given redox potential of the substrate

$$(J_1)_{opt} = tg\left(\frac{\alpha}{2}\right) \cos \alpha \, L_{22} X_2 z \tag{45}$$

$$(J_2)_{opt} = \cos \alpha \, L_{22} X_2 \tag{46}$$

These functions as well as $(X_1)_{opt}$ given by Eq. (37), $(J_1/J_2)_{opt}$ given by Eq. (38), and η_{opt} given by Eq. (33) are plotted in Figure 5 versus the degree of coupling. Clearly $(J_1)_{opt}$ passes through a maximum. Maximization of (Eq. 45) with respect to q yields the degree of coupling q_{max} where $(J_1)_{opt}$ assumes the maximum value $(J_1)_{max}$

$$q_{max} = F^{-\frac{1}{2}} \tag{48}$$

where $F = (\sqrt{5} + 1)/2$. This is a remarkable result insofar as it allows the expression of all thermodynamic quantities of interest in terms of the golden number F (abbreviated F because it is also the limit of the Fibonacci series). Remembering a few identities of the golden number

$$\frac{F}{F-1} = F^2 = F + 1 \tag{49}$$

we obtain after a few turnabouts the results collected in Table 1. In this table the thermodynamic parameters of interest are listed according to the general form

Table 1. Maximal parameters of oxidative phosphorylation. Parameters are listed in this table for n up to 6 according to the form defined in Eq. (50). See text for further details.
$F = (\sqrt{5}+1)/2$

Parameter $= F^{-\frac{n}{2}}$ constant

n	Parameter	Constant
1	q_{max}	-
2	$(J_2)_{max}$	$(L_{22}X_2)^{-1}$
3	$(J_1/J_2)_{max}$	-
4	$(\dot{S}_{sh})q_{max}$	$L_{22}X_2{}^2$
5	$(J_1)_{max}$	$(zL_{22}X_2)^{-1}$
6	η_{max}	-

$$(Param)_{max} = F^{-\frac{n}{2}} \text{ constant} \tag{50}$$

Every exponent n from 1 through 6 has a corresponding thermodynamic parameter.

The degree of coupling $q_{max} = 0.78$ appears rather low when compared to the values obtained from isolated mitochondria. Let us therefore consider an alternative problem. As can be seen from Figure 5, the optimal phosphate potential has a maximum at $q = 1$ where the net rate of ATP synthesis vanishes. Hence the maximum of the optimal output $(J_1X_1)_{opt}$ must occur at a value of q intermediate between q_{max} and 1. This degree of coupling can be obtained by maximizing

$$(J_1X_1)_{opt} = - \text{tg}^2\left(\frac{\alpha}{2}\right) \cos \alpha \ L_{22}X_2{}^2 \tag{51}$$

with respect to q. The result is the degree of coupling that allows the mitochondria to produce a maximal output power at optimal efficiency. To avoid confusion with the maximal values calculated above and with the maximal output defined by Kedem and Caplan (11), let us call these values ideal. For q_{id} we obtain

$$q_{id} = \sqrt{2(\sqrt{2}-1)}$$

Table 2 is a list of numerical values for some thermodynamic quantities of interest for the cases of maximal rate (maximal values) and maximal output (ideal values) under the constraint of optimal efficiency. From these values it appears that the ideal values agree rather closely with experimentally determined values.

It is now tempting to speculate that mitochondria have evolved to a state that allows them to produce a maximal output at optimal efficiency. It is important to stress that this question cannot be solved solely on the basis of isolated mitochondria. The only information that can be obtained from isolated mitochondria is their degree of coupling.

The aim of many laboratories has been, and is still, to isolate mitochondria such that they have the highest possible degree of coupling. It is possible that during these isolation procedures the degree of

Table 2. Ideal and maximal values. Numerical factors for several thermodynamic
parameters are listed in this table according to the form: Parameter = numerical
factor • constant. This allows an easy calculation for the thermodynamic quantities of interest. For example: What are the ideal phosphate potential $(X_1)_{id}$ and
P/O ratio $(J_1/J_2)_{id}$ with NADH as an oxidizable substrate? $X_2 = 50$ kcal, $z = 3$
yields $(X_1)_{id} = 10.73$ kcal; $(J_1/J_2)_{id} = 1.93$. For the meaning of the symbols and
for further details see text

Parameter	Numerical factor		Constant
	Ideal	Maximal	
q	0.910	0.786	–
J_1	0.267	0.300	$z\,L_{22}X_2$
X_1	– 0.644	– 0.486	X_2/z
J_1X_1	– 0.172	– 0.146	$L_{22}X_2^{\,2}$
J_2	0.414	0.618	$L_{22}X_2$
J_1/J_2	0.644	0.486	z
η	0.414	0.236	–
$\dot S$	0.243	0.472	$L_{22}X_2^{\,2}$
r	5.83	2.62	–
α	65.53	51.83	(degrees)

coupling of the mitochondria is improved by removal of free fatty acids
through addition of defatted bovine serum albumin, addition of complexing agents for divalent cations, etc.

Therefore the degree of coupling measured in vitro no longer necessarily reflects the degree of coupling of the same mitochondria in vivo.
From this we conclude that the question of whether mitochondria in the
cell are of the ideal rather than of the maximal type can only be decided by in vivo measurements or by modifying the isolation procedures
for mitochondria in such a way that, when using the procedures, no artificial improvement of q results.

In this context it is interesting to discuss briefly the effect of a
fluctuating environment on the efficiency of oxidative phosphorylation.
It is probable that in vivo not all mitochondria within a cell have
exactly the same degree of coupling since natural uncouplers such as
long-chain free fatty acids, Na^+ ions in the case of muscle cells, etc.,
show local concentration fluctuations within the cytosol. Hence the
overall degree of coupling measured in vivo or in vitro always represents an average value. It is from this average value of q that the optimal efficiency η_{opt} is calculated using Eq. (33). This calculation
is not necessarily correct. Within the framework of an ensemble theory
we recently calculated the effect of a fluctuating q on the ensemble
average value $<\eta_{opt}>$. Due to the strong convexity properties of η_{opt}
as shown in Figure 5, it follows from Jensen's inequality that

$$E\,\eta_{opt}\,(q) \geq \eta_{opt}\,(Eq) \tag{52}$$

where E denotes the mathematical expectation.

This matter is too complicated to be discussed here in detail and will
be published elsewhere (J.W. Stucki and W. Horsthemke, in preparation),

but the final result, (52), is highly nontrivial. It states that the *efficiency of oxidative phosphorylation is increased by fluctuations of the degree of coupling*. Under realistic assumptions this increase can be estimated to be as high as 20%.

To conclude this discussion we finally compare our results in Table 2 with the *adenylate energy charge* as introduced by Atkinson (21):

$$\text{energy charge} = \frac{1}{2}\frac{2\text{ATP} + \text{ADP}}{\text{ATP} + \text{ADP} + \text{AMP}} \tag{53}$$

Atkinson suggested that this ratio is maintained constant at a value of 0.9 in the cell by some regulatory mechanism. This value corresponds to a phosphate potential of 11 kcal/mol, and from Table 2 we notice that this corresponds closely to the ideal phosphate potential. Hence, if it is assumed that the production of ATP by the mitochondria and its utilization by various processes are the dominating energy flux in the cell, then the regulatory mechanism mentioned by Atkinson might well be explained on the basis of an ideal degree of coupling of the mitochondria operating in an optimal environment. The fact that the value 0.9 for the energy charge has been measured in vivo in several studies, favors the hypothesis of mitochondria being of the ideal rather than of the maximal type.

Entropy Production of Oxidative Phosphorylation

The foregoing paragraph was based on the fundamental assumption that mitochondria operate near optimal efficiency in the cellular environment. This assumption is certainly reasonable, but its justification is for the moment based entirely on teleologic arguments. The available experimental evidence to support this hypothesis is by no means sufficient and the observed agreement of the adenylate energy charge and the optimal efficiency hypothesis might be a coincidence. However, the formal apparatus established in the foregoing sections should allow further exploration of this hypothesis by suitable experiments.

Apart from an experimental confirmation, the biochemist should also be able to understand and to prove this hypothesis on the basis of the laws of macroscopic physics, notably thermodynamics.

This final theoretical section is not meant to give a final answer to the problem of whether mitochondria in the cell operate at optimal efficiency. Its aim is rater to clarify and delimit the problems to be solved. For the sake of simplicity we still maintain the linearity assumption between forces and flows as defined in Eq. (3). It migth well turn out that the problem of optimal efficiency cannot be tackled successfully with linear nonequilibrium thermodynamics. Nevertheless, the linear approach can serve as a guideline for future work.

In the foregoing sections we encountered three different distinguished steady states: static head, level flow, and the state of optimal efficiency. The entropy production at these states expressed in our parametric form is

$$(\dot{S})_{sh} = \cos^2\alpha \; L_{22}X_2^2 \tag{54}$$

$$(\dot{S})_{opt} = 2\frac{\cos^2\alpha}{1 + \cos\alpha} L_{22}X_2^2 \tag{55}$$

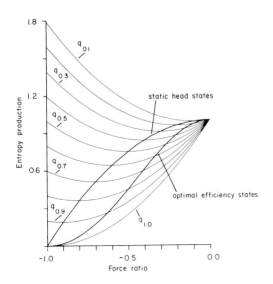

Fig. 6. Entropy production as a function of the force ratio. *Thin lines* represent plots of Eq. (58) for values of $q = 0.1$ up to $q = 1.0$ as shown in the Figure. $z = L_{22} = X_2 = 1$. *Thick lines* represent the loci of the static head states: $\dot{S}_{sh} = 1 - z^2 X_1^2/X_2^2$ and of the optimal efficiency states: $\dot{S}_{opt} = (1 - z^2 X_1^2/X_2^2) / (1 + z^2 X_1^2/X_2^2)$

$$(\dot{S})_{lf} = L_{22} X_2^2 \qquad (56)$$

These relations yield the inequality

$$(\dot{S})_{sh} < (\dot{S})_{opt} < (\dot{S})_{lf} \qquad (57)$$

This inequality indicates the major problem with which we are faced: entropy production at optimal efficiency is not minimal. Hence this state cannot be declared to be a natural unconstrained steady state of the system by the theorem of minimal entropy production. Using Eqs. (9a and 9b), (10), and (12) we obtain for the total entropy production (Eq. 8) in terms of the force ratio

$$\dot{S} = \left(z^2 \frac{X_1^2}{X_2^2} + 2qz \frac{X_1}{X_2} + 1 \right) L_{22} X_2 \qquad (58)$$

The minimum of this function occurs at $X_1/X_2 = -q/z$, i.e., at static head. This is also apparent from Figure 6, where $\dot{S}$ is plotted as a function of the force ratio for various values of q. This result shows that the static head is the state of minimal entropy production. By taking the theorem of minimal entropy production as a global evolution and stability criterion for the linear system, we conclude that mitochondria will invariably evolve toward state 4 when no constraints are imposed on the phosphate potential from the outside world. This theoretical prediction is so amply documented by experimental data that it might sound trivial. Hence isolated unconstrained mitochondria will never freely evolve toward the state of optimal efficiency but always remain within a finite distance from this state unless $q = 0$ or $q = 1$.

A calculation of the maximum of this distance

$$\Delta\dot{S} = (\dot{S})_{opt} - (\dot{S})_{sh} = \cos^2\alpha \; tg^2\left(\frac{\alpha}{2}\right) (\dot{S})_{lf} \qquad (59)$$

with respect to q yields the result that this maximum occurs at q_{max}. The maximal value is

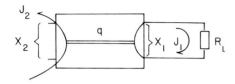

Fig. 7. Input-output relationship of oxidative phosphorylation in mitochondria in vivo. For the meaning of symbols see text

$$\dot{S}_{max} = F^{-5} (\dot{S})_{1f} \tag{60}$$

We note further that at q_{max} also the differences $(X_1)_{opt} - (X_1)_{sh}$ and $(J_1)_{opt} - (J_1)_{sh}$ attain a maximum.

These results underline once more that the state of optimal entropy production is not characterized by some minimal properties of entropy production as defined by Eq. (8). Of course, this does not contradict the assumption that mitochondria work at a state of optimal efficiency in the cellular environment. These results only point to the fundamental limitation in the adopted scheme (9a, 9b). This model is entirely limited to the description of isolated mitochondria and tells nothing about the real situation in the cell. The problem of whether mitochondria operate at optimal efficiency in the cell must be treated by a model that also considers the cellular environment.

To gain qualitative insight into the situation of the mitochondria in the cellular environment, we have to slightly extend the model we used thus far (Fig. 1). Let us summarize the entropy production of all ATP-utilizing reactions in the cell by one single term Av, the overall affinity of the ATP-utilizing reactions A times the velocity of the ATP-utilization by these processes v. With this, the cellular entropy production can be written

$$\dot{S}_{Cell} = \dot{S}_{Mit} + Av = J_1 X_1 + J_2 X_2 + Av \tag{61}$$

The general scheme for this equation is given in Figure 7, where R_L represents the overall load resistance of the cellular ATP-utilizing reactions. With this definition and within the linear domain of the adopted formalism we have

$$Av = \frac{X_1^2}{R_L} \tag{62}$$

For the sake of simplicity we introduce the notation $x = zX_1/X_2$ for the force ratio. With this the cellular entropy production becomes

$$\dot{S}_{Cell} = \left\{ x^2 \left(1 + \frac{1}{R_L L_{11}} \right) + 2qx + 1 \right\} L_{22} X_2^2 \tag{63}$$

Taking the derivative of Eq. (63) with respect to x and substituting $x = x_{opt}$ at $d\dot{S}_{cell}/dx = 0$ yields

$$(R_L L_{11})^{-1} = R_{phos}/R_L = \sqrt{1 - q^2} \tag{64}$$

where R_{phos} is the internal resistance of phosphorylation.

Hence

$$\dot{S}_{Cell} = \left\{ x^2 (1 + \sqrt{1 - q^2}) + 2qx + 1 \right\} L_{22} X_2^2 \tag{65}$$

The crucial relation (64), which forces $\dot{S}_{cell}$ to be minimal at x_{opt}, is the impedance matching for constant input power as introduced by Kedem and Caplan (11). At the beginning of this section we were interested in the circumstances under which the mitochondria in the cell would evolve toward the state of optimal efficiency as their natural steady state. We are now in the position to answer this question:

Theorem

Suppose that the resistance of phosphorylation R_{phos} is matched to the load resistance R_L of the ATP-utilizing reactions in the cell according to relation (64). Under these conditions the natural steady state of the mitochondria within the cellular environment is the state of optimal efficiency of oxidative phosphorylation.

This theorem can be established by virtue of the theorem of minimal entropy production at the steady state which, in the linear domain of thermodynamics, is a general evolutionary criterion as well as a stability criterion (12 - 14). The necessary and sufficient condition of impedance matching (64) should be amenable to experimental verification. Note that the results in Section 5 were obtained under the assumption that mitochondria operate at optimal efficiency in the cellular environment. The above theorem describes the conditions under which this assumption is fulfilled. Hence the values q_{max} and q_{id} have their special physical meaning mentioned in the preceding section only if the criterion of impedance matching as stated above is met.

To conclude this section we will now develop a geometric interpretation of the entropy production at different states. Let us rewrite the parametrization of $\dot{S}_{sh}$ and $\dot{S}_{opt}$ introduced in Eqs. (54 - 56) in terms of q by using the normalization $\dot{S}_{1f} = L_{22}X_2^2 = 1$

$$(\dot{S})_{sh} = 1 - q^2 \tag{66}$$

$$(\dot{S})_{opt} = \frac{2(1 - q^2)}{1 + \sqrt{1 - q^2}} \tag{67}$$

$$(\dot{S})_{1f} = 1 \tag{68}$$

Now we split the entropy production at optimal efficiency $\dot{S}_{opt}$ into two terms: a term $\dot{S}_b$ representing the basal entropy production due to the input power $(J_2X_2)_{opt}$ and a second term $\dot{S}_w$ representing the entropy gained by the output power $(J_1X_1)_{opt}$ (see Eq. 51) at optimal efficiency

$$\dot{S}_{opt} = \dot{S}_b + \dot{S}_w \tag{69}$$

Then

$$\dot{S}_b - \sqrt{1 - q^2} = \cos \alpha \tag{70}$$

and

$$\dot{S}_w = - \frac{\sqrt{1 - q^2}(1 - \sqrt{1 - q^2})}{1 + \sqrt{1 - q^2}} - \cos \alpha \, tg^2\left(\frac{\alpha}{2}\right) \tag{71}$$

This latter expression can also be rewritten as

$$\dot{S}_w = - \dot{S}_b \, \eta_{opt} \tag{72}$$

which is a restatement of definition (29).

Let us consider again the difference $\Delta\dot{S}$ introduced in Eq. (59), which can now be expressed in terms of $\dot{S}_b$ and $\dot{S}_w$

284

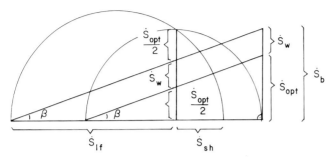

Fig. 8. Geometric construction of $\dot{S}_w$ and $\dot{S}_b$. Given $\dot{S}_{1f}$ and $\dot{S}_{sh}$. $\dot{S}_b$ are obtained as the geometric mean (Eq. 75). Addition and subtraction of $\dot{S}_b$ from $\dot{S}_{1f}$ yields the proportion (74), from which $\dot{S}_w$ is constructed. Note that $\dot{S}_{1f} = 1$ due to normalization Eq. (68). From this construction it is apparent that $\dot{S}_{opt} < \dot{S}_b$ since $\dot{S}_w < 0$. For further details see text

$$\Delta \dot{S} = - \dot{S}_b \dot{S}_w \tag{73}$$

A straightforward maximization of Eq. (72) shows that $\dot{S}_w$ assumes a maximum at q_{id}, whereas the product $|\dot{S}_b \dot{S}_w|$ has its maximal value at q_{max} as stated in the foregoing (Eq. 60).

These remarkable properties give a physical meaning to the two distinguished degrees of coupling q_{id} and q_{max} in terms of the mitochondrial entropy production: *at q_{max} the entropy production due to respiration times the entropy production due to phosphorylation is maximal, whereas at q_{id} only the entropy production due to phosphorylation is maximal.*

Equations (70 - 71) can be used to visualize directly the extreme properties of $\dot{S}_w$ by a geometric construction. Dividing Eq. (71) by Eq. (70) yields the fundamental proportion

$$\frac{\dot{S}_w}{\dot{S}_b} = \frac{1 - \sqrt{1 - q^2}}{1 + \sqrt{1 - q^2}} = \frac{1 - \dot{S}_b}{1 + \dot{S}_b} = tg^2 \left(\frac{\alpha}{2}\right) = n_{opt} \tag{74}$$

By interpreting Eq. (70) as a geometric mean of $\dot{S}_{sh}$ and $\dot{S}_{1f}$

$$\dot{S}_b = \sqrt{\dot{S}_{sh} \dot{S}_{1f}} \tag{75}$$

we can use Eq. (74) to construct $\dot{S}_w$ and $\dot{S}_b$ geometrically, as explained in Figure 8. The angle β introduced in this figure is given by

$$tg\ \beta = \frac{\dot{S}_b}{1 + \dot{S}_b} = \frac{\cos \alpha}{1 + \cos \alpha} \tag{76}$$

This relation takes the values

$$tg\ \beta_{id} = \frac{\sqrt{2} - 1}{\sqrt{2}} \quad \text{at } q_{id} \tag{77}$$

$$tg\ \beta_{max} = \frac{F - 1}{F} \quad \text{at } q_{max} \tag{78}$$

From this it may be noted that $\sqrt{2}$ plays a role in the ideal situation similar to that of F in the maximal situation. The geometric construc-

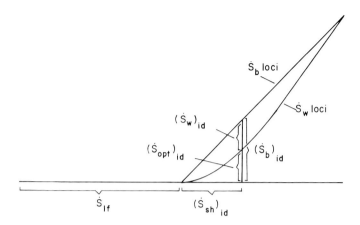

$\dot{S}_b$ loci

$\dot{S}_w$ loci

$(\dot{S}_w)_{id}$

$(\dot{S}_{opt})_{id}$

$(\dot{S}_b)_{id}$

$\dot{S}_{lf}$

$(\dot{S}_{sh})_{id}$

<u>Fig. 9.</u> Geometric construction of the $\dot{S}_b$ and $\dot{S}_w$ loci for $0 \leqslant q \leqslant 1$. The construc-
tion shown in Figure 8 was carried out for all values $0 \leqslant \dot{S}_{sh} \leqslant \dot{S}_{lf}$ corresponding to
$0 \leqslant q \leqslant 1$. The resulting loci of $\dot{S}_b$ and $\dot{S}_w$ are represented by the two corresponding
curves. The physical meaning of the distances is indicated for the ideal case where
$\dot{S}_w$ is maximal

tion of $\dot{S}_w$ and $\dot{S}_b$ for all values of $0 < q < 1$ as depicted in Figure 9
allows a direct visualization of the maximum of $\dot{S}_w$ at q_{id}. Everything
is geometry!

Concluding Remarks

This theory offers some new insights into the problem of efficiency
of mitochondrial energy conversion. Of special importance is the non-
intuitive result that a degree of coupling of oxidative phosphorylation close to unity is not adequate for the mitochondria operating in
vivo. The extensive experimental work with isolated mitochondria suggested that the machinery of oxidative phosphorylation should be designed such as to yield the highest possible phosphate potential. The
results in Section 5 show that such high potentials necessarily imply
a slow rate of ATP production. It seems obvious that the main task of
the mitochondria in vivo is to maintain a high flow of ATP synthesis
rather than to maintain a high phosphate potential with an essentially
vanishing net flow of ATP. This latter situation is typical for isolated mitochondria (state 4). Hence it seems highly probable that mitochondria were designed to maximize the output power of the phosphorylating machinery in vivo. This, however, would imply a degree of coupling of only 0.91 or even lower, in contrast to a value close to one
as heuristically assumed from studies with isolated mitochondria.

Another nontrivial result of this theory is that the optimal operating
conditions for oxidative phosphorylation necessitate a specially matched
environment. It seems possible to decide, by suitable relaxation measurements, whether the condition (64) of impedance matching is full-
filled under in vivo conditions. If there is impedance matching, then
the mitochondria operate indeed at optimal efficiency as shown in the
theorem in Section 6. The experimental verification of impedance matching is of immediate practical interest for our theory since this matching is one of the fundamental assumptions made in Section 5.

The present theory is surely limited by the introduced simplifications. Especially the representation of all ATP-utilizing processes by one lumped circuit only might turn out to be a gross oversimplification of the matter. Hence the theory has to be extended toward a more realistic description of the real situation by considering more than just one ATP-utilizing reaction in the cytosol. Furthermore, it may be necessary to introduce deviations from linearity between flows and forces by introducing higher order terms in Eq. (3), as well as to consider the time dependence of the phenomenologic coefficients L_{ij} in future versions of this theory. However, although there may be considerable discrepancies between the results of this paper and the experimental observations, it is to be hoped that these data generally agree at least on a qualitative basis.

Summary

A theory of the efficiency of mitochondrial energy conversion, in particular oxidative phosphorylation, has been developed within the framework of linear nonequilibrium thermodynamics. The efficiency of oxidative phosphorylation is critically dependent on the degree of coupling (q) of oxidative phosphorylation and on the force ratio, i.e., the ratio of the clamped phosphate potential and the redox potential of the substrate. Efficiency is zero at both maximal (static head) and zero (level flow) force ratios and passes through a maximum (optimal efficiency) in between these values. Experimental observations indicate that mitochondria in vivo operate at optimal efficiency. Under the conditions of optimal efficiency, there are two distinct degrees of coupling: (1) q_{max} where the *rate* of ATP synthesis is maximal (maximal case), and (2) q_{id} where the *output power* of oxidative phosphorylation, i.e., the developed phosphate potential times the rate of ATP production, is maximal (ideal case). The experimentally observed value of 0.9 for the adenylate energy charge in vivo corresponds exactly to the ideal case. This indicates that oxidative phosphorylation has evolved to a state where the output power rather than the rate of ATP production is maximal. On the basis of the principle of minimal entropy production at steady state, the following theorem has been proved: Mitochondria reach a state of optimal efficiency whenever the resistance of phosphorylation (R_{phos}) matches the overall resistance (R_{load}) of the cellular ATP-utilizing reactions, according to the relation

$$R_{phos}/R_{load} = \sqrt{1 - q^2} .$$

Acknowledgments: This work has been supported by several grants from the Swiss National Science Foundation. The author is particularly indebted to Profs. G. Nicolis, and H. Reuter to Drs. W. Horsthemke, R. Lefever, and D. Walz for stimulating discussions and to Prof. I. Prigogine for his continued interest and support of this work.

References

1. Ernster, L., Nordenbrand, K.: In: Dynamics of Energy Transducing Membranes. Ernster, L., Estabrook, R.W., Slater, E.C. (eds.) BBA Library. Amsterdam: Elsevier Scientific Publ. Co., 1974, Vol. 13, p. 283
2. Rottenberg, H., Caplan, S.R., Essig, A.: In: Membranes and Ion Transport. Bittar, E.E. (ed.). London: Wiley-Interscience, 1970, Vol. 1, p. 165

3. Caplan, S.R.: In: Current Topics in Bioenergetics. Sanadi, D.R. (ed.). New York: Academic Press, 1971, Vol. 4, p. 1
4. Stucki, J.W.: Eur. J. Biochem. $\underline{68}$, 551 (1976)
5. Carafoli, E., Rossi, C.S., Lehninger, A.L.: Biochem. Biophys. Res. Commun. $\underline{19}$, 609 (1965)
6. Lehninger, A.L., Carafoli, E., Rossi, C.S.: Adv. Enzymol. $\underline{29}$, 259 (1967)
7. Stucki, J.W., Ineichen, E.A.: Eur. J. Biochem. $\underline{48}$, 365 (1974)
8. Stucki, J.W., Walter, P.: Eur. J. Biochem. $\underline{30}$, 60 (1972)
9. Prigogine, I.: Introduction to Thermodynamics of Irreversible Processes. New York: Interscience, 1967
10. Katchalsky, A., Curran, P.F.: Nonequilibrium Thermodynamics in Biophysics. Cambridge, Massachusetts: Harvard University Press, 1965
11. Kedem, O., Caplan, S.R.: Trans. Faraday Soc. $\underline{61}$, 1897 (1965)
12. Stucki, J.W.: Progr. Biophys. Mol. Biol., $\underline{33}$, 99 (1978)
13. Nicolis, G., Prigogine, I.: Self-Organization in Nonequilibrium Systems. New York: John Wiley, 1977
14. Glansdorff, P., Prigogine, I.: Thermodynamic Theory of Structure, Stability and Fluctuations. New York: John Wiley, 1971
15. Rottenberg, H.: In: Progress in Surface and Membrane Science. Danielli, J.F., Cadenhead, A., Rosenberg, M.D. (eds.). New York: Academic Press, 1978, in press
16. Owen, Ch.S., Wilson, D.F.: Arch. Biochem. Biophys. $\underline{161}$, 581 (1974)
17. Erecińska, M., Veech, R.L., Wilson, D.F.: Arch. Biochem. Biophys. $\underline{160}$, 412 (1974)
18. Caplan, S.R., Essig, A.: Proc. Natl. Acad. Sci. U.S.A. $\underline{64}$, 211 (1969)
19. Chance, B., Williams, G.R.: Adv. Enzymol. $\underline{17}$, 92 (1956)
20. Davis, E.J., Lumeng, L.: J. Biol. Chem. $\underline{250}$, 2275 (1975)
21. Atkinson, D.E.: Biochemistry $\underline{7}$, 4030 (1968)

The journal of

Membrane
Biology

An international
journal for studies
on the structure,
function and genesis
of biomembranes

Title No. 232

During the past decade work on the membrane has expanded
so widely that is has come to the forefront of biological research.
Many formerly disparate disciplines now find a common
meeting ground in their study of the biomembrane. THE
JOURNAL OF MEMBRANE BIOLOGY integrates the
diverse aspects of membrane biology and serves as a stimulus
among the several disciplines with membrane research.

Springer-Verlag
New York Inc.

A sample copy as well as subscription and back-volume information available upon request

Please address:
Springer-Verlag New York Inc.
175 Fifth Avenue, New York, NY 10010, USA
or:
Springer-Verlag, Promotion Department
P. O. Box 10 52 80, D-6900 Heidelberg 1

Membrane Transport in Biology

Editors:
G. Giebisch, D. C. Tosteson,
H. H. Ussing

Membranes regulate the transport and distribution of the components of living cells. These processes underlie a wide variety of cellular functions. This Handbook brings together contributions from a large number of investigators and provides a comprehensive view of the present state of knowledge in this field. Volume I treats the general properties of biological membranes, some aspects of physical chemistry relevant to transport phenomena, and the movement of substances across lipid bilayers. Volume II considers transport across single biological membranes, both the plasma membranes separating the inside and outside of cells and the membranes surrounding intracellular organelles. This volume illuminates how transport across single membranes underlies such important biological functions as excitation and propagation in nerve and muscle cells and the initiation of shape changes in contractile cells. Volume III deals with transport across isolated epithelia, while Volume IV treats organs specialized for transport, notably the kidney and the gastro-intestinal tract. Therofore, the theme units Volumes III and IV is transport across at least two single biological membranes arranged in series. The difference in the transport properties between two such membranes accomplishes the net movement of substances from one to the other side of an epithelium or layer of cells, that is, absortion and secretion.

Springer-Verlag
Berlin
Heidelberg
New York

Volume 1
Concepts and Models

Editor: D. C. Tosteson

With contribution by O. S. Anderson, J. E. Hall, D. J. Hanahan, U. V. Lassen, P. K. Lauf, R. J. Lefkowitz, E. Racker, B. E. Rasmussen, S. A. Rudolph, F. A. Sauer, C. W. Slayman, G. Stark, O. Sten-Knudson, H. H. Ussing

1978. 108 figures. Approx. 450 pages
ISBN 3-540-08687-0

Volume 2
Transport Across Single Biological Membranes

Editor: T. C. Tosteson

With contribution by L. Beaugé, A. Herold, G. Inesi, V. L. Lew, R. I. Macey, L. J. Mullins, B. Sarkadi, A. Scarpa, D. C. Tosteson, R. A. Venosa, D. Walker, G. Weissmann

178. 103 figures. Approx. 450 pages
ISBN 3-540-08780-X

Volume 3
Transport Across Multi-Membrane Systems

Editor: G. Giebisch

With contribution by M. A. Bisson, D. Erlij, A. L. Finn, J. Gutknecht, D. F. Hastings, J. Hess Thaysen, A. Leaf, E. A. C. MacRobbie, S. H. P. Madrell, A. Martinez-Palomo, J. L. Oschman, H. H. Ussing, E. M. Wright, J. A. Zadunaisky, K. Zerahn

1978. 97 figures, 26 tables. Approx. 600 pages
ISBN 3-540-08596-3

Volume 4
Transport Organs

Editor: G. Giebisch

With contributions by T. E. Andreoli, E. Boulpaep, M. Claret, F. E. Curry, J. H. Dirks, J. G. Forte, G. Giebisch, C. W. Gottschalk, R. N. Khuri, R. Kinne, W. E. Lassiter, E. W. van Lennep, T. E. Machen, G. Malnic, D. W. Powell, G. F. Quamme, E. M. Renkin, L. Reuss, J. A. Schafer, S. Schultz, I. Schulz, R. A. L. Sutton, K. Ullrich, E. E. Windhager, J. A. Young

1978. 219 figures. Approx. 600 pages
ISBN 3-540-08895-4